Mathematical Methods and Quantum Mathematics for Economics and Finance

Belal Ehsan Baaquie

Mathematical Methods and Quantum Mathematics for Economics and Finance

Springer

Belal Ehsan Baaquie
International Centre for Education
in Islamic Finance
Kuala Lumpur, Malaysia

ISBN 978-981-15-6610-3 ISBN 978-981-15-6611-0 (eBook)
https://doi.org/10.1007/978-981-15-6611-0

This Springer imprint is published by the registered company Springer Nature Singapore Pte Ltd.
The registered company address is: 152 Beach Road, #21-01/04 Gateway East, Singapore 189721, Singapore

The three great branches of mathematics are, in historical order, Geometry, Algebra, and Analysis. Geometry we owe essentially to Greek civilization, algebra is of Indo-Arab origin, and analysis (or calculus) was the creation of Newton and Leibniz, ushering in the modern era.

Sir Michael Atiyah (Fields Medalist 1966. Quoted in Mathematics as Metaphor by Yuri I. Manin, American Mathematical Society (2007)).

Preface

Modern economics and finance increasingly employs mathematical techniques as an essential aspect of its formalism as well as in its applications. Given the pace of development of economics and finance, a short and up to date introduction to mathematics has become a necessary pre-requisite for all graduate students, even for those not specializing in quantitative finance.

This book is an introductory graduate text for students of economics and finance. The bedrock of the mathematics required is linear algebra, calculus and probability theory, and which is the main focus of the book—with many examples and problems being chosen from economics and finance. The topics covered could be useful for anyone seeking an introduction to higher mathematics.

The book is divided into five Parts. The material in Parts One to Four can be covered in 14 weeks. The book starts by focusing on the key subjects of linear algebra and calculus before going on to more advanced topics such as differential equations, probability theory and stochastic calculus. Part Four of the book ends with an introduction to the mathematical under-pinnings of stochastic calculus by focusing on a derivation of the Black-Scholes and Merton equations.

Part Five on Quantum Mathematics is an introduction to the mathematical techniques drawn from quantum mechanics, and introduces the reader to key ideas of quantum mathematics such as state space, Hamiltonian, Lagrangian and path integration. Research and applications of quantum mathematics to economics and finance is an emerging field [12], and 'quantum finance' is becoming part of mainstream finance [4, 5].

Part Five introduces readers to advanced mathematical techniques that could be employed in more complex formulations of topics in economics and finance.

Kuala Lumpur, Malaysia — Belal Ehsan Baaquie

Acknowledgements I thank Frederick H.Willeboordse for his consistent support and Muhammad Mahmudul Karim for his valuable input in the preparation of the book. I thank Mohamed Eskandar Shah Mohd Rasid for his precious support and encouragement for writing this book.

This book is dedicated to my wife Najma and to my family members Tazkiah, Farah and Arzish. Their love, affection, support and encouragement have made this book possible. For this and much more, I am eternally indebted to all of them.

Synopsis

Abstract

A brief summary is given of the five parts of the book, highlighting the logic for the choice of topics covered in the various chapters. Starting from introductory topics, the subsequent chapters cover linear algebra, calculus and probability theory, laying the foundation for analyzing more advanced topics, including quantum mathematics.

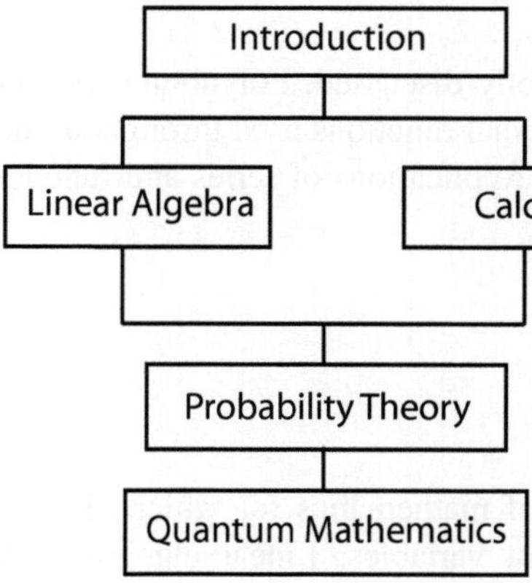

The figure shows the flow chart of the major parts of the book. The introduction is a review of well-known results. The three parts—Linear Algebra, Calculus and Probability Theory—are the foundation of the book and need to be mastered by students and practitioners of economics and finance, as well as those specializing in financial engineering. These three parts form the core knowledge required for the rest of the book.

Quantum Mathematics is included with the objective of extending and expanding the mathematics of economics and finance by showing a deep and fundamental connection between the mathematical formalism of the two subjects.

A pre-requisite for studying Quantum Mathematics is to have a basic grounding in Linear Algebra, Calculus and Probability. Quantum Mathematics is meant for students who intend to specialize in—as well as for specialists (from other fields) interested in—quantitative finance and economics.

The subjects covered in this book are customized to the needs of readers who have little or no background in mathematics. The material discussed in the first four Parts of the book prepares readers to follow the advanced topics covered in Part Four; it is hoped that Part Five will be accessible to those who master the subjects discussed in the first four Parts. To help the students grapple with and grasp the book's material, each Chapter of the first four Parts have problem sets.

The following is a list of useful references on mathematical methods for economics and finance: [1, 2, 3, 15, 17, 18, 19, 20, 23, 27, 28, 35, 38, 39]. These books are not referenced in the main text as they cover the material of the first four parts.

References to material on more advanced topics like the Merton equation and topics on quantum mathematics, not covered in the standard references, are included in the text.

I. Introduction

Series and functions are briefly discussed. Polynomial equations and finite series lead to the infinite series. Special functions used throughout the book, including the exponential, are introduced. Applications of series and functions in economics and finance are discussed.

II. Linear Algebra

Linear algebra is a branch of mathematics for which all the equations are linear functions of the independent variables. Linear algebra is one of the pillars of modern mathematics, and of science and engineering.

Topics as diverse as geometry and functional analysis are based on linear algebra, and it is fundamental in modern presentations of geometry, including for defining basic objects such as lines, planes and rotations. For nonlinear systems, for which the fundamental equations require quadratic and higher power of the independent variables, the equations are often linearized and linear algebra provides an approximation of the nonlinear equations.

Linear algebra is introduced in the book using ideas of vectors and matrices; it is shown that simultaneous linear equations lead to the analysis of matrices and determinants. Tensor product for matrices is defined and the analysis of the determinant shows its geometrical significance. Square matrices are studied in some detail, including the spectral decomposition of symmetric and Hermitian matrices.

III. Calculus

Calculus is a vast subject and this part provides a brief introduction to the essential ideas of calculus. Integral and differential calculus as well as differential equations are at the foundation of modern mathematics, science and engineering. Calculus is indispensable in the study of economics and finance. An intuitive introduction is given to calculus, with a view to its applications in various branches of economics and finance.

Integration is introduced as the logical extension of convergent infinite series. Definite and indefinite integrals are defined and the result of integrating a few special cases is derived from first principles. Gaussian integration is studied in great detail as it is required in many of the advanced topics.

Differentiation is seen as the inverse of integration, and the rules for differentiation are derived. A discussion of optimization in the context of economics is discussed and the Hessian matrix for the maxima or minima of multiple variables is derived. Constrained optimization is illustrated using examples from economics. Ordinary differential equations, both first and second order, are studied with focus on linear differential equations.

IV. Probability Theory

The theory of probability is the mathematical discipline that analyzes the nature of uncertainty and provides a precise definition for it. There are many views on what is the best description of uncertainty and it is an open question whether all types and forms of uncertainty are within the ambit of probability theory. In economics and finance, the future prices of financial instruments are uncertain as are the prices of real estate, commodities and so on; probability theory plays an essential role in modeling this uncertain future.

Probability theory is based on the concept of random variables, and various types of discrete and continuous random variables are studied. The binomial option model based on the binomial random variable is analysed in some detail to illustrate the utility of probability theory. The general properties of probability distributions are introduced, and fundamental concepts such as the joint, conditional and marginal probability are discussed. The central limit theorem for independent random variables is derived from first principles.

Concepts from probability theory and calculus are used to define stochastic processes, and Ito calculus is briefly discussed. The important special cases of the

Black-Schloes model for option pricing and the Merton model from corporate finance are studied in great detail to allow the reader to have an insight into the mathematics of stochastic processes.

V. Quantum Mathematics

Quantum Mathematics arises from the uncertainty and indeterminacy that is fundamental to quantum mechanics [6]—and provides a complete set of mathematical ideas for describing, explaining and modeling uncertain and random phenomena, be it classical or quantum. The concepts and theoretical underpinnings of quantum mechanics lead to a whole set of new mathematical ideas and have given rise to the subject of *quantum mathematics*. In particular, the theoretical framework of quantum mechanics provides the mathematical tools for studying classical randomness that occurs in economics and finance [12].

Path integration, also called functional integration, is a natural generalization of integral calculus and is essentially the integral calculus of infinitely many independent variables. There is however a fundamental feature of path integration that sets it apart from functional integration in general, and which is the role played by the Hamiltonian in the formalism. All the path integrals discussed in this book have an underlying linear structure that is encoded in the Hamiltonian operator and its linear vector state space. It is this combination of the path integral and its underlying linear vector space and Hamiltonian that provides a powerful and flexible mathematical machinery that can address a vast variety and range of diverse problems. Topics from finance have been chosen that demonstrate the interplay of a system's path integral, state space and Hamiltonian.

The Hamiltonian operator and the mathematical formalism of path integration makes it eminently suitable for describing classical randomness that is a recurrent theme in economics and finance. The material covered provides an introduction to the foundations of path integration and is a primer to the techniques and methods employed in the study of quantum finance, as formulated in [4] and [5].

Many problems arising in finance and economics can be addressed by quantum mathematics. This part of the book studies the mathematical aspect of path integrals and Hamiltonians as applied to finance. All the topics and subjects have been specifically chosen to illustrate the application of quantum mathematics to finance.[1]

The synthesis of liner algebra with calculus leads to functional analysis, and which is the mathematical basis of quantum mathematics. The Dirac notation for functional analysis is introduced. The spectral decomposition of linear algebra is extended to define Fourier transformations. Functional differentiation and functional integration are defined, and illustrated using specific examples.

[1]Applications of quantum mathematics to economics are discussed in [11] and are excluded here for the sake of brevity.

The concept of Hamiltonian and path integrals are discussed in the context of finance; the Black-Schloes and Merton models are analyzed to illustrate the application of these concepts to finance. The martingale condition is derived using the Hamiltonian and the path integral is used to evaluate the pricing kernel for both the Black-Schloes and Merton models. Quantum fields are introduced to study coupon bonds and interest rates, with a linear quantum field being used to study coupon bonds and a nonlinear quantum field used for studying the Libor model of interest rates.

Contents

Part I Introduction

Part II Linear Algebra

About the Author

Belal Ehsan Baaquie is a Professor at The International Centre for Education in Islamic Finance, Malaysia. His previous position was Professor at the National University of Singapore. He studied theoretical physics at the California Institute of Technology (BS) and Cornell University (PhD), specializing in quantum field theory. He has worked extensively in applying quantum mathematics to problems in finance and economics. His research interests cover many fields, and he has written over ten books on science and finance, including The Theoretical Foundations of Quantum Mechanics (Springer, 2013). (Author page: https://www.amazon.com/-/e/B07GCC7WQ4).

Part I
Introduction

Chapter 1
Series and Functions

Abstract The basic rules of algebra are reviewed. The finite and infinite series, including the expansion of a function, are introduced and some applications to finance and economics are discussed.

1.1 Introduction

Quadratic and higher order polynomial are analyzed and the roots of polynomial equations are discussed. Functions of a real variable are defined; the exponential, logarithm and associated functions are illustrated by the compounding and discounting of future cash flows. Demand and supply functions, as well as a few payoff functions from option theory, are discussed.

1.2 Basic Algebra

Integers, real and complex numbers are briefly reviewed in the Appendix. Numbers have definite values, such as 3, $\sqrt{2}$ and so on. However, there are many phenomenon that need to be described by a quantity that takes values in a range and is said to have a variable value; examples of variables are the prices of a commodity at different times, the temperature at different locations and so on.

Variables, denoted by $x, y, z, \ldots$, hence have no fixed value, but rather takes value in a set consisting of many different numbers, which can be discrete or continuous. For example, the number of customers is a discrete quantity, whereas the price p of a commodity is always positive and takes continuous values in the set of positive real numbers; in other words, $p \in \mathbb{R}_+ = \{0, \ldots, +\infty\}$.

Let $x \in \mathbb{R}$ be a real variable. It follows the rules of algebra

$$x + x = 2x; \quad x^a x^b = x^{a+b}; \quad \frac{1}{x} = x^{-1}; \quad xx^{-1} = 1, \ldots$$

B. E. Baaquie, *Mathematical Methods and Quantum Mathematics for Economics and Finance*, https://doi.org/10.1007/978-981-15-6611-0_1

For example

$$(x - a)(x + a) = x^2 - a^2$$

Using the rules of algebra one can form functions of x, denoted by $f(x)$. Some of the most important functions are the following.

- $f(x) = c_0 + c_1 x + c_2 x^2 + \cdots + c_n x^n$: polynomial. The coefficients for polynomial $f(x)$ can be assigned numerical values; an example is

$$-9x^4 + 0.1x^3 + 15x - 0.9$$

- $f(x) = \sin(x);\ \ \cos(x);\ \ e^x; \ldots$: Transcendental functions
- $f(x) = \sum_{n=0}^{\infty} c_n x^n$: infinite series

1.2.1 *Binomial Expansion*

The Binomial theorem states that

$$\begin{aligned}(x + y)^n &= x^n + nx^{n-1}y + \cdots + \frac{n(n-1)\cdots(n-k+1)}{k!}x^{n-k}y^k \cdots + y^n \\ &= \sum_{k=0}^{N} \frac{n!}{(n-k)!k!}x^{N-k}y^k\end{aligned}$$

where k factorial is given by

$$k! = 1.2. \cdots (k-1) \cdot k; \quad 0! \equiv 1$$

For example

$$(x + y)^2 = x^2 + y^2 + 2xy; \quad (1 + x)^3 = 1 + 3x + 3x^2 + x^3$$

A proof of the binomial expansion is given in Sect. 8.3.2 and Noteworthy 8.2.

1.3 Quadratic Polynomial

One of the most important polynomial is the quadratic form given by

$$f(x) = ax^2 + bx + c; \quad a, b, c: \text{ real}$$

The function $f(x)$ is zero for special values of x called the *roots* of the $f(x)$; denote the roots by x_+, x_-. This yields the following

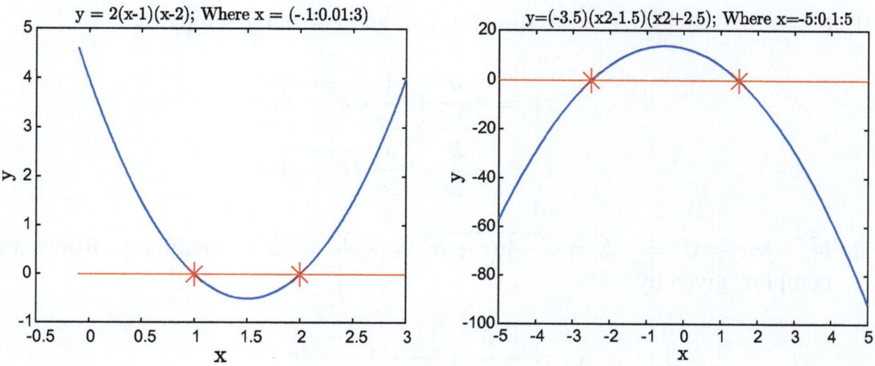

Fig. 1.1 Quadratic function with location of the roots. Published with permission of © Belal E. Baaquie 2020. All Rights Reserved

$$f(x) = a(x - x_+)(x - x_-)$$

To find x_+, x_-, consider the following simplification (Fig.1.1).

$$\begin{aligned} f(x) &= a(x^2 + \frac{b}{a}x) + c \\ &= a(x + \frac{b}{2a})^2 - \frac{1}{4a}(b^2 - 4ac) = a\Big[(x + \frac{b}{2a})^2 - (\frac{1}{2a})^2\Delta^2\Big] \end{aligned}$$

where

$$\Delta = \sqrt{b^2 - 4ac}: \text{ Discriminant}$$

Using $x^2 - a^2 = (x - a)(x + a)$ yields

$$\begin{aligned} f(x) &= a\Big[(x + \frac{b}{2a})^2 - (\frac{1}{2a})^2\Delta^2\Big] = a\Big(x + \frac{b}{2a} - \frac{1}{2a}\Delta\Big)\Big(x + \frac{b}{2a} + \frac{1}{2a}\Delta\Big) \\ &= a(x - x_+)(x - x_-) \end{aligned}$$

Hence, the roots of $f(x)$ are given by

$$x_+ = -\frac{b}{2a} + \frac{1}{2a}\Delta$$
$$x_- = -\frac{b}{2a} - \frac{1}{2a}\Delta$$

The discriminant has three cases.

1. $b^2 - 4ac > 0 \Rightarrow \Delta$: real. Roots x_+, x_- are real and given by

$$x_+ = -\frac{b}{2a} + \frac{1}{2a}\sqrt{b^2 - 4ac}$$
$$x_- = -\frac{b}{2a} - \frac{1}{2a}\sqrt{b^2 - 4ac}$$

2. $b^2 - 4ac < 0 \Rightarrow \Delta = \sqrt{-4ac + b^2} = i\sqrt{4ac - b^2}$: imaginary. Roots are complex, given by

$$x_+ = -\frac{b}{2a} + \frac{i}{2a}\sqrt{4ac - b^2}$$
$$x_- = -\frac{b}{2a} - \frac{i}{2a}\sqrt{4ac - b^2} = x_+^*$$

3. $b^2 - 4ac = 0 \Rightarrow \Delta = 0$: real. Roots are real and given by

$$x_+ = x_- = -\frac{b}{2a}$$

The complex roots come in complex conjugate pairs; the reason being that the starting was a quadratic equation with real coefficients a, b, c. Let $x_+ = z$; the conjugate roots yield the following

$$f(x) = a(x - z)(x - z^*) = a\Big(x^2 - (z + z^*)x + zz^*\Big)$$

The coefficients are real since $z + z^*$ and zz^* are both real; in particular, for $z = p + iq$, with p, q: real, we have

$$f(x) = x^2 - 2px + (p^2 + q^2) : \text{ real quadratic polynomial}$$

1.3.1 Higher Order Polynomial

For an nth order real polynomial $f(x) = c_0 + c_1 x + c_2 x^2 + \cdots + c_n x^n$ there are n polynomial roots; the polynomial is given by

$$f(x) = c_n(x - x_1)(x - x_2)\cdots(x - x_n)$$

The complex polynomial roots come in conjugate pairs for the reason given above, and yield the following.

- For n: even, all roots are either real or come in conjugate pairs.

- For n: odd, there is always a single real root, and the rest of the roots are either real or come in conjugate pairs.
- The extension of the real numbers to the complex numbers is sufficient to provide a solution to all polynomial equations.

1.4 Finite Series

The functions of variable x can be defined by the means of finite and infinite series.

Consider the finite series

$$f(x) = c_0 + c_1 x^1 + c_2 x^2 \cdots + c_N x^N \cdots = \sum_{n=0}^{N} c_n x^n$$

Example. The NPV (Net Present Value) for a firm with cash flows CF_n for the next N years can be written as

$$\text{NPV} = \frac{CF_1}{1+r} + \frac{CF_2}{(1+r)^2} + \cdots \frac{CF_N}{(1+r)^N} - \text{Outlay}$$

To get the internal rate of return IRR, define $x = 1/(1 + IRR)$; then the IRR is given by solving for the roots of the polynomial equation

$$f(x) = CF_1 x + CF_2 x^2 + \cdots CF_N x^N - \text{Outlay} = 0$$

The IRR is usually found numerically by solving the equation above.

The NPV for a firm giving *constant returns* for the next N years can be written as

$$\begin{aligned}\text{NPV} &= \frac{c}{1+r} + \frac{c}{(1+r)^2} + \cdots \frac{c}{(1+r)^N} - \text{Outlay} \\ &= \sum_{n=1}^{N} \frac{c}{(1+r)^n} - \text{Outlay}\end{aligned}$$

The series for NPV is a case of the geometric series, defined by

$$S_N = a + a^2 + \cdots + a^N = \sum_{n=1}^{N} a^n$$

The geometric series can be exactly summed using the following procedure. Note

$$aS_N = a^2 + a^3 + \cdots + a^{N+1}$$

Subtracting S_N from aS_N yields

$$(a-1)S_N = -a + a^{N+1}$$

and hence

$$S_N = a\frac{1-a^N}{1-a} \tag{1.4.1}$$

NPV can be written using Eq. 1.4.4 and $a = 1/(1+r)$.

1.4.1 Example: NPV

The cash flow for the nth year is given by CF_n. Let the company function for N years. For fixed future cash flows given by $c = CF_n, n = 1, \cdots N$ the net revenue of the company is given by

$$R = \sum_{n=1}^{N} \frac{c}{(1+r)^n} = ca\frac{1-a^N}{1-a}; \quad a = \frac{1}{1+r}$$

Hence, revenue R is given by

$$R = \frac{c}{r}\Big(1 - (\frac{1}{1+r})^N\Big)$$

For $r \to 0$, the revenue has the expected limit

$$R \to \frac{c}{r}\Big(1-(1-Nr)\Big) = Nc$$

Suppose the initial cost of project B is 100,000. The project is expected to provide 30,000 for 4 years. The discounting rate is 3.2%. What is the present day valuation of the project? For $r = 3.2\%$ and $N = 4$

$$NPV = -\text{Outlay} + \frac{CF_1}{(1+r)^1} + \frac{CF_2}{(1+r)^2} + \frac{CF_3}{(1+r)^3} + \frac{CF_4}{(1+r)^4}$$
$$NPV = -100000 + \frac{30000}{1.0032^1} + \frac{30000}{1.032^2} + \frac{30000}{1.032^3} + \frac{30000}{1.032^4} = 10981.67$$

1.4.2 Example: Coupon Bond

To further understand the geometric series consider a coupon bond, denoted by $\mathcal{B}(t)$, that pays fixed coupons at future times T_n years—and returning the principal L when the bond matures at future time $T_N = T$ years. See Fig. 1.2. Suppose the bond is issued at $t = 0$, with the issuer having an internal rate of return given by irr; the value of the bond is then

$$\mathcal{B}(t) = \frac{\alpha}{1+irr} + \frac{\alpha}{(1+irr)^2} + \cdots \frac{\alpha}{(1+irr)^N} + \frac{1}{(1+irr)^N} L \quad (1.4.2)$$

The price of the coupon bond is the sum is the sum of the discounted cash flows in the future. $\mathcal{B}(t)$ is written in terms of yield to maturity $y(t) = y$ by the following equation

$$\mathcal{B}(t) = \frac{\alpha}{1+y} + \frac{\alpha}{(1+y)^2} + \cdots \frac{\alpha}{(1+y)^N} + \frac{1}{(1+y)^N} L \quad (1.4.3)$$

$$= L\left[\frac{c}{1+y} + \frac{c}{(1+y)^2} + \cdots \frac{c}{(1+y)^N} + \frac{1}{(1+y)^N}\right] \quad (1.4.4)$$

where $\alpha = cL$.

Using Eq. 1.4.3 and $a = 1/(1+y)$ yields

$$\mathcal{B}(t) = L\left[\frac{c}{1+y} \cdot \frac{1 - \frac{1}{(1+y)^N}}{1 - \frac{1}{(1+y)}} + \frac{1}{(1+r)^N}\right]$$

$$= L\left[\frac{c}{y} \cdot \{1 - \frac{1}{(1+y)^N}\} + \frac{1}{(1+y)^N}\right] \quad (1.4.5)$$

$$\Rightarrow \mathcal{B}(t) = \frac{c}{y} L + \frac{y-c}{y} \cdot \frac{L}{(1+y)^N}$$

The par value of a bond is when the bond's value is equal to the principal. When the yield to maturity is equal to the coupon rate, given by $y = c$, then, from Eq. 1.4.5

$$\mathcal{B}(t) = L \text{ for } y = c: \text{ Par value}$$

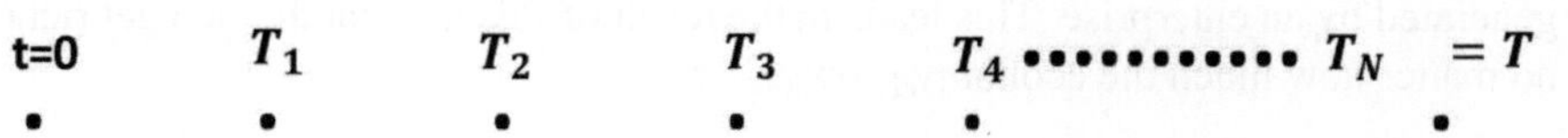

Fig. 1.2 The future time, denoted by $T_1, T_2, \cdots, T_N$ when coupons are paid. Published with permission of © Belal E. Baaquie 2020. All Rights Reserved

The below and above par values of the bond are given by

$$\mathcal{B}(t) = \begin{cases} \frac{c}{y}L + \frac{y-c}{y} \cdot \frac{L}{(1+y)^N} < L \text{ for } y > c: & \text{below par} \\ \frac{c}{y}L + \frac{y-c}{y} \cdot \frac{L}{(1+y)^N} > L \text{ for } y < c: & \text{above par} \end{cases}$$

1.4.3 Example: Valuation of a Firm

A firm's value today is determined by all the future dividends D_n (cash flows) that it can generate. The discounting of D_n is determined by discount factor r. The value of the firm, denoted by V, is given by

$$\begin{aligned} V &= \frac{D_1}{1+r} + \frac{D_2}{(1+r)^2} + \cdots \frac{D_N}{(1+r)^N} + \ldots \\ &= \sum_{n=1}^{\infty} \frac{D_n}{(1+r)^n} \end{aligned}$$

Suppose the firm grows at the rate of g and the dividends are given by

$$D_n = D_0(1+g)^n$$

Then the value of the firm is given by

$$V = D_0 \sum_{n=1}^{\infty} \frac{(1+g)^n}{(1+r)^n}$$

Summing this geometric series using Eq. 1.6.1 yields, for $a = (1+g)/(1+r)$, the following

$$V = D_0 \frac{a}{1-a} = D_0 \frac{1+g}{r-g}$$

Note under the present economic system

$$r > g$$

In other words, the return on accumulated capital is always greater than the wealth generated by an enterprise. This leads to the result of Piketty, that the rich get richer no matter how much the economy grows.

1.4.4 *Example: 1/3*

Consider the following finite decimal expansion; from Eq. 1.4.3

$$\underbrace{0.33333333333333330000\cdots}_{N\text{ non-zero entries}} = 3\{\frac{1}{10} + \frac{1}{10^2}\cdots + \frac{1}{10^{N+1}}\}$$
$$= \frac{3}{10}\frac{1-(\frac{1}{10})^N}{1-\frac{1}{10}} = \frac{1}{3}(1-(\frac{1}{10})^N)$$
$$\Rightarrow \lim_{N\to\infty}\underbrace{0.33333333333333330000\cdots}_{N\text{ non-zero entries}} = 0.333333333333\cdots \to \frac{1}{3}$$

Note that for any finite N, the decimal expansion is not equal to 1/3, differing from it by a factor of 10^{-N}. Only in the limit that $N \to \infty$ does the decimal expansion become equal to 1/3.

1.5 Infinite Series

A function can have an expansion in an infinite series, known as the Taylor and Laurent series. In particular

$$f(x) = \sum_{n=0}^{\infty} c_n x^n : \text{ Taylor series}$$

and

$$f(x) = \sum_{n=-\infty}^{\infty} c_n x^n : \text{ Laurent series}$$

What is the general condition that an infinite series converges? Consider, using notation $c_n x^n = u_n$, the following infinite series

$$\sum_{n=0}^{\infty} u_n$$

Conditional convergence means that that

$$\lim_{N\to\infty}\sum_{n=0}^{N} u_n \to \text{finite}$$

Infinite series are unlike finite series. Consider for example the infinite series with alternating signs

$$1-\frac{1}{2}+\frac{1}{3}-\frac{1}{4}+\frac{1}{5}\cdots=\sum_{n=1}^{\infty}\frac{(-1)^{n+1}}{n}=\ln(2)$$

On summing the series as given above, the series apparently converges to $\ln(2)$. However, it was shown by Riemann that the series does not have any unique value and can have any value, depending on how the series is rearranged. If one groups the terms differently by moving around infinitely many terms in the series, one has[1]

$$1-\frac{1}{2}-\frac{1}{4}+\frac{1}{3}-\frac{1}{6}-\frac{1}{8}+\frac{1}{5}-\frac{1}{10}-\frac{1}{12}\cdots$$

the sum is composed of three terms

$$\frac{1}{2k-1}-\frac{1}{2(2k-1)}-\frac{1}{4k}=\frac{1}{2(2k-1)}-\frac{1}{4k}$$

The series can, hence, be written out as

$$\frac{1}{2}\Big(1-\frac{1}{2}+\frac{1}{3}-\frac{1}{4}+\frac{1}{5}\cdots\Big)=\frac{1}{2}\ln(2)$$

What went wrong? Infinite series that are conditionally convergent can take any value, and the mistake made above was that of *re-arranging infinitely* many terms in the series. This is not permitted due to the conditional convergence.

An infinite series that is absolutely convergent is defined by the requirement that its absolute sum *converges*

$$\lim_{N\to\infty}\sum_{n=0}^{N}|u_n|\to \text{finite}$$

Note for all real numbers x, the absolute value $|x|$ is defined as follows

$$|x|=\sqrt{x^2}=\begin{cases} x & x\geq 0\\ -x & x<0\end{cases}$$

For complex number z

$$|z|=\sqrt{z^*z}$$

[1] https://en.wikipedia.org/wiki/Riemann_series_theorem#Changing_the_sum.

An absolutely convergent series has a *unique* value, and one can rearrange all the terms without changing the value of the series. For the series above, it is not absolutely convergent since

$$\sum_{n=1}^{N}\left|\frac{(-1)^{n+1}}{n}\right| = \sum_{n=1}^{N}\frac{1}{n} \simeq \ln(N) \to \infty$$

1.6 Cauchy Convergence

Consider the infinite series

$$\lim_{N\to\infty}\sum_{n=0}^{N} u_n$$

Consider only the case of absolute convergence for which conditionally convergent series are excluded. Hence the following series is analyzed

$$\lim_{N\to\infty}\sum_{n=0}^{N} |u_n| \to \text{finite}$$

Clearly, to converge, the terms u_n must get smaller and smaller as $n \to \infty$. But how fast? The Cauchy criterion for convergence requires that, for an arbitrarily number denoted by ϵ, there exists an N that depends on ϵ, given by $N = N(\epsilon)$, such that, for $m, n > N$

$$\Big|S_m - S_n\Big| < \epsilon : \;\; m, n > N(\epsilon)$$

where

$$S_m = \sum_{j=0}^{m} u_j$$

Let $m = n + p$; then Cauchy convergence yields

$$\Big|u_{n+1} + \cdots + u_{n+p}\Big| < \epsilon : \;\; p, n > N(\epsilon)$$

1.6.1 Example: Geometric Series

Consider the geometric finite series

$$\tilde{S}_n = \sum_{k=1}^{m} a^n = a + a^2 \cdots a^n = a\frac{1-a^{n+1}}{1-a}$$

Let $m > n > N$; replace a by $|a|$ and consider

$$\Big|S_m - S_n\Big| = \frac{|a|^{n+1}}{1-|a|}(|1-|a|^{m-n}|)$$

For the series to converge, we need

$$|a|^{m-n} \to 0 \;\Rightarrow\; |a| < 1$$

To determine how large is N consider

$$\Big|S_m - S_n\Big| \to |a|^N = \epsilon : \;\; m > n > N(\epsilon)$$

Hence

$$N\ln(|a|) = \ln(\epsilon) \;\Rightarrow\; N = \frac{\ln(\epsilon)}{\ln(|a|)} \to +\infty$$

since $\ln(|a|) < 0$.

In general, a can be positive or negative, and the condition that the infinite geometric series have a finite value is given by

$$|a| < 1$$

For example, if $a = 1$, the infinite series is divergent since

$$S = 1 + 1 + 1 + 1 + \cdots = N \to \infty$$

In the limit of $N \to \infty$, for $|a| < 1$ we have

$$|a| < 1 \;\Rightarrow\; \lim_{N\to\infty} a^N \to 0$$

Hence, from Eq. 1.4.1

$$\tilde{S} = \frac{a}{1-a} \tag{1.6.1}$$

Sometimes, the geometric series includes a term independent of a and is written as

$$S = \sum_{n=0}^{\infty} a^n = 1 + a + a^2 \cdots a^N \cdots$$

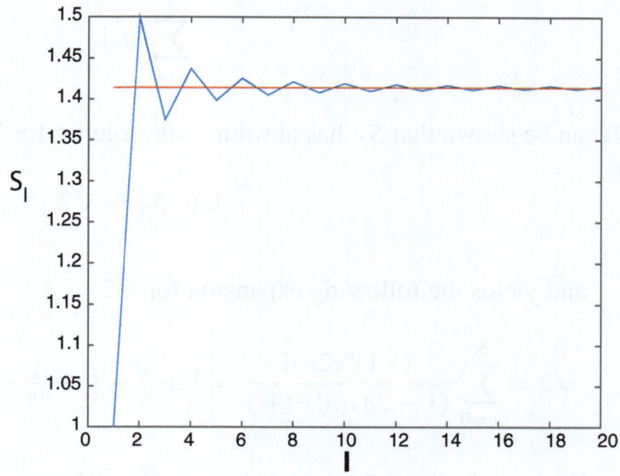

Fig. 1.3 The partial sum S_N converges to $\sqrt{2}$. Published with permission of © Belal E. Baaquie 2020. All Rights Reserved

For this case

$$S = 1 + \tilde{S} = \frac{1}{1-a}$$

1.7 Expansion of $\sqrt{1+x}$

The number $\sqrt{2}$ is analyzed to see how it is related to the rational numbers. Consider the general problem of Taylor expansion of the function $\sqrt{1+x}$; if the limit of $x \to 1$ yields a convergent Taylor expansion, $\sqrt{2}$ can be represented by the expansion.[2]

$$\sqrt{1+x} = \sum_{n=0}^{\infty} \frac{(-1)^n (2n)!}{(1-2n)(n!)^2(4^n)} x^n = 1 + \tfrac{1}{2}x - \tfrac{1}{8}x^2 + \tfrac{1}{16}x^3 - \tfrac{5}{128}x^4 + \cdots$$

The infinite series of rational numbers converges to $\sqrt{2}$; the convergence is oscillatory, the partial sum sometimes being greater and sometimes less than $\sqrt{2}$. To examine how the infinite series converges to $\sqrt{2}$, define the partial sum by

$$S_N = \sum_{n=0}^{N} \frac{(-1)^n (2n)!}{(1-2n)(n!)^2(4^n)} x^n = \sum_{n=0}^{N} u_n$$

For Cauchy convergence consider

[2]The Taylors series is given by https://en.wikipedia.org/wiki/Squareroot.

$$\sum_{n=0}^{N} |u_n|$$

It can be shown that S_N has absolute convergence for $|x| \le 1$. Hence

$$\lim_{N\to\infty} S_N = \sqrt{2}$$

and yields the following expansion for $\sqrt{2}$

$$\sqrt{2} = \sum_{n=0}^{\infty} \frac{(-1)^n (2n)!}{(1-2n)(n!)^2(4^n)} = 1 + \tfrac{1}{2} - \tfrac{1}{8} + \tfrac{1}{16} - \tfrac{5}{128} + \cdots = 1.41421\cdots$$

Figure 1.3 shows S_N converging to $\sqrt{2}$ as N is increased. In the 'neighborhood' of $\sqrt{2}$, there are infinitely many rational numbers. The 'neighborhood' of a number in calculus is denoted by the infinitesimal ϵ.

1.8 Problems

1. Solve the following quadratic equations

$$x^2 + 4x - 21 = 0; \quad 3x^2 + 6x + 10 = 0; \quad -3x^2 + 12x = -1$$

2. Suppose the initial cost of project A is 100,000. The project is expected to provide 40,000 for 3 years. The discounting rate is 4.011%. What is your evaluation of the project?

3. If you had to choose between project A and project B, which one would you choose and why?

4. Suppose that the current dividend payout of a company is \$1.80 and the rate of return required is 15% while the constant growth rate is 5%. What is the current price of the company's stock?
   Hint:

$$V = D_0 \frac{1+g}{r-g}$$

5. Consider the following polynomial

$$f(x) = a_1 x^n + a_n x^{n-1} + \cdots + a_n$$

   How many roots does $f(x)$ have? Write $f(x)$ in terms of its roots.

6. Consider the price of a coupon bond given by

$$\mathcal{B} = \frac{\alpha}{1+y} + \frac{\alpha}{(1+y)^2} + \cdots + \frac{\alpha}{(1+y)^N} + \frac{L}{(1+y)^N}$$

What is the par value of the coupon α?

7. Find the sum of the infinite series

$$S = 1 + a + a^2 + \cdots + a^n + \cdots$$

8. Consider the quadratic equation

$$ax^2 + bx + c = 0$$

Under what condition does the quadratic equation have solutions that are imaginary? Why do imaginary solutions come in complex conjugate pairs?

9. Consider the partial sum

$$S_N = \sum_{k=0}^{N} \frac{(2k+1)!}{2^{3k+1}(k!)^2} \to \sqrt{2}$$

Show that S_N is Cauchy convergent.

10. Consider the expansion of $\sqrt{2}$ given by

$$\sqrt{2} = \sum_{n=0}^{\infty} \frac{(-1)^n (2n)!}{(1-2n)(n!)^2(4^n)} = 1 + \tfrac{1}{2} - \tfrac{1}{8} + \tfrac{1}{16} - \tfrac{5}{128} + \ldots$$

(a) How many terms of the infinite series are required for an accuracy of 10^{-4}?
(b) How many terms of the infinite series are required for an accuracy of 10^{-6}?

1.9 Functions

Let $f(x)$ be a function of real variable x. The function is a mapping from the set to which x belongs—called the domain and denoted by $\mathcal{D}$—to another set $\mathcal{R}$, called the range of the function; see Fig. 1.4. The mapping is expressed as follows

$$f : \mathcal{D} \to \mathcal{R}; \;\; y = f(x) : \; x \in \mathcal{D}, \; y \in \mathcal{R}$$

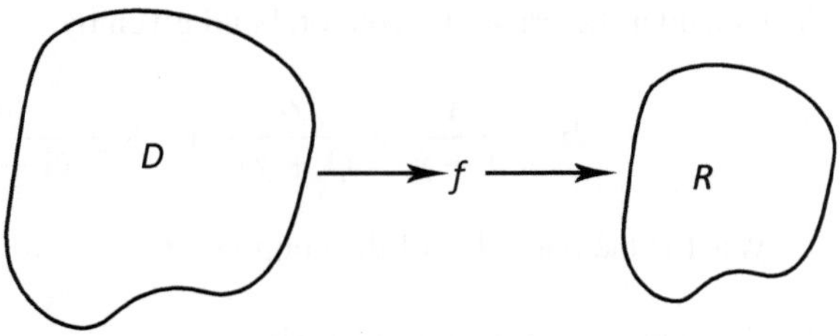

Fig. 1.4 Domain (D) of x and range R of $y = f(x)$. Published with permission of © Belal E. Baaquie 2020. All Rights Reserved

Functions satisfy many of rules of algebra, some of them are the following.

- $(f \pm g)(x) = f(x) \pm g(x)$: point-wise addition/subtraction
- $(fg)(x) = f(x)g(x)$: point-wise multiplication
- $(f/g)(x) = f(x)/g(x)$: point-wise division
- $(f \odot g)(x) = f(g(x))$: function of a function

Example: Let $f(x) = \exp x;\ g(x) = x^2$; then

- $(f \pm g)(x) = \exp x \pm x^2$
- $(fg)(x) = x^2 \exp x$
- $(f/g)(x) = x^{-2} \exp x$
- $(f \odot g)(x) = f(g(x)) = f(x^2) = \exp\{x^2\}$

The polynomial is one of the most commonly used function

$$f(x) = c_0 + c_1 x + c_2 x^2 + \cdots + c_n x^n$$

The coefficients for polynomial $f(x)$ can be assigned numerical values. A number of functions of the variable x, mentioned earlier, have an infinite series expansion. Note that all functions do not always have a series expansion.

1.10 Exponential Function

The exponential is one of the most important function. The exponential function is of central importance in economics. In finance, the discounting of future cash flows and compounding of cash deposits, is ubiquitous; for continuous time, the concept of exponential plays a key role.

Consider a simple interest of 100% per year on a cash deposit of value M. After one year, the depositor gets $M + M = 2M$ (Fig. 1.5).

Suppose the depositor is given the choice of having a half-yearly payout, which he then again deposits. The depositor then gets $M + \frac{1}{2}M$ after 1/2 year and then again deposits to get—after a year—

$$M(1 + \frac{1}{2}) + M(1 + \frac{1}{2})\frac{1}{2} = M(1 + \frac{1}{2})^2$$

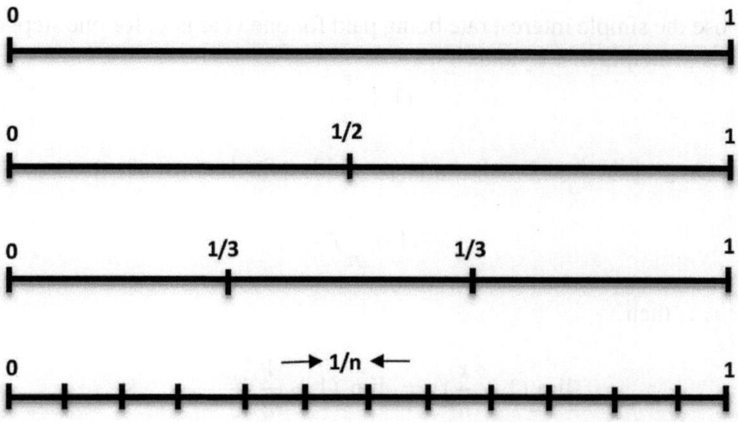

Fig. 1.5 Compounding fixed deposit. Published with permission of © Belal E. Baaquie 2020. All Rights Reserved

One can then carry on and deposit for 1/3 of a year and compound the initial deposit three times and obtain $M(1+\frac{1}{3})^3$; for n payments per year, the amount received is

$$M(1+\frac{1}{n})^n$$

n	1	2	3	$\cdots$	10^3	10^4
Amount received	2.0	2.25	2.37	$\cdots$	2.717	2.718

From Table above one can see that the depositor will never receive $3M$ no matter how many times the deposit is compounded. Let $M = 1$ as it plays no role in the definition of the exponential.

Taking the limit of $n \to \infty$ means that the money is deposited for a small interval $1/n$, which becomes ϵ. Taking the limit of $n \to \infty$ yields a finite number, known as the exponential and denoted by the symbol of e.

$$\lim_{n\to\infty}(1+\frac{1}{n})^n = e = 2.71828\cdots$$

The exponential e is an irrational, and in fact, a transcendental number.[3]

[3]A transcendental number is not the roote of any polynomial.

Suppose the simple interest rate being paid for one year is x; for one step, interest rate x plus principla of \$1 yields

$$(1 + \frac{x}{n})$$

Compounding of the interest after n steps yields a final return of

$$(1 + \frac{x}{n})^n$$

Let $n = mx$; then

$$\begin{aligned}
\lim_{n\to\infty}(1 + \frac{x}{n})^n &= \lim_{m\to\infty}(1 + \frac{1}{m})^{mx} \\
&= \Big(\lim_{m\to\infty}(1 + \frac{1}{m})^m\Big)^x = e^x \\
\Rightarrow \lim_{n\to\infty}(1 + \frac{x}{n})^n &= e^x
\end{aligned} \qquad (1.10.1)$$

The following notation is used for the exponential

$$e^x \equiv \exp(x)$$

Using the Binomial theorem, it can be shown from Eq. 1.10.1 that

$$e^x = \sum_{n=0}^{\infty} \frac{x^n}{n!} \qquad (1.10.2)$$

Noteworthy 1.1: Exponential function and Binomial expansion

The binomial theorem states that

$$(1 + x)^n = \sum_{k=0}^{n} \binom{n}{k} x^k$$

Using the binomial theorem, the expansion in Eq. 1.10.1 is given by the following

$$\lim_{n\to\infty}(1 + \frac{x}{n})^n = \lim_{n\to\infty} \sum_{k=0}^{n} \binom{n}{k} (\frac{x}{n})^k \qquad (1.10.3)$$

Note

$$\lim_{n\to\infty} \binom{n}{k} = \lim_{n\to\infty} \frac{n!}{(n-k)!k!} = \lim_{n\to\infty} \frac{n(n-1)\cdots(n+1-k)}{k!} \to \frac{n^k}{k!}$$

Hence, from Eq. 1.10.3

$$\lim_{n\to\infty}(1+\frac{x}{n})^n = \lim_{n\to\infty}\sum_{k=0}^{n}\frac{n^k}{k!}(\frac{x}{n})^k \to \sum_{k=0}^{\infty}\frac{1}{k!}x^k = e^x$$

The exponential function, shown in Fig. 1.6, has the important property that, for real x

$$e^x > 0: \ \ x \in [-\infty, +\infty]$$

From Eq. 1.10.2, e^x increases at a faster rate than any power of x. For small x,

$$e^x = \sum_{n=0}^{\infty}\frac{x^n}{n!} \approx 1 + x + \frac{x^2}{2!}\cdots \tag{1.10.4}$$

Special values are

$$\exp(-\infty) = 0; \ \ \exp(0) = 1; \ \ \exp(+\infty) = +\infty$$

The proof of Eq. 1.10.1, using Eq. 1.10.2, is the following

$$\lim_{N\to\infty}(1+\frac{x}{N})^N = \lim_{N\to\infty}\Big(\exp\{\frac{x}{N} + O(1/N^2)\}\Big)^N \to e^x$$

Separating out the even and odd powers of x yields

$$e^x = \cosh x + \sinh x$$

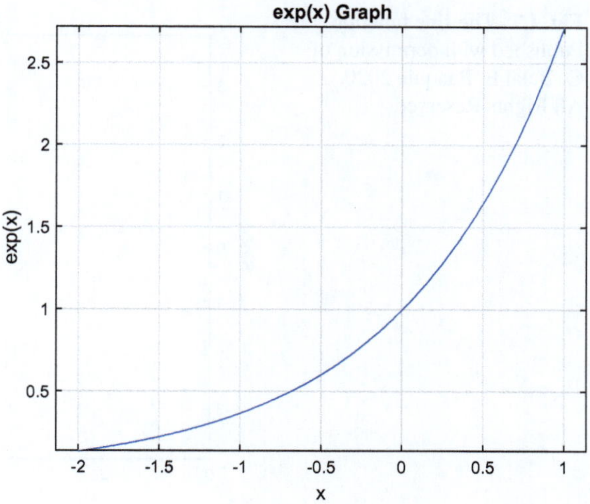

Fig. 1.6 The exponential function. Published with permission of © Belal E. Baaquie 2020. All Rights Reserved

where

$$\cosh x = \sum_{n=0}^{\infty} \frac{x^{2n}}{(2n)!}; \quad \sinh x = \sum_{n=0}^{\infty} \frac{x^{2n+1}}{(2n+1)!}$$

For complex variable $z = x + i\theta$, the exponential function yields

$$e^z = e^x e^{i\theta}; \quad e^{i\theta} = \cos\theta + i\sin\theta$$

where

$$\cos\theta = \sum_{n=0}^{\infty} (-1)^n \frac{\theta^{2n}}{(2n)!}; \quad \sin\theta = \sum_{n=0}^{\infty} (-1)^n \frac{\theta^{2n+1}}{(2n+1)!}$$

The sine and cosine function have the important representation given by

$$\cos\theta = \frac{1}{2}(e^{i\theta} + e^{-i\theta}); \quad \sin\theta = \frac{1}{2i}(e^{i\theta} - e^{-i\theta}) \tag{1.10.5}$$

The sine and cosine functions have the key property of being *periodic*, that is, they repeat their values as the argument of the functions go through one cycle.

$$\cos(\theta) = \cos(\theta + 2\pi); \quad \sin(\theta) = \sin(\theta + 2\pi) \quad \Rightarrow \quad e^{i\theta} = e^{i(\theta+2\pi)} \tag{1.10.6}$$

Note from Eq. 1.10.6 that

$$e^{2\pi i} = 1; \quad e^{\pi i} = -1$$

The sine function is shown in Fig. 1.7 and has the periodicity given above. To prove the periodicity of the sine and cosine functions, it is easiest to use their definition

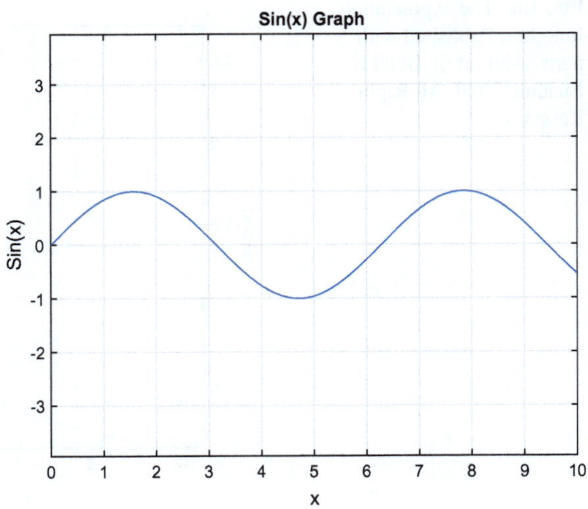

Fig. 1.7 The sine function. Published with permission of © Belal E. Baaquie 2020. All Rights Reserved

from trigonometry. To prove the periodicity from the series expansion is difficult since one needs to define π for the infinite series.

1.10.1 Parametric Representation

A function of the variable x can be expressed indirectly, using an auxiliary variable called a parameter. Consider the function of two variables x, y given by

$$r^2 = f(x, y) = x^2 + y^2$$

This function can be represented in a parametric form using the auxiliary variable θ by the following

$$x = r\cos(\theta); \quad y = r\sin(\theta)$$

since, from Eq. 1.10.5

$$\cos^2(\theta) + \sin^2(\theta) = 1$$

Eliminating θ yields r as a function of x, y.

1.10.2 Continuous Compounding

Consider putting money M in a bank for a time ϵ (can be finite or infinitesimal) and earning a simple interest rate of r; then after ϵ time interval the money is given by

$$M(1) = (1 + \epsilon r)M$$

Suppose one rolls over the money N times—after every ϵ time interval; one then obtains

$$M(N) = (1 + \epsilon r)^N M$$

One can consider the limit when one deposits the money for an instant and takes it back and rolls it and so on for a total time interval of T. The number of times N that the money is rolled over is given by

$$T = N\epsilon$$

Let us take the limit of $\epsilon \to 0$; holding T fixed means that

$$N = \lim_{\epsilon \to 0} \frac{T}{\epsilon} \to \infty$$

Hence, the money in the bank after time interval T, denoted by $M(T)$, is given by

$$M(T) = \lim_{N\to\infty}\left(1+\frac{rT}{N}\right)^N M$$

From Eq. 1.10.1

$$M(T) = e^{rT}M \tag{1.10.7}$$

Equation 1.10.7 yields the compounded value of M at fixed interest rate r.

The value of a coupon bond, with fixed coupons, from Eq. 1.4.5, is given by

$$\mathcal{B}(t) = L\left[\frac{c}{y}\cdot\{1-\frac{1}{(1+y)^N}\}+\frac{1}{(1+y)^N}\right] \tag{1.10.8}$$

Bond valuation for continuous time can be written in terms of the exponential function.

Suppose the bond matures at some fixed future time T, with the time left for maturity equal to $T-t$. In N payments, the total amount paid in coupons, until the maturity of the bond, is given by Nc; instead of coupons being paid at discrete times, one can equivalently consider the coupons being paid at some rate d for each interval of time $\epsilon = (T-t)/N$, and shown in Fig. 1.8; similarly, let the yield to maturity y be defined for continuous time by continuous discounting given by r. Then

$$Nc = d(T-t) \ \Rightarrow\ c = d\frac{(T-t)}{N};\ y = r\frac{(T-t)}{N}$$

Hence, from Eq. 1.10.8, taking the limit of $N\to\infty$ yields

$$\begin{aligned}\mathcal{B}(t) &= L\frac{c}{y}\cdot\{1-\frac{1}{(1+y)^N}\}+\frac{L}{(1+y)^N}\\ &= L\frac{d}{r}\cdot\{1-\frac{1}{(1+\frac{r(T-t)}{N})^N}\}+\frac{L}{(1+\frac{r(T-t)}{N})^N}\\ \Rightarrow\ \mathcal{B}(t) &= L\frac{d}{r}\cdot\{1-e^{-r(T-t)}\}+Le^{-r(T-t)}\end{aligned}$$

The result for the value of the bond has a simple and intuitive explanation. The first term is the value of the bond due to the coupons, and its contribution to the value of the bond falls as the bond matures; the second term is the discounted present day value of the principal that is to be paid back when the bond matures. The bond has a par value for $d = r$.

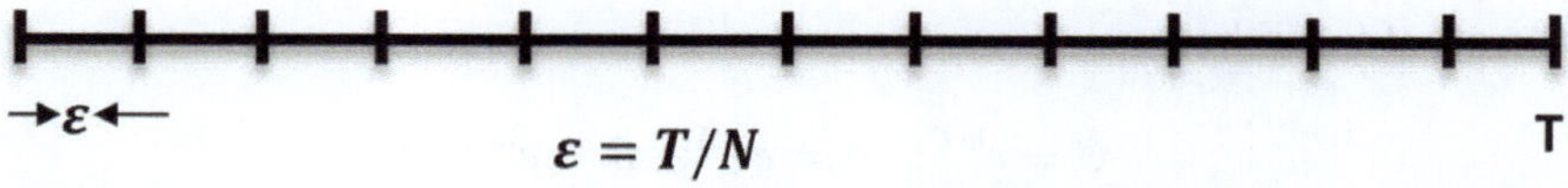

Fig. 1.8 Coupon bond valuation for continuous time. Published with permission of © Belal E. Baaquie 2020. All Rights Reserved

1.10.3 Logarithm Function

The *inverse* of the exponential function is the logarithmic function given by

$$\ln(x); \quad \exp(\ln(x)) = x$$

The logarithmic function, shown in Fig. 1.9, has the following values limit given by

$$\lim_{x\to 0} \ln(x) \to -\infty; \quad \ln(1) = 0$$

The logarithmic function has the following properties

$$\ln(x^a) = a\ln(x) \;\; \Rightarrow \;\; x^a = \exp\{a\ln(x)\} = \Big(\exp\{\ln(x)\}\Big)^a = x^a$$

Let

$$C = AB \tag{1.10.9}$$

Fig. 1.9 The logarithmic function. Published with permission of © Belal E. Baaquie 2020. All Rights Reserved

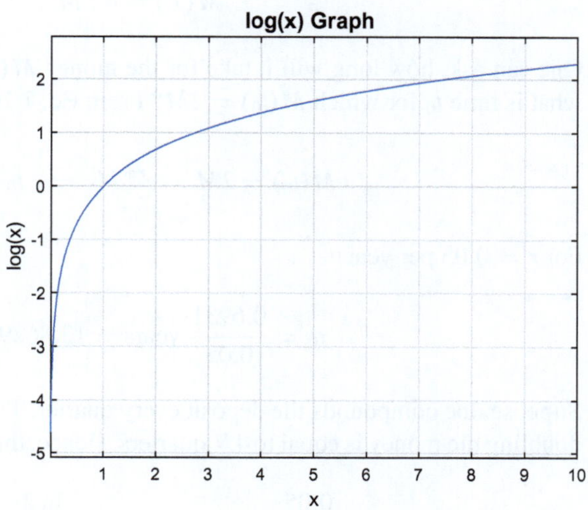

In terms of logarithms

$$C = e^{\ln C}; \quad A = e^{\ln A}; \quad B = e^{\ln B}$$

Hence, from Eq. 1.10.9

$$AB = C \Rightarrow e^{\ln A + \ln B} = e^{\ln C} \Rightarrow \ln(AB) = \ln A + \ln B$$

Note

$$\ln(\frac{x}{x}) = 0 = \ln(x) + \ln(\frac{1}{x}) \Rightarrow \ln(\frac{1}{x}) = -\ln(x)$$

Hence

$$\ln(\frac{x}{y}) = \ln(x) - \ln(y)$$

For $|x| < 1$, the series expansion yields

$$\ln(1+x) = x - \frac{x^2}{2} + \frac{x^3}{3} \cdots + (-1)^{(n-1)}\frac{x^n}{n} \cdots$$

and

$$\ln(1-x) = -x - \frac{x^2}{2} - \frac{x^3}{3} \cdots - \frac{x^n}{n} \cdots$$

1.10.3.1 Example

$$M(T) = e^{rT} M$$

One can ask, how long will it take for the money $M(T)$ to double. In other words, what is time t_0 for which $M(t_0) = 2M$? From Eq. 1.10.7, we have

$$M(t_0) = 2M = e^{rt_0} M \Rightarrow t_0 = \frac{\ln 2}{r}$$

For $r = 0.05$ per year

$$t_0 = \frac{0.6931}{0.05} \text{ years} = 13.8629 \text{ years}$$

Suppose one compounds the deposit every quarter. Then the number of years N for doubling the money is equal to $4N$ quarters. Hence, the discrete compounding yields

$$2 = (1 + \frac{0.05}{4})^{4N} \Rightarrow N = \frac{\ln 2}{4\ln(1.0125)} = 13.95\text{years}$$

Time doubling for the discrete case is longer than continuous compounding. We can use the expansion for logarithms and approximate

$$\ln(1 + 0.0125) \simeq 0.0125$$

Then

$$N \simeq \frac{\ln 2}{0.05} = 13.8620$$

Approximate value of logarithm is closer to the continuum value.

1.11 Supply and Demand Functions

Demand and supply are the basis of many ideas in microeconomics. There are many functions that can be used to model, and we choose the exponential function to model supply and demand. The utility function $\mathcal{U}$ describes the demand for an amount (quantity) q_i of a commodity to be consumed by a consumer.

Let the price of the commodity be p_i; then microeconomic theory states that that a consumers consumption is based on maximizing his/her utility, subject to the budget constraint. Hence

$$\text{maximize}_q\ (\mathcal{U}[q]) \ \text{ subject to } \sum_{i=1}^{N} p_i q_i = M : \text{ constraint } \Rightarrow \ q_i = q_i(p, M)$$

For a utility function that depends on only one commodity, the maximization is unnecessary since the quantity consumed is given by[4]

$$q = \frac{M}{p}$$

Note the price of a commodity is always positive $p_i > 0$; hence the price of the commodity can be represented by

$$p_i = p_{i0} e^{x_i} > 0$$

For a single commodity

$$p = p_0 e^x \ \Rightarrow \ q = \frac{M}{p_0} e^{-x} \tag{1.11.1}$$

[4]For more than one commodity, the maximization becomes a nontrivial problem; we will return to this later.

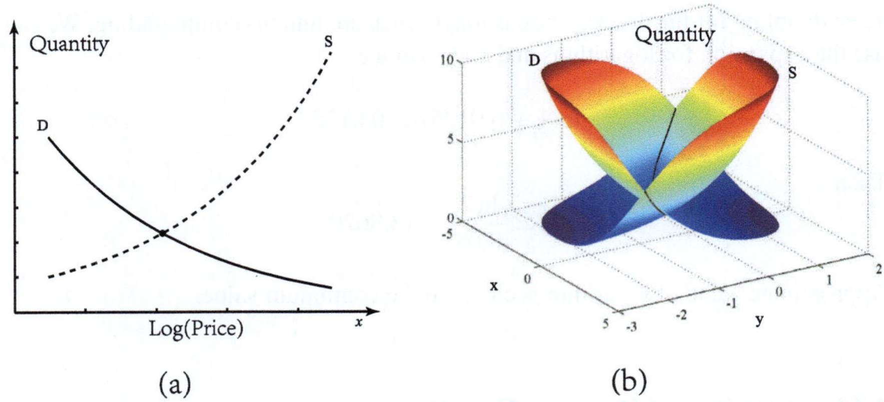

Fig. 1.10 The supply and demand of quantity versus logarithm of price, for **a** one and **b** two commodities, respectively. For large p_i, the demand price goes as $\mathcal{D}_i \simeq e^{-a_i x_i}$ and the supply prices goes as $\mathcal{S}_i \simeq e^{b_i x_i}$. Published with permission of © Belal E. Baaquie 2020. All Rights Reserved

Suppose the utility function is a dimensionless function given by (Fig. 1.10).

$$\mathcal{U}[q] = (\frac{q}{q_0})^a$$

where q_0 is some unit for the quantity q.

The demand function $\mathcal{D}[p]$ is a function of price p and is equal to the utility, once the maximization—subject to the budget constraint—has yielded quantity as a function of price. An example of this is given in Eq. 1.11.1. Hence, the demand function is given by

$$\mathcal{D}[p] = \mathcal{U}[q(p)] = \left(\frac{1}{q_0}\frac{M}{p_0}e^{-x}\right)^a = de^{-ax}; \;\; d = \left(\frac{M}{q_0 p_0}\right)^a$$

Suppose the supply function $\mathcal{S}[p]$ is given by

$$\mathcal{S}[p] = \tilde{s}p^b = se^{bx}; \;\; s = \tilde{s}p_0^b$$

In microeconomics, the market price $\tilde{p}$ is fixed by the requirement of supply being equal to demand, which yields

$$\mathcal{D}[\tilde{p}] = \mathcal{S}[\tilde{p}] \;\; \Rightarrow \;\; de^{-a\tilde{x}} = se^{b\tilde{x}}$$

Hence, the market price of a commodity is given by

$$e^{(a+b)\tilde{x}} = \frac{d}{s} \;\; \Rightarrow \;\; \tilde{p} = p_0 e^{\tilde{x}} = p_0(\frac{d}{s})^{1/(a+b)}$$

1.12 Option Theory Payoff

Option theory is discussed in Chaps. 9 and 11. In this Section, functions of the underlying security are analyzed using the payoff functions of options.

The buyer of a call option has the right, but not an obligation, to buy an underlying security S at a fixed future time T for a fixed strike price. Let $S(T)$ be the value of a security at a future time T when the option matures, and let K be a strike price, denoted by K. If the security $S(T)$ at future time T is less than K, the buyer does not exercise the option; whereas if $S(T)$ is bigger than K, the buyer exercises the option.

The payoff of the option, called the *vanilla option*, is given by

$$\mathcal{P}(\text{call}) \equiv g(S) = (S - K)_+ = \begin{cases} S - K \,;\, S > K \\ 0 \quad\quad\;\; ;\, S < K \end{cases} \tag{1.12.1}$$

A put option gives the buyer the right, but not an obligation, to sell the security at a pre-fixed strike price is the future. The payoff for the put option is given by

$$\mathcal{P}(\text{put}) \equiv h(S) = (K - S)_+ = \begin{cases} K - S \,;\, K > S \\ 0 \quad\quad\;\; ;\, K < S \end{cases} \tag{1.12.2}$$

The call and put option are shown in Fig. 1.11a, b, respectively.

The digital (binary) option has payoff $\mathcal{P}$ given by

$$\mathcal{P}(\text{digital}) = \begin{cases} 1 \,;\, S > K \\ 0 \,;\, S < K \end{cases} \tag{1.12.3}$$

The evolution of the security $S = S(t)$ is stochastic, with its value at payoff $S(T)$ being a random variable. For a path independent option, the option price today, at time t, denoted by $C(S, t, K, T, \cdots)$, is given by the expectation value of the payoff

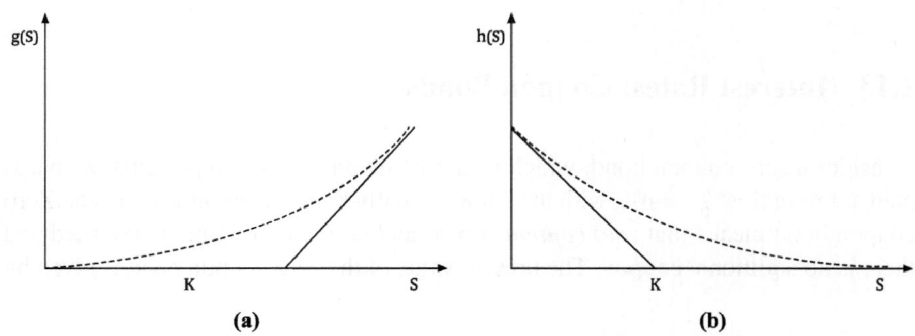

Fig. 1.11 **a** The call and **b** the put option. The dashed lines are the value of the call and put option before maturity. Published with permission of © Belal E. Baaquie 2020. All Rights Reserved

function, discounted to the present time. For spot interest rate r, the price of the option is given by

$$C(S, T-t, K, T) = e^{-r(T-t)} E\Big[\mathcal{P}(S(T), K, \cdots)\Big]$$

A *compound option* is one for which the payoff is defined in terms of another option on the same security. Consider a call option on a call option. Let the price of the call on call option be denoted by $C(S, T_2 - t, K_2)$, having a strike price K_2 and maturing at some future time $T_2 > t$. A compound option depends on two strike prices $k_1$and k_2, as well as on two times of exercise T_1, T_2.

The holder of the call on call option has the right to buy a *new call option* at time T_1 and at strike price K_1; the new call option has maturity T_2 and strike price K_2. The payoff function of the call on call option, at time T_1 and for security price $S(T_1)$, is given by

$$\begin{aligned}\mathcal{P}(\text{compound})(T_1) &= \Big[C(S(T_1), T_2 - T_1, K_2) - K_1\Big]_+ \\ &= \begin{cases} C(S(T_1), T_2 - T_1, K_2) - K_1 & ; S(T_1) > S_* \\ 0 & ; S(T_1) < S_* \end{cases}\end{aligned}$$

where S_* is defined by

$$C(S_*, T_2 - T_1, K_2) = K_1$$

The payoff function for a call on call option at time T_2 is given by

$$\mathcal{P}(\text{compound})(T_2) = \begin{cases} S(T_2) - K_2 & ; S(T_2) > K_2 \text{ and } S(T_1) > S_* \\ 0 & ; \text{otherwise} \end{cases}$$

There are four basic types of compound options: Call on Call, Call on Put, Put on Put and Put on Call. The compound option is a path dependent option, depending on the value of the security S at two times T_1, T_2.

1.13 Interest Rates; Coupon Bonds

Consider a zero coupon bond, which is an instrument that has a principal L that is paid at future time $T = N\ell$, with the tenor ℓ usually being six months or a year. Zero coupon bond means that *zero coupon* is paid and only the principal is returned and there is no additional coupon. The present value of the zero coupon bond is given by

$$B(t, T) = \frac{1}{(1 + r_N(t))^N} \; : \; \text{Zero Coupon (Discounting) Bond}$$

where $r_N(t)$ is discounting rate at time t for a payment made at time$T = N\ell$ in the future. In general, $r_n(t)$ is the rate of discounting, at time t, of the future payment at time T_n- and is called the zero coupon yield curve (ZCYC).

Consider a coupon bond, with principal L, paying fixed coupons c_n at times $T_n = n\ell$, and maturing at some time T in the future. The value of the coupon bond, at time t, is given by

$$\mathcal{B}(t) = \sum_{n=1}^{N} \frac{c_n}{(1 + r_n(t))^n} + \frac{L}{(1 + r_N(t))^N} = \sum_{n=1}^{N} c_n B(t, T_n) + L B(t, T)$$

A coupon bond is a portfolio of zero coupon bonds with different maturities, with coupons c_n being paid every time one of the zero coupon bond $B(t, T_n)$ matures.

To compare coupon bonds with different maturities, coupon payments and principal L, one finds a yield to maturity $y = y(t)$ for the coupon bond so that

$$\mathcal{B}(t) = \sum_{n=1}^{N} \frac{c_n}{(1 + y)^n} + \frac{L}{(1 + y)^N}$$

The yield to maturity of vastly different coupon bonds can be compared by comparing their yields to maturity y. The bond that has a larger y has a higher rate of discounting and in effect is paying a higher rate of interest compared to a bond with a lower y.

For a sukuk, one should not have fixed coupons as this would amount to an interest rate instrument.[5] Instead, one expects the coupons c_n to depend on time $c_n(t)$, and to be stochastic; the price of a sukuk is given by averaging over the fluctuations of the coupons and yields

$$\mathcal{S}(t) = \sum_{n=1}^{N} \frac{E[c_n(t)]}{(1 + r_n(t))^n} + \frac{L}{(1 + r_n(t))^N}$$

The discounting $r_n(t)$ is done using the *expected rate of return* for the firm issuing the sukuk.

Interest rate swaps are defined to be a contract in which one party pays the floating Libor $L(t, T_n)$ interest rate and receives interest payments at a fixed rate R. A swaplet has one cash flow, like a zero coupon bond; for the exchange of fixed versus floating rate at a future time T_n, its present value is given by

$$\text{Swaplet}_n(t) = \frac{1}{(1 + r_n(t))^n}(R - L(t, T_{n-1})) \lesseqgtr 0$$

[5] Sukuks based on payments coming from rental incomes have fixed coupons equal to the rental payments.

The value of a swap – consisting of N payments – is given by summing over individual swaplets and yields

$$\mathrm{Swap}(t) = \sum_{n=1}^{N} \frac{1}{(1 + r_n(t))^n} (R - L(t, T_{n-1})) \lesseqgtr 0$$

A swaption is an option on the interest swap; for the option maturing at future time T, the payoff function is given by

$$\mathcal{P} = \left[\mathrm{Swap}(T)\right]_+$$

1.14 Problems

1. Suppose that $f(x) = x^3$ and $g(x) = x^4 - 2$. Evaluate the following.
   $(f + g)(x)$
   $(fg)(x)$
   $(f/g)(x)$
   $(f \circ g)(x)$
2. Suppose you deposit your money in a bank on a continuous compounding basis. If the interest rate is 4%, how long will it take for the money

   (a) to double.
   (b) to triple.

3. Suppose you deposit your money in a bank on a discrete compounding basis that compounds semi-annually. If the interest rate is 4%, how long will it take for the money

   (a) to double.
   (b) to triple.

   Of the previous two cases (continuous vs. discrete compounding) which takes longer to double and triple? Explain why (mathematically).
4. What's the relationship between the frequency of compounding (i.e. annually, semi-annually, quarterly, etc.) and time? Explain why (mathematically).
5. Suppose you have Stock A, B and C, which are priced at \$10, \$20 and \$30, respectively. What is the needed capital if you want to construct a portfolio that consists of 2, 4 and 6 stocks of A, B and C, respectively? Can you construct a different portfolio of the three stocks with the exact capital amount?
6. Suppose you raised an additional \$600 and you want to invest the whole amount in one stock only (either Stock A, B or C). How many stocks can you purchase of each stock?
7. Calculate the monthly payment CF on a 30 year mortgage loan of \$300,000 at a mortgage interest rate of 4.7% per year.

8. A web development freelancer charges \$100 for an assessment of web problems plus \$50 an hour for each hour of work. Express the cost y of a freelancer as a function of the number of hours x the job involves.
9. A tennis racket producing company has a fixed cost of \$500,000 and a marginal cost of \$100 per racket.

   (a) Express the company's cost y as a function of the number of x rackets produced.
   (b) Estimate the cost for $x = 5,000$.

10. Write the payoff function, and draw a figure showing the function, for the following options.

    - Bear spread. Let $K_1 < K_2$. Buy a call option with strike price K_2 and sell a call option with strike price K_1.
    - Bull spread. Buy a call option with strike price K_1 and sell a call option with strike price K_2.
    - Straddle. Sell a call and put option with same maturity and same strike price.
    - Strangle. Let $K_1 < K_2$. Sell a put option with strike price K_1 and sell a call option with strike price K_2.

Part II
Linear Algebra

Chapter 2
Simultaneous Linear Equations

Abstract Linear algebra is introduced using simultaneous equations. Linear transformations are used to introduce the idea of vectors and matrices. The concept of a finite dimensional linear vector space is introduced for describing the properties of vectors. Basis vectors and rules for vector addition and scalar products are defined. Linear transformation are interpreted as the action of matrices on vectors and which yields the defining equations of how vectors are transformed under the action of matrices. The generalization of vectors to N-dimensions is made.

2.1 Introduction

Main topics to be covered in this Chapter are the following.

- Vectors
- Linear transformations
- N-dimensional simultaneous equations.

2.2 Two Commodities

Consider two commodities. The quantity of demand for first commodity is denoted by Q_{d1} and the quantity of supply for first commodity is denoted by Q_{s1}, and similarly by Q_{d2}, Q_{s2} for the second commodity. Suppose the supply and demand of both commodities depends linearly on their price, denoted by P_1, P_2; then

$$Q_{d1} = a_1 + b_1 P_1 + c_1 P_2$$
$$Q_{s1} = p_1 + q_1 P_1 + r_1 P_2$$

and

$$Q_{d2} = a_2 + b_2 P_1 + c_2 P_2$$
$$Q_{s2} = p_2 + q_2 P_1 + r_2 P_2$$

B. E. Baaquie, *Mathematical Methods and Quantum Mathematics for Economics and Finance*, https://doi.org/10.1007/978-981-15-6611-0_2

The general equilibrium model states that in equilibrium, the prices of the two commodities take values such that the quantity of demand and supply of the commodities are equal; hence

$$Q_{d1} = Q_{s1}; \quad Q_{d2} = Q_{s2}$$

and from equations above

$$(a_1 - p_1) + (b_1 - q_1)P_1 + (c_1 - r_1)P_2 = 0$$
$$(a_2 - p_2) + (b_2 - q_2)P_1 + (c_2 - r_2)P_2 = 0$$

Define the new parameters

$$a = b_1 - q_1; \quad c = c_1 - r_1; \quad b = a_2 - p_2; \quad d = c_2 - r_2$$

and

$$h = -(a_1 - p_1); \quad j = -(a_2 - p_2)$$

Then the equations for the equilibrium prices are given by

$$aP_1 + cP_2 = h \tag{2.2.1}$$
$$bP_1 + dP_2 = j \tag{2.2.2}$$

Equations 2.2.1 and 2.2.2 state that the price earned by the sellers of the two commodities is given by P_1, P_2. One needs to simultaneously solve both the equations above to obtain the equilibrium commodity prices. One can think of the simultaneous equations as the transformation of commodity prices P_1, P_2 into h, j.

To solve for the two simultaneous equations, multiply Eq. 2.2.1 by b and Eq. 2.2.2 by a; subtracting the two equations yields

$$(cb - ad)P_2 = bh - aj \;\Rightarrow\; P_2 = \frac{bh - aj}{cb - ad}$$

Substituting the value of P_2 in Eq. 2.2.1 yields

$$aP_1 + c\frac{bh - aj}{cb - ad} = h \;\Rightarrow\; P_1 = \frac{cj - hd}{cb - ad}$$

Consider the numerical example of

$$a = 1; b = 3; c = 3; d = 0; h = 3; j = 6$$

Then

$$P_1 = 2 \ ; \ P_2 = \frac{1}{3} \tag{2.2.3}$$

The equilibrium prices of N commodities is a direct generalization of Eqs. 2.2.1 and 2.2.2. Let the N commodity prices be denoted by $P_1, \ldots, P_N$. Generalizing the notation for N linear equations yields

$$\begin{aligned} a_{11}P_1 + \cdots + a_{1N}P_N &= h_1 \\ &\cdots \\ a_{i1}P_1 + \cdots + a_{iN}P_N &= h_i \\ &\cdots \\ a_{N1}P_1 + \cdots + a_{NN}P_N &= h_N \end{aligned} \tag{2.2.4}$$

Equation 2.2.4 is a system of N simultaneous equations and is the generalization of two simultaneous linear Eqs. 2.2.1 and 2.2.2. These N simultaneous linear equations have to be solved for obtaining the equilibrium prices $P_1, \ldots, P_N$ of N commodities. The simple process of substitution and elimination used for solving for two commodity prices P_1, P_2 has to be generalized to obtain commodity prices $P_1, \ldots, P_N$.

Vectors and linear transformations are mathematical concepts of great generality. To contextualize their study in economics and finance, vectors and linear transformations are analyzed from the point of view of solving the simultaneous equations required for evaluating equilibrium commodity prices. An N-dimensional vector is how the N commodity prices $P_1, \cdots, P_N$ are organized; the linear transformation is the transformation, given in Eq. 2.2.4, of the prices into the N-dimensional vector $h_1, \ldots, h_N$.

2.3 Vectors

A vector is a collection of numbers; an example of a 2-dimensional vector is the price of two commodities P_1, P_2. A vector can be geometrically thought of as an arrow in space. An arrow has both length and direction. An arrow is a straight line from its base to its tip. How does one mathematically represent the arrow?

For simplicity, consider a two dimensional space, denoted by E_2, and consider an xy-grid, with sides of unit length. Let $\mathbf{u}$ be a vector that can point in any direction, and is shown in Fig. 2.1. A vector, such as $\mathbf{u}$, will be denoted by a boldface letter to differentiate it from a single number.

Let the arrow's base be fixed at the origin and with its tip lying on one of the grid points; one needs to specify how to get from the base of the arrow to its tip. Since one can move in two directions, one way of doing this is to specify how many steps one needs to move in the x-direction, say 3 steps and how many directions one needs to move in the y direction, say -2 steps—the minus to indicate one is moving below the x-axis.

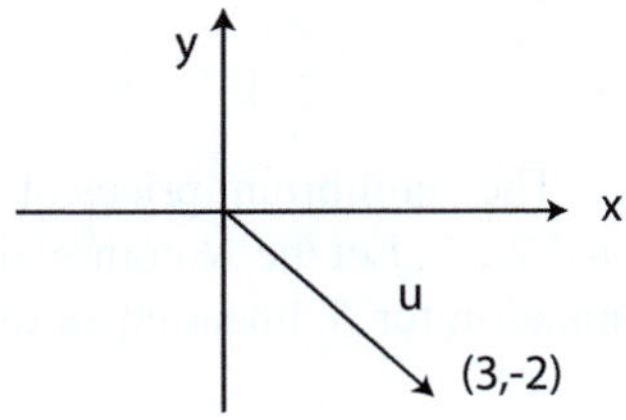

Fig. 2.1 Vector **u**. Published with permission of © Belal E. Baaquie 2020. All Rights Reserved

Since the directions x and y are different ('orthogonal'), the numbers representing the steps in the x and y directions cannot be written side by side. The notation for 'column vector', denoted by $\mathbf{u} \in E_2$, is given by

$$\mathbf{u} = \begin{bmatrix} 3 \\ -2 \end{bmatrix}$$

The entry 3 is called the x-component of the vector and -2 the y-component. An example of a vector is the solution for equilibrium prices given in Eq. 2.2.3 and yields the vector

$$\mathbf{P} = \begin{bmatrix} 2 \\ \frac{1}{3} \end{bmatrix}$$

Note there are four quadrants to the plane, shown in Fig. 2.2, and given by the following.

- Quadrant I: $x, y \geq 0$

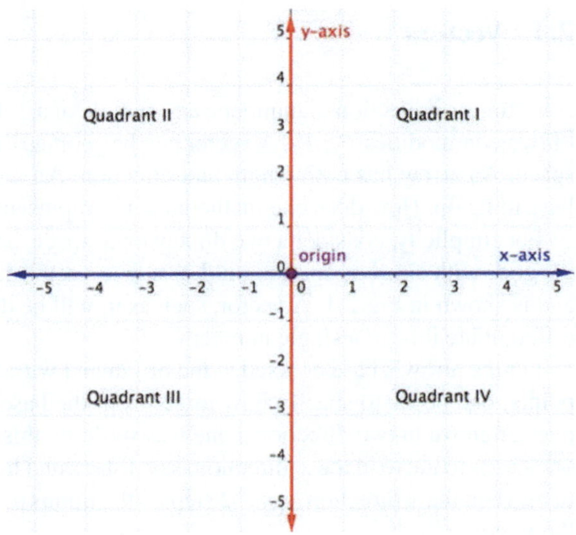

Fig. 2.2 The four quadrants of the plane. Published with permission of © Belal E. Baaquie 2020. All Rights Reserved

- Quadrant II: $x \leq 0;\ y \geq 0$
- Quadrant III: $x, y \leq 0$
- Quadrant IV: $x \geq 0;\ y \leq 0$.

The price vector $\mathbf{P}$ is special since both P_1, P_2 must be positive. Hence, a price vector can only point in the *first quadrant*.

A 'row vector' is called the transpose of $\mathbf{u}$ and denoted by $\mathbf{u}^T$. The transpose of $\mathbf{u}$ is written as

$$\mathbf{u}^T = \begin{bmatrix} 3 & -2 \end{bmatrix}$$

The transpose is written with brackets because 2 and 3 appearing inside the bracket is not a number, but an ordered pair (array).

More generally, a vector can have arbitrary components and is represented by

$$\mathbf{u} = \begin{bmatrix} a \\ b \end{bmatrix};\ \mathbf{u}^T = \begin{bmatrix} a & b \end{bmatrix}$$

For two commodities

$$\mathbf{P} = \begin{bmatrix} P_1 \\ P_2 \end{bmatrix};\ \mathbf{P}^T = \begin{bmatrix} P_1 & P_2 \end{bmatrix}$$

Mathematically speaking, a vector is an ordered set. A vector need not necessarily be a geometrical object and can consist of any two quantities. For example, a vector representing grain and its price can take the form

$$\mathbf{u} = \begin{bmatrix} a\ \text{kg} \\ b\ \text{Dollars} \end{bmatrix}$$

Consider the price vector $\mathbf{P}$. Suppose there are two markets for the two commodities that are widely separated; then one expects that the prices may be different in the two markets for reasons such as transportation, site of production and so on. Let the two price vectors be denoted by $\mathbf{P}$ and $\mathbf{P}'$. The average price is then

$$P_{1\text{Avg}} = \frac{1}{2}(P_1 + P_1');\ \ P_{2\text{Avg}} = \frac{1}{2}(P_2 + P_2')$$

In vector notation

$$\mathbf{P}_{\text{Avg}} = \frac{1}{2}(\mathbf{P} + \mathbf{P}') = \frac{1}{2}\mathbf{P} + \frac{1}{2}\mathbf{P}'$$

The example of the price vector shows that there two rudimentary and fundamental operations that one can perform with vectors, which are the following.

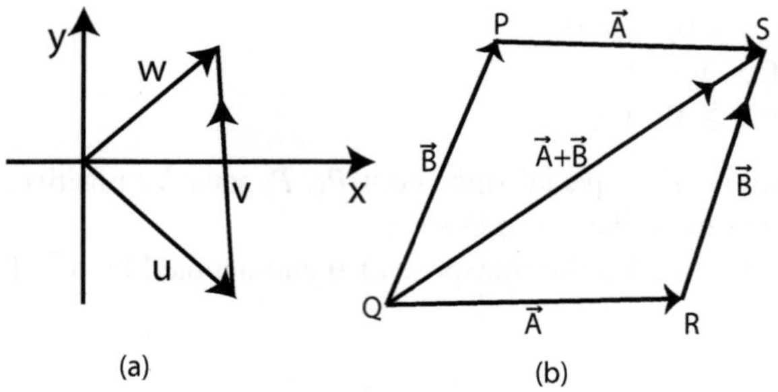

Fig. 2.3 Two ways of adding vectors. **a** Adding vectors **u** and **w** resulting in vector **v**, in which vector **v** does not start at the origin. **b** Adding vectors **A** and **B** resulting in vector **A+B**; all three vectors in one representation (PQR) start at the origin. Published with permission of © Belal E. Baaquie 2020. All Rights Reserved

- **Vector addition**: From two vectors **u**, **v**, one can form a third vector which is their vector sum and given by

$$\mathbf{w} = \mathbf{u} + \mathbf{v}$$

One can think of the vector **w** as the result of starting with vector **u**, with its base at the origin, and then placing the vector **v** on the tip of vector **u**; the resulting vector **w** starts at the origin, which recall is the origin of vector **u**, and ends at the tip of vector **v**. See Fig. 2.3a.

Figure 2.3b has an important identity encoded in it. Note that vector **A+B** can be constructed in two different ways, one by adding the vectors that lie along QPS, and another way by adding the vectors that lie along QRS. This is possible because the two vectors labeled by **A** (one along QR and another along PS) are *parallel* and so are the two vectors labeled by **B** (and lying along QP and another along RS). Parallel means that *all* the components of the two vectors that are parallel are equal.

Figure 2.3b diagram shows that vectors can be freely moved parallel to themselves anywhere in the plane, or in general, in any linear vector space. The origin has no significance for a linear vector space and is chosen for convinience, similar to choosing the orthogonal axis. Vectors can be freely moved around the plane; however, for convenience vectors will usually be considered to be starting at the origin.[1]

In component notation, let

$$\mathbf{u} = \begin{bmatrix} a \\ b \end{bmatrix}; \quad \mathbf{v} = \begin{bmatrix} c \\ d \end{bmatrix}$$

[1]In curved spaces, one cannot move around vectors and the concept of parallel transport is required, which depends on the path taken for moving a vector from one point to another. The change of the vector undergoing parallel transport depends on the curvature of the space.

Vector addition, which was described geometrically, is expressed by adding the coordinates in the x and y-directions separately and yields

$$\mathbf{w} = \mathbf{u} + \mathbf{v} = \begin{bmatrix} a + c \\ b + d \end{bmatrix} \tag{2.3.1}$$

Vector addition is associative in the sense that

$$(\mathbf{u} + \mathbf{v}) + \mathbf{w} = \mathbf{u} + (\mathbf{v} + \mathbf{w}) = \mathbf{u} + \mathbf{v} + \mathbf{w}$$

- **Scalar multiplication**: for a real or complex number λ, one has a new vector $\mathbf{w}$ given by

$$\mathbf{w} = \lambda \mathbf{v}$$

By scalar multiplication, the length of a given vector is changed. If $\lambda = 2$, the vector $\mathbf{w}$ has twice the length of $\mathbf{v}$ and points in the same direction and for $\lambda = -2$, the vector $\mathbf{w}$ has twice the length of $\mathbf{v}$ but points in the *opposite direction*.

Consider again the price vector **P**; one can be either be long on the sale of the two commodities or can be short. Going long means that one pays, and hence one is less than what one started off by an amount -**P**; going short means that one receives an amount **P**; hence multiplying by a negative number means making a purchase, or going long. If one buys and sells the same commodity, the result is zero expenditure and is expressed by $\mathbf{P} - \mathbf{P} = 0$.

In general, from the rule of scalar multiplication it follows that one can multiply by -1; hence there exists a *zero vector* $\varnothing$ such that

$$(-1)\mathbf{v} = -\mathbf{v} \;\Rightarrow\; \mathbf{v} + (-\mathbf{v}) = \varnothing = \begin{bmatrix} 0 \\ 0 \end{bmatrix}$$

The points of E_2 can each be uniquely associated with a vector from the origin to that point. The vectors of E_2 obey the rules of vector addition and multiplication by a scalar. Under these vector operations, the result are vectors that also are elements of E_2. Viewed in light of vector algebra, E_2 is a two dimensional linear *vector space*.

2.4 Basis Vectors

A typical price vector is given by

$$\mathbf{P} = \begin{bmatrix} 2 \\ \frac{1}{3} \end{bmatrix}$$

Using the rules for vector addition and scalar multiplication given in Eq. 2.3.1 there two underlying vectors given by

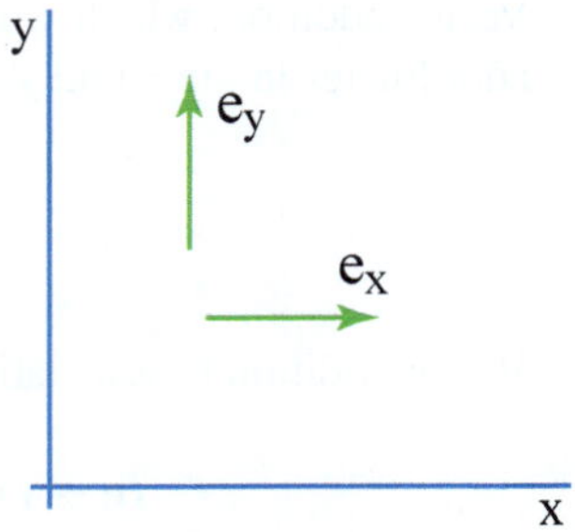

Fig. 2.4 Basis vectors for the two-dimensional state space. Published with permission of © Belal E. Baaquie 2020. All Rights Reserved

$$\mathbf{P} = \begin{bmatrix} 2 \\ \frac{1}{3} \end{bmatrix} = 2\begin{bmatrix} 1 \\ 0 \end{bmatrix} + \frac{1}{3}\begin{bmatrix} 0 \\ 1 \end{bmatrix}$$

The basis state vectors, as shown in Fig. 2.4, and their dual basis vectors have the following representation

$$\mathbf{e}_x = \begin{bmatrix} 1 \\ 0 \end{bmatrix};\ e_y = \begin{bmatrix} 0 \\ 1 \end{bmatrix} \qquad (2.4.1)$$
$$\mathbf{e}_x^T = \begin{bmatrix} 1\ 0 \end{bmatrix};\ \mathbf{e}_y^T = \begin{bmatrix} 0\ 1 \end{bmatrix}$$

Vectors $\mathbf{e}_x, \mathbf{e}_y$ are the basis vectors of the two-dimensional vector space. What is the significance of these basis vectors? Note the fact that if

$$a\mathbf{e}_x + b\mathbf{e}_y = 0 \ \Rightarrow \ a, b = 0$$

In other words, linear independence implies that

$$a = 0 = b$$

What this means is that $\mathbf{e}_x, \mathbf{e}_y$ are *linearly independent*—essentially they are independent of each other. Conversely, vectors $\mathbf{u}, \mathbf{v}$ are said to be *linearly dependent* if there are nonzero numbers a, b such that

$$a\mathbf{u} + b\mathbf{v} = 0;\ \ a, b \neq 0$$

In E_2 this equation is trivial but in higher dimensions it has more content.

2.4.1 Scalar Product

Given a price vector **P**, how can one recover the price of the individual commodities, given by P_1, P_2? For this one needs to utilize what is called the scalar product of two vectors. Since a vector has many components, the rule for the scalar product of two vector has to be defined.[2]

The scalar product or *inner product* of two vectors is defined in such a manner so that it is independent of the basis vectors used for representing the vectors. For vectors **u**, **v**, the definition is given by

$$\mathbf{u}^T \cdot \mathbf{v} = \begin{bmatrix} a \; b \end{bmatrix} \cdot \begin{bmatrix} c \\ d \end{bmatrix} = ac + bd$$

For the basis states

$$\mathbf{e}_x^T \cdot \mathbf{e}_x = \begin{bmatrix} 1 \; 0 \end{bmatrix} \cdot \begin{bmatrix} 1 \\ 0 \end{bmatrix} = 1; \; \mathbf{e}_y^T \cdot \mathbf{e}_y = \begin{bmatrix} 0 \; 1 \end{bmatrix} \cdot \begin{bmatrix} 0 \\ 1 \end{bmatrix} = 1 \tag{2.4.2}$$

Furthermore

$$\mathbf{e}_y^T \cdot \mathbf{e}_x = \begin{bmatrix} 0 \; 1 \end{bmatrix} \cdot \begin{bmatrix} 1 \\ 0 \end{bmatrix} = 0 : \text{ Orthogonal} \tag{2.4.3}$$

For the price vector, one has

$$\mathbf{e}_x^T \cdot \mathbf{P} = \begin{bmatrix} 1 \; 0 \end{bmatrix} \cdot \begin{bmatrix} P_1 \\ P_2 \end{bmatrix} = P_1; \; \mathbf{e}_y^T \cdot \mathbf{P} = \begin{bmatrix} 0 \; 1 \end{bmatrix} \cdot \begin{bmatrix} P_1 \\ P_2 \end{bmatrix} = P_2$$

The price vector can hence be written in terms of the basis vectors as follows

$$\mathbf{P} = P_1 \mathbf{e}_x + P_2 \mathbf{e}_y$$

The linear span of two vectors **P** and $\mathbf{P}'$ is defined by the vector sum $a\mathbf{P} + b\mathbf{P}'$; if **P** and $\mathbf{P}'$ are linearly independent, then by varying $a, b \in \mathbb{R}$ the linear span can be seen to cover all four first quadrants, even though $\mathbf{P}_1$ and $\mathbf{P}_2$ are ≥ 0.

A very important consequence of the linear independence is that *any* two dimensional vector **u** can be expressed as a vector sum of the basis vectors, with appropriate scalings. In other words, all vectors $u \in E_2$ have the representation

$$\mathbf{u} = x\mathbf{e}_x + y\mathbf{e}_y$$

By varying x, y through all the real numbers $\mathbb{R}$, every vector in E_2 can be reached. The *linear span* of the basis vectors $\mathbf{e}_x, \mathbf{e}_y$ covers all of E_2.

[2]The tensor product of two vectors is discussed later.

Consider vectors **u**, **v**, **w** and scalars a, b. The scalar product is a *linear operation* given by

$$\mathbf{u}^T \cdot (a\mathbf{v} + b\mathbf{w}) = a\mathbf{u}^T \cdot \mathbf{v} + b\mathbf{u}^T \cdot \mathbf{w} \tag{2.4.4}$$

The definition of a scalar product yields the length of a vector **u** by the following

$$|\mathbf{u}|^2 \equiv \mathbf{u}^T \cdot \mathbf{u} = \begin{bmatrix} a \ b \end{bmatrix} \cdot \begin{bmatrix} a \\ b \end{bmatrix} = a^2 + b^2 \ \Rightarrow \ |\mathbf{u}| = \sqrt{a^2 + b^2}$$

Note the length of the basis vectors, from Eq. 2.4.3 is 1, called normal, since

$$\mathbf{e}_x^T \cdot \mathbf{e}_x = 1 = \mathbf{e}_y^T \cdot \mathbf{e}_y$$

The basis states for this reason are called orthonormal, meaning basis states that are orthogonal and normal.

Any pair of vectors can be brought to start from the origin; two vectors of any higher dimension span a plane, as shown in Fig. 2.5. Since vectors are straight lines in a plane (spanned by them) one may ask: what is the angle between two vectors? Let the angle between the two vectors by θ, as shown in Fig. 2.5.

The scalar product provides the following definition

$$\mathbf{u}^T \cdot \mathbf{v} = |\mathbf{u}||\mathbf{v}| \cos(\theta)$$

From the Fig. 2.5 one can interpret the scalar product as the projection of one of the vectors, in the Fig. 2.5 vector **u** in the direction of vector **v**; the length of the projection is $|u| \cos(\theta)$.

Hence, we see that the basis vectors are orthogonal, with $\theta = \pi/2$; orthogonality of two vectors is a sufficient (but not necessary) condition for them to be linearly independent.

In components, using the linearity of the scalar product as given in Eq. 2.4.4 and the orthogonormality of the basis vectors given in Eqs. 2.4.2 and 2.4.3 yields

$$\mathbf{u}^T \cdot \mathbf{v} = (a\mathbf{e}_x^T + b\mathbf{e}_y^T) \cdot (c\mathbf{e}_x + d\mathbf{e}_y) = ac\mathbf{e}_x^T \cdot \mathbf{e}_x + bd\mathbf{e}_y^T \cdot \mathbf{e}_y = ac + bd$$

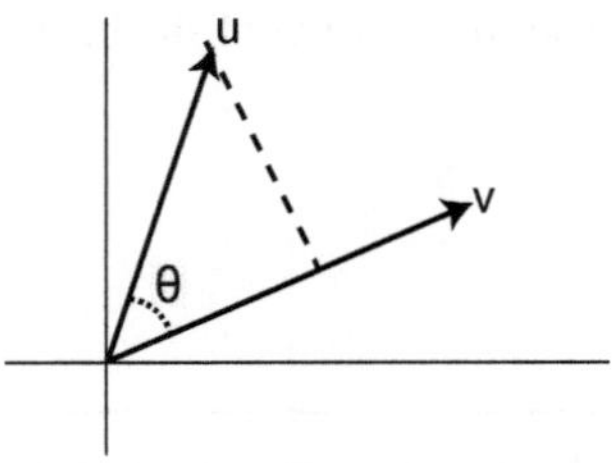

Fig. 2.5 Angle θ between two vectors **u**, **v**. Published with permission of © Belal E. Baaquie 2020. All Rights Reserved

2.5 Linear Transformations; Matrices

Recall from Eqs. 2.2.1 and 2.2.2 that

$$aP_1 + cP_2 = h$$
$$bP_1 + dP_2 = j$$

The right hand side of Eqs. 2.2.1 and 2.2.2 is also a vector given by

$$\mathbf{J} = \begin{bmatrix} h \\ j \end{bmatrix}$$

The defining equations are written in terms of P_1, P_2, which are the components of the price vector **P**. How can this set of two equations be written directly in terms of the vector **P**?

Functions are a mapping from a variable x to another variable $f(x)$. Similar to a function, Eqs. 2.2.1 and 2.2.2 yield a mapping of vector **P** to vector **J**, given by

$$\mathcal{M}: \ \mathcal{M}[\mathbf{P}] = \mathbf{J}$$

How can we write the mapping $\mathcal{M}$ of one vector to another? There are many ways of mapping vectors, and the simplest mapping between two vectors is what is called a *linear transformation*, of which Eqs. 2.2.1 and 2.2.2 is an example.

Consider a linear transformation L of a vector into another vector. The term linear is used as L acts only on the vector **v** and not its nonlinear functions that may, for example, depend on $\mathbf{v}^T \cdot \mathbf{v}$ and so on.

The action of L on **u** is to transform the vector into another vector **v**, as shown in Fig. 2.6.

$$L(\mathbf{u}) = \mathbf{v}$$

As shown in Fig. 2.6a, one can rotate a vector as well as change its length, a shown in Fig. 2.6b.

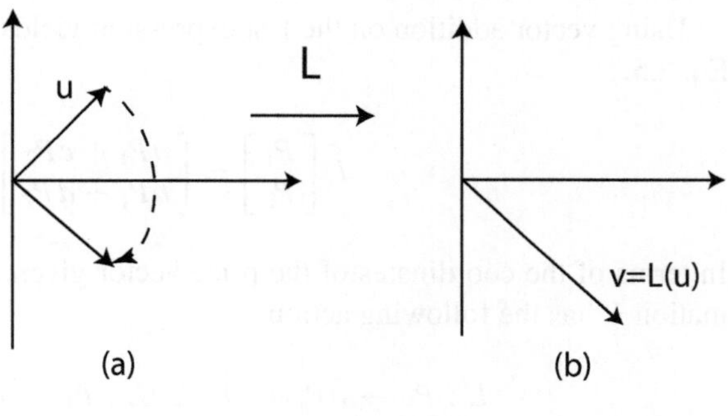

Fig. 2.6 Linear transformation on vector u resulting in vector v. Published with permission of © Belal E. Baaquie 2020. All Rights Reserved

For vectors **u**,**v** and scalars a, b, let $\mathbf{w} = a\mathbf{u} + b\mathbf{v}$. The linear transformation L has the following *defining* property

$$L(\mathbf{w}) = L(a\mathbf{u} + b\mathbf{v}) = aL(\mathbf{u}) + bL(\mathbf{v}) \tag{2.5.1}$$

Writing Eqs. 2.2.1 and 2.2.2 as a vector equation yields

$$\begin{bmatrix} aP_1 + cP_2 \\ bP_1 + dP_2 \end{bmatrix} = \begin{bmatrix} h \\ j \end{bmatrix} = P_1 \begin{bmatrix} a \\ b \end{bmatrix} + P_2 \begin{bmatrix} c \\ d \end{bmatrix} \tag{2.5.2}$$

Consider the two dimensional price vector

$$\mathbf{P} = P_1\mathbf{e}_x + P_2\mathbf{e}_y = P_1 \begin{bmatrix} 1 \\ 0 \end{bmatrix} + P_2 \begin{bmatrix} 0 \\ 1 \end{bmatrix} \tag{2.5.3}$$

and hence

$$L(\mathbf{P}) = \mathbf{J} = P_1 \begin{bmatrix} a \\ b \end{bmatrix} + P_2 \begin{bmatrix} c \\ d \end{bmatrix} \tag{2.5.4}$$

Acting directly on the vector **P** using the linearity of L given in Eq. 2.5.1 yields, from Eq. 2.5.4

$$L(\mathbf{P}) = L(P_1\mathbf{e}_x + P_2\mathbf{e}_y) = P_1L(\mathbf{e}_x) + P_2L(\mathbf{e}_y) = P_1 \begin{bmatrix} a \\ b \end{bmatrix} + P_2 \begin{bmatrix} c \\ d \end{bmatrix} \tag{2.5.5}$$

The *defining equation* for L is its action on the basis states; Eqs. 2.5.5 and 2.5.3 yield the following

$$L(\mathbf{e}_x) = L \begin{bmatrix} 1 \\ 0 \end{bmatrix} = \begin{bmatrix} a \\ b \end{bmatrix}; \quad L(\mathbf{e}_y) = L \begin{bmatrix} 0 \\ 1 \end{bmatrix} = \begin{bmatrix} c \\ d \end{bmatrix} \tag{2.5.6}$$

The action of a linear transformation L on every vector is uniquely specified by defining how the transformation acts on the basis vectors.

Using vector addition on the last expression yields the expected answer given in Eq. 2.5.2

$$L \begin{bmatrix} P_1 \\ P_2 \end{bmatrix} = \begin{bmatrix} aP_1 + cP_2 \\ bP_1 + dP_2 \end{bmatrix} \tag{2.5.7}$$

In terms of the coordinates of the price vector given by P_1, P_2, the linear transformation L has the following action

$$L : P_1 \to aP_1 + cP_2 \ ; \quad L : P_2 \to bP_1 + dP_2 \tag{2.5.8}$$

Since a complete knowledge of L has been derived, can L be directly be specified without having to refer to its actions on any specific vector $\mathbf{P}$? The coefficients a, b, c, d need to be organized into a single entity that defines L so that the action of L on any vector can be produced. L is clearly not a vector since it has more coefficients than are allowed for a two dimensional vector. A *new* mathematical object, called a *matrix*, is defined by organizing all the four coefficients into one structure

$$L = \begin{bmatrix} a & c \\ b & d \end{bmatrix}$$

Note the columns are precisely the mapping of the basis vectors.

From Eq. 2.5.7 the action of the matrix on a vector is defined by the following

$$L\begin{bmatrix} P_1 \\ P_2 \end{bmatrix} = \begin{bmatrix} a & c \\ b & d \end{bmatrix}\begin{bmatrix} P_1 \\ P_2 \end{bmatrix} = \begin{bmatrix} aP_1 + cP_2 \\ bP_1 + dP_2 \end{bmatrix} \tag{2.5.9}$$

Equation 2.5.9 encodes the generic rule of matrix multiplication, which is that the row from the left multiplies the column on the right. The result was obtained ealier for the scalar product of two vectors but this rule is true in general, and is discussed in the (next) Chap. 3 on matrices.

Hence, in matrix notation, Eqs. 2.2.1 and 2.2.2 are re-written as

$$L[\mathbf{P}] = \mathbf{J} \ \Rightarrow \ \begin{bmatrix} a & c \\ b & d \end{bmatrix}\begin{bmatrix} P_1 \\ P_2 \end{bmatrix} = \begin{bmatrix} h \\ j \end{bmatrix}$$

2.6 E_N: N-Dimensional Linear Vector Space

The discussion for two dimensional vector space E_2 generalizes to N-dimensions. Let $\mathbf{e}_i,\ i = 1, 2 \ldots, N$ be N orthogonal basis states with unit norm. Then

$$\mathbf{e}_i^T \cdot \mathbf{e}_j = 0 : \ i \neq j; \ \ \mathbf{e}_i^T \cdot \mathbf{e}_i = 1 \tag{2.6.1}$$

A realization of the basis vectors is given by

$$\mathbf{e}_n = \begin{bmatrix} 0 \\ \vdots \\ 1 \\ 0 \\ \vdots \\ 0 \end{bmatrix} \ ; \ \mathbf{e}_n^T = \underbrace{\begin{bmatrix} 0 \cdots 1 \, 0 \cdots \end{bmatrix}}_{n\text{-th position}} \tag{2.6.2}$$

A compact way of writing Eq. 2.6.1 is the following

$$\mathbf{e}_m^T \cdot \mathbf{e}_n = \delta_{n-m} \equiv \begin{cases} 1 & n = m \\ 0 & n \neq m \end{cases} : m, n = 1, \ldots, N \tag{2.6.3}$$

Note δ_{n-m} is the Kronecker delta function and has only integer arguments.

The basis states are linearly independent since

$$\sum_{i=1}^{N} a_i \mathbf{e}_i = 0 \ \Rightarrow \ a_i = 0 \text{ for all } i$$

The linear span of $\mathbf{e}_i$ is all linear combinations of the basis vectors and covers the entire space E_N. To represent an arbitrary vector $\mathbf{x}$ in E_N, the following notation is used

$$\mathbf{x} = \begin{bmatrix} x(1) \\ \vdots \\ x(i) \\ \vdots \\ x(N) \end{bmatrix} ; \ \mathbf{e}_i^T \cdot \mathbf{x} = x(i) \tag{2.6.4}$$

In terms of the basis vectors, $\mathbf{x}$ can be represented by

$$\mathbf{x} = \sum_{i=1}^{N} x(i) \mathbf{e}_i$$

The scalar product for vectors x, y, using Eq. 2.6.1, is given by

$$\mathbf{x}^T \cdot \mathbf{y} = \sum_{i,j=1}^{N} x(i) y(j) (\mathbf{e}_i^T \cdot \mathbf{e}_j) = \sum_{i,j=1}^{N} x(i) y(j) \delta_{i-j} = \sum_{i=1}^{N} x(i) y(i)$$

The angle θ between two vectors $\mathbf{x}, \mathbf{y}$ in the N-dimensional vector space is defined, as one would expect, by the following

$$\mathbf{x}^T \cdot \mathbf{y} = |\mathbf{x}||\mathbf{y}| \cos(\theta); \ |\mathbf{x}| = \sqrt{\sum_i x(i)^2}; \ |\mathbf{y}| = \sqrt{\sum_i y(i)^2}$$

2.7 Linear Transformations of E_N

The results for the two dimensional vector space are generalized to the N-dimensional case. From Eq. 2.5.1 for any arbitrary vector the action of the linear transformation is given by

$$L(\mathbf{x}) = L\left(\sum_{i=1}^{N} x(i)\mathbf{e}_i\right) = \sum_{i=1}^{N} x(i)L(\mathbf{e}_i)$$

Hence, as mentioned earlier, it is sufficient to know $L(\mathbf{e}_i)$ to fully define L. The effect of L on the basis vectors can be expressed as linear combinations of the basis vectors themselves; recall the two dimensional case, as given in Eq. 2.5.6, can be written as the following

$$L(\mathbf{e}_x) = a\mathbf{e}_x + b\mathbf{e}_y; \quad L(\mathbf{e}_y) = c\mathbf{e}_x + d\mathbf{e}_y \tag{2.7.1}$$

Similarly, the action of L on the basis vectors of E_N yields the following

$$L(\mathbf{e}_i) = \sum_{j=1}^{N} A_{ji}\mathbf{e}_j = \sum_{j=1}^{N} A_{ij}^T\mathbf{e}_j \;\Rightarrow\; A_{ij}^T = A_{ji}$$

Note the ordering of the indexes for A_{ji}. Generalizing the result given in Eq. 2.5.5 yields the following

$$\begin{aligned} L(\mathbf{x}) &= L(\sum_i x(i)\mathbf{e}_i) = \sum_i x(i)L(\mathbf{e}_i) \\ &= \sum_{ij} x(i)A_{ji}\mathbf{e}_j = \sum_j \Big[\sum_i x(i)A_{ji}\Big]\mathbf{e}_j \\ \Rightarrow\; L(\mathbf{x}) &\equiv \sum_j y(j)\mathbf{e}_j \end{aligned} \tag{2.7.2}$$

where the generalization of Eq. 2.5.8 is given by

$$y(j) = \sum_i A_{ji}x(i)$$

Let the price vector for N commodities be given by

$$\mathbf{P} = \begin{bmatrix} P_1 \\ \vdots \\ P_i \\ \vdots \\ P_N \end{bmatrix}; \; \mathbf{J} = \begin{bmatrix} J_1 \\ \vdots \\ J_i \\ \vdots \\ J_N \end{bmatrix} \tag{2.7.3}$$

and let a linear transformation be given by

$$L = \begin{bmatrix} a_{11} & a_{12} & .. & a_{1N} \\ a_{21} & a_{22} & .. & a_{2N} \\ .... & & & \\ a_{i1} & a_{i2} & .. & a_{iN} \\ .... & & & \\ a_{N1} & a_{N2} & .. & a_{NN} \end{bmatrix}$$

The linear equation given in Eq. 2.2.4 can be written in matrix notation in the following manner.

$$L[\mathbf{P}] = \mathbf{J}$$

In terms of the components, the linear transformation and the vectors are given by the following equation

$$\begin{bmatrix} a_{11} & a_{12} & .. & a_{1N} \\ a_{21} & a_{22} & .. & a_{2N} \\ .... & & & \\ a_{i1} & a_{i2} & .. & a_{iN} \\ .... & & & \\ a_{N1} & a_{N2} & .. & a_{NN} \end{bmatrix} \begin{bmatrix} P_1 \\ \vdots \\ P_i \\ \vdots \\ P_N \end{bmatrix} = \begin{bmatrix} J_1 \\ \vdots \\ J_i \\ \vdots \\ J_N \end{bmatrix} \tag{2.7.4}$$

To solve for N unknown prices given by $\mathbf{P}$, the linear equation in Eq. 2.7.4 has N equations and is sufficient to yield a unique solution. Hence, in principle, similar to the process of substitution and elimination used for evaluating the prices P_1, P_2 in Eq. 2.2.3, one can obtain the solution for $\mathbf{P}$. However, as one can imagine, the algebra for N unknown prices will become increasingly difficult to handle.

A far more efficient method for solving for the unknown prices is the following

$$\mathbf{P} = L^{-1}\mathbf{J}$$

To implement the above solution, one needs to define L^{-1}, and this is the subject of linear algebra which is discussed in the next few chapters.

2.8 Problems

1. The macroeconomic equations for GDP Y and consumption C are

$$Y = C + I + G; \quad C = a + bY$$

where G is Government expenditure and I is the net yearly investment.
(a) Write these equations using vectors and matrix for the unknown Y and C.
(b) Solve the equations for Y and C.

2. A retailer sells 700 USB's, 400 bulbs, and 200 clocks players each week. The selling price of USB is \$4, Bulb \$6, and clocks is \$150. The cost to the shop is \$3.25 for a USB, \$4.75 for a Bulb, and \$125 for a clock. Arrange the goods and prices into a vector. Find the total weekly profit.
3. Consider vectors

$$\mathbf{u} = \begin{bmatrix} 3 \\ -2 \end{bmatrix}; \quad \mathbf{w} = \begin{bmatrix} 1 \\ 0 \end{bmatrix}$$

- Form the scalar product $\mathbf{u}^T \cdot \mathbf{w}$.
- What is the angle between $\mathbf{u}^T$ and $\mathbf{w}$?
- Are the two vectors linearly independent?
- Draw the vectors in E_2.

4. Consider the vector and matrix

$$\mathbf{u} = \begin{bmatrix} 3 \\ -2 \\ 5 \end{bmatrix}; \quad M = \begin{bmatrix} 2\,3\,5 \\ 3\,1\,2 \\ 1\,4\,3 \end{bmatrix}$$

- Evaluate $\mathbf{u}^T M$, $M\mathbf{u}$ and $\mathbf{u}^T M\mathbf{u}$.
- Evaluate M^2.

5. Consider the matrices

$$A = \begin{bmatrix} -9 & 2/3 \\ -4 & 5 \end{bmatrix}; \quad B = \begin{bmatrix} -2 & 7 \\ 3 & -2 \end{bmatrix}$$

- Find the matrix C where

$$C = -6A + 5B$$

- Find AB and BA.
- Show that $(AB)C = A(BC)$.

6. Consider matrices

$$M = \begin{bmatrix} 2\,3\,5 \\ 3\,1\,2 \\ 1\,4\,3 \end{bmatrix}; \quad N = \begin{bmatrix} 2\,3 \\ 3\,1 \\ 1\,4 \end{bmatrix}; \quad L = \begin{bmatrix} 2\,3\,5 \\ 3\,1\,2 \end{bmatrix}$$

- Evaluate MN and LM.

7. Find the linear transformation L that transforms

$$\mathbf{u} = \begin{bmatrix} 1 \\ 0 \end{bmatrix}; \quad \mathbf{w} = \begin{bmatrix} 0 \\ 1 \end{bmatrix}$$

into

$$\mathbf{u}' = \begin{bmatrix} 3 \\ -2 \end{bmatrix}; \ \mathbf{w}' = \begin{bmatrix} 0 \\ 2 \end{bmatrix}$$

8. Find the result of applying the linear transformation L obtained in Problem 5 to

$$\mathbf{u}' = \begin{bmatrix} 3 \\ -2 \end{bmatrix}; \ \mathbf{w}' = \begin{bmatrix} 0 \\ 2 \end{bmatrix}$$

Optional: What is the linear span of the vectors generated by repeatedly applying L.

Chapter 3
Matrices

Abstract Matrices are analyzed from first principles. The rules of matrix addition and multiplication are shown to follow from the definition of the transformation of a vectors by a matrix. The properties of 2×2 matrices are used as an exemplars for the general features of matrices. The properties of determinants are discussed, and an explanation based on geometry is given for the significance of a determinant being zero. Central ideas of tenor (outer) product, eigenvalues and eigenvectors are discussed. A derivation is given of the spectral decomposition, using 2×2 matrices as an example.

3.1 Introduction

Recall in Chap. 2, the concept of the matrix was introduced to represent linear transformations on vectors. In particular, for transforming two dimensional price vectors, the matrix was given by

$$L = \begin{bmatrix} a & c \\ b & d \end{bmatrix}$$

Consider the following example for transforming the price vector for four commodities. A company sells four goods to two customers for 3 days; the customers buy the *same* fixed quantity of commodities every day. Let price matrix be P and the quantity sold be Q. For three days, the daily prices are given in rows of P and the daily quantities sold are given in columns of Q.

P, Q have the following representation

$$P = \underbrace{\begin{bmatrix} 1 & 2 & 3 & 7 \\ 4 & 6 & 5 & 8 \\ 9 & 10 & 11 & 12 \end{bmatrix}}_{\text{commodities}} \quad Q = \underbrace{\begin{bmatrix} 20 & 35 \\ 30 & 45 \\ 40 & 55 \\ 50 & 65 \end{bmatrix}}_{\text{customers}} \quad \Rightarrow \quad P \cdot Q = \underbrace{\begin{bmatrix} a_{11} & a_{12} \\ a_{21} & a_{22} \\ a_{31} & a_{32} \end{bmatrix}}_{\text{customers}}$$

B. E. Baaquie, *Mathematical Methods and Quantum Mathematics for Economics and Finance*, https://doi.org/10.1007/978-981-15-6611-0_3

One would like to know how much was paid by the two customers for each of the three days; the first customer pays the amounts of a_{11}, a_{21}, a_{31} and second customer pays the amounts of a_{12}, a_{22}, a_{32}. To compute the revenues paid by the respective customers, one needs the rules for matrix multiplication, denoted by $\cdot$, and is discussed in Sect. 3.2.

Consider the most general case that a trader sells N commodities and daily prices of these commodities are recorded; denote the price of commodity n and on day m by p_{mn}, and which is organized in a $M \times N$ matrix denoted by P. Let the quantity q_{nl} of commodity n be sold to the lth customer for M days, and which is organized in a $N \times L$ matrix denoted by Q. There are L total customers. After M days the trader would obtain the following lists.

$$P = \underbrace{\begin{bmatrix} p_{11} & p_{12} & \cdots & p_{1N} \\ p_{21} & p_{22} & \cdots & p_{2N} \\ \cdots & & & \\ p_{M1} & p_{M2} & \cdots & p_{MN} \end{bmatrix}_{MN}}_{\text{commodities}} \; ; \; Q = \underbrace{\begin{bmatrix} q_{11} & q_{12} & \cdots & q_{1L} \\ q_{21} & q_{22} & \cdots & q_{2N} \\ \cdots & & & \\ q_{N1} & q_{N2} & \cdots & q_{NL} \end{bmatrix}_{NL}}_{\text{customers}}$$

To foreground the following discussion, matrix multiplying P and Q will yield a $M \times L$ matrix of revenues accruing to the seller from each of the customer and given by

$$P_{MN} \cdot Q_{NL} = [PQ]_{ML}$$

Matrices are mathematical objects that organize a list such as daily commodity prices, and can be used to then analyze these lists of numbers. In general, a matrix is an array that consists of M-rows, that are arranged horizontally, and N columns, arranged vertically. A typical $M \times N$ matrix A is given below.

$$A = \begin{bmatrix} a_{11} & a_{12} & \cdots & a_{1N} \\ a_{21} & a_{22} & \cdots & a_{2N} \\ \cdots & & & \\ a_{M1} & a_{M2} & \cdots & a_{MN} \end{bmatrix}$$

The individual entries of the matrix, denoted by a_{mn}, are the elements of the matrix. The elements can be real or complex numbers. For matrix A, there are in total MN elements.

A particular row for matrix A consists of N elements; for example, the second row is given by

$$\begin{bmatrix} a_{21} & a_{22} & \cdots & a_{2N} \end{bmatrix}$$

One can think of a row as the transpose of a vector. Similarly, each column is given by M elements and can be thought of as vector of length M. The second column for matrix A is given by

$$\begin{bmatrix} a_{12} \\ a_{22} \\ \cdots\cdot \\ a_{M2} \end{bmatrix}$$

A matrix that is $1 \times M$, and denoted by the column vector $\mathbf{v}$, is given by

$$\mathbf{v} = \begin{bmatrix} v(1) \\ v(2) \\ \cdots\cdot \\ v(M) \end{bmatrix}$$

One can see that $\mathbf{v}$ is equivalent to a M-dimensional vector and all vectors, in fact, are special cases of matrices.

The following are some of the properties of matrices. Let A, B be $M \times N$ matrices. Then

- $$(A + B)\mathbf{v} = A\mathbf{v} + B\mathbf{v}$$

  In components

  $$[(A + B)\mathbf{v}]_i = \sum_{j=1}^{N}(A + B)_{ij}v(j) = \sum_{j=1}^{N} A_{ij}v(j) + \sum_{j=1}^{N} B_{ij}v(j)$$

  Since $\mathbf{v}$ is arbitrary, matrix addition is done element by element

  $$(A + B) = A + B \;\Rightarrow\; (A + B)_{ij} = A_{ij} + B_{ij}$$

  One has the null matrix

  $$(A - A) = A - A = 0 = \begin{bmatrix} 0 & 0 & .. & 0 \\ 0 & 0 & .. & 0 \\ & \ddots & & \\ 0 & 0 & .. & 0 \end{bmatrix}$$

- The result of multiplying A by a scalar λ results in λA, which in components is given by

  $$(\lambda A)_{ij} = \lambda A_{ij}$$

3.2 Matrix Multiplication

The multiplication of two matrices, if one views them in isolation, seem quite arbitrary and counter-intuitive. To see how the multiplication rules arise very naturally, the two dimensional case is carried out in full detail. Unlike most books—where the multiplication rules are ad hoc and just postulated—these rules, it will be seen, are a necessary consequence of the action of the matrices on the vectors of the underlying vector space.

From Eq. 2.5.9, the matrix effects the following transformation on a vector

$$L\begin{bmatrix} x \\ y \end{bmatrix} = \begin{bmatrix} a & c \\ b & d \end{bmatrix}\begin{bmatrix} x \\ y \end{bmatrix} = \begin{bmatrix} ax + cy \\ bx + dy \end{bmatrix}$$

The transformation of a vector by a matrix fixes the rule for the multiplication of two matrices.

Consider the following two dimensional matrices given by

$$L' = \begin{bmatrix} a' & c' \\ b' & d' \end{bmatrix}; \quad L = \begin{bmatrix} a & c \\ b & d \end{bmatrix}$$

We want to obtain matrix M, which is the result of the multiplication of these matrices, and given by

$$M = L'L = \begin{bmatrix} a' & c' \\ b' & d' \end{bmatrix} \times \begin{bmatrix} a & c \\ b & d \end{bmatrix} \tag{3.2.1}$$

The symbol $\times$ has been put in to indicate that we need to define how these matrices are multiplied. Usually the symbol $\times$ is assumed and not written explicitly.

The definition of the action of linear transformations given in Eq. 2.5.7 yields

$$L'L\begin{bmatrix} x \\ y \end{bmatrix} = L' \times \begin{bmatrix} a & c \\ b & d \end{bmatrix}\begin{bmatrix} x \\ y \end{bmatrix} = \begin{bmatrix} a' & c' \\ b' & d' \end{bmatrix}\begin{bmatrix} ax + cy \\ bx + dy \end{bmatrix} \tag{3.2.2}$$

Resolving the vector in Eq. 3.2.2 into its basis states yields

$$L'L\begin{bmatrix} x \\ y \end{bmatrix} = \begin{bmatrix} a' & c' \\ b' & d' \end{bmatrix}\left\{(ax + cy)\begin{bmatrix} 1 \\ 0 \end{bmatrix} + (bx + dy)\begin{bmatrix} 0 \\ 1 \end{bmatrix}\right\} \tag{3.2.3}$$

Using the result given in Eq. 2.5.6, the definition of the linear transformation yields, from Eq. 3.2.3

$$
\begin{aligned}
L'L\begin{bmatrix} x \\ y \end{bmatrix} &= (ax+cy)\begin{bmatrix} a' \\ b' \end{bmatrix} + (bx+dy)\begin{bmatrix} c' \\ d' \end{bmatrix} \\
&= \begin{bmatrix} a'(ax+cy)+c'(bx+dy) \\ b'(ax+cy)+d'(bx+dy) \end{bmatrix} \\
&= \begin{bmatrix} (a'a+c'b)x+(a'c+c'd)y \\ (b'a+d'b)x+(b'c+d'd)y \end{bmatrix}
\end{aligned}
\tag{3.2.4}
$$

Using the result given in Eq. 2.5.9, the result obtained above in Eq. 3.2.4 can be written as

$$
L'L\begin{bmatrix} x \\ y \end{bmatrix} = \begin{bmatrix} a'a+c'b & a'c+c'd \\ b'a+d'b & b'c+d'd \end{bmatrix}\begin{bmatrix} x \\ y \end{bmatrix}
\tag{3.2.5}
$$

Hence, from Eqs. 3.2.1 and 3.2.5, the rule for matrix multiplication is given by

$$
M = L'L = \begin{bmatrix} a' & c' \\ b' & d' \end{bmatrix}\begin{bmatrix} a & c \\ b & d \end{bmatrix} = \begin{bmatrix} a'a+c'b & a'c+c'd \\ b'a+d'b & b'c+d'd \end{bmatrix}
\tag{3.2.6}
$$

The rule for matrix multiplication is that each row on the left, one by one multiplies, each column on the right, one by one. This is the same rule that was obtained earlier in multiplying a matrix into a vector. The result depends on the dimensions of the matrices. The rules obtained for the 2×2 matrices generalize to the case of multiplying any two matrices.

Let A be a square $M \times M$ matrix and B be a $M \times L$ matrix, as given below

$$
A = \begin{bmatrix} a_{11} & a_{12} & .. & a_{1M} \\ a_{21} & a_{22} & .. & a_{2M} \\ .... & & & \\ a_{M1} & a_{M2} & .. & a_{MM} \end{bmatrix}; \quad B = \begin{bmatrix} b_{11} & b_{12} & .. & b_{1L} \\ b_{21} & b_{22} & .. & b_{2L} \\ .... & & & \\ b_{M1} & b_{N2} & .. & b_{ML} \end{bmatrix}
$$

Matrix multiplication requires the number of columns in A be equal to the number of rows in B. The general result for AB yields a $M \times L$ matrix, and is given by

$$
AB = \begin{bmatrix} \sum a_{1i}b_{i1} & \sum a_{1i}b_{i2} & .. & \sum a_{1i}b_{iL} \\ \sum a_{2i}b_{i1} & \sum a_{2i}b_{i1} & .. & \sum a_{2i}b_{iL} \\ .... & & & \\ \sum a_{Mi}b_{i1} & \sum a_{2i}b_{i1} & .. & \sum a_{Mi}b_{iL} \end{bmatrix}; \quad \sum \equiv \sum_{i=1}^{M}
$$

The unit matrix $\mathbb{I}$ leaves the matrix unchanged on multiplication

$$
A\mathbb{I} = \mathbb{I}A; \quad \mathbb{I} = \begin{bmatrix} 1 & 0 & .. & 0 \\ 0 & 1 & .. & 0 \\ \cdots & & & \\ 0 & 0 & .. & 1 \end{bmatrix}
\tag{3.2.7}
$$

In the example discussed earlier what is the for three days expenditure E of the two customers? The matrices P, Q have been organized precisely to have E given by the matrix multiplication PQ. Hence

$$E = PQ = \underbrace{\begin{bmatrix} 1 & 2 & 3 & 7 \\ 4 & 6 & 5 & 8 \\ 9 & 10 & 11 & 12 \end{bmatrix}}_{\text{commodities}} \underbrace{\begin{bmatrix} 20 & 35 \\ 30 & 45 \\ 40 & 55 \\ 50 & 65 \end{bmatrix}}_{\text{customers}} = \underbrace{\begin{bmatrix} 550 & 745 \\ 860 & 1025 \\ 1520 & 2150 \end{bmatrix}}_{\text{revenue}}$$

For example, on the first day, customer 1 has to pay the following

$$1 \times 20 + 2 \times 30 + 3 \times 40 + 7 \times 50 = 550$$

The example illustrates that for accounting and other problems where there are lists of entries that need to be processed, matrix multiplication arises quite naturally.

The two dimensional column vector is a 2×1 matrix, since the vector has two rows and one column. A 2×2 matrix on the left can multiply a 2×1 matrix on the right since the number of number of columns of the matrix on the left must match the number of rows on the right. Multiplying yields our earlier result

$$\begin{bmatrix} a & c \\ b & d \end{bmatrix} \begin{bmatrix} x \\ y \end{bmatrix} = \begin{bmatrix} ax + cy \\ bx + dy \end{bmatrix}$$

The scalar product is also a multiplication of a $1 \times N$ matrix with a $N \times 1$ and yields a 1×1, which is simply a number, given by

$$\begin{bmatrix} a(1) \; a(2) \; .. \; a(N) \end{bmatrix} \begin{bmatrix} v(1) \\ v(2) \\ .... \\ v(N) \end{bmatrix} = a(1)v(1) + a(2)v(2) \cdots a(N)v(N)$$

Consider the $M \times L$ matrix B; it acts on a $L \times 1$ (vector) $\mathbf{v}$ and yields a matrix (vector) $M \times 1$, given by $\mathbf{v}$. In matrix notation, the action of the matrix B on $\mathbf{v}$ is represented as follows ($\sum \equiv \sum_{i=1}^{L}$)

$$B \begin{bmatrix} v(1) \\ v(2) \\ .... \\ v(L) \end{bmatrix} = \begin{bmatrix} b_{11} & b_{12} & .. & b_{1L} \\ b_{21} & b_{22} & .. & b_{2L} \\ .... & & & \\ b_{M1} & b_{N2} & .. & b_{ML} \end{bmatrix} \begin{bmatrix} v(1) \\ v(2) \\ .... \\ v(L) \end{bmatrix} = \begin{bmatrix} \sum b_{1i} v(i) \\ \sum b_{2i} v(i) \\ .... \\ \sum b_{Mi} v(i) \end{bmatrix}$$

In vector notation

$$B\mathbf{v} = \mathbf{w} \;\; \Rightarrow \;\; w(i) = \sum_{j=1}^{L} b_{ij} v(j); \;\; i = 1, \ldots, M$$

In other words, B maps vectors a L dimensional vector $\mathbf{v}$ to a M dimensional vector $\mathbf{w}$. In terms of vector space, B maps linear vector space E_L to vector space E_M, and yields

$$B : E_L \rightarrow E_M$$

3.3 Properties of $N \times N$ Matrices

The rest of the discussion will treat only $N \times N$ *square matrices*, as these matrices play a specially important role in many problems. Let A, B, C be $N \times N$ square matrices and let $\mathbf{v}$ be an N-dimensional vector.

- Matrix multiplication is *not commutative* and is given by

$$(AB)_{ij} = \sum_k A_{ik} B_{kj} \quad \Rightarrow \quad AB \neq BA$$

- Matrix multiplication is associative with

$$(AB)C = A(BC) = ABC$$

To illustrate the non-commutativity of matrices, consider two matrices

$$A = \begin{bmatrix} 1 & -3 \\ 2 & 5 \end{bmatrix}; \; B = \begin{bmatrix} 1 & 4 \\ -2 & 3 \end{bmatrix}$$

Then

$$AB = \begin{bmatrix} 7 & -5 \\ -8 & 23 \end{bmatrix} \neq BA = \begin{bmatrix} 9 & 17 \\ 4 & 21 \end{bmatrix} \quad \Rightarrow \quad AB - BA \neq 0$$

The non-commutativity of matrices has far reaching consequences.

3.4 System of Linear Equations

Consider two vectors

$$\mathbf{v} = \begin{bmatrix} v(1) \\ v(2) \\ \dots \\ v(N) \end{bmatrix}; \; \mathbf{b} = \begin{bmatrix} b(1) \\ b(2) \\ \dots \\ b(N) \end{bmatrix}$$

Suppose the vector satisfies the system of linear equations given by

$$\begin{aligned}a_{11}v(1)+a_{12}v(2)..+a_{1N}v(N)&=b(1)\\ a_{21}v(1)+a_{22}v(2)..+a_{2N}v(N)&=b(2)\\ &....\\ a_{N1}v(1)+a_{N2}v(2)..+a_{NN}v(N)&=b(N)\end{aligned}$$

The system of linear equations can be written in a matrix form using the rules for the multiplication of a matrix into a vector. Define matrix A by

$$A=\begin{bmatrix}a_{11} & a_{12} & .. & a_{1N}\\ a_{21} & a_{22} & .. & a_{2N}\\ .... & & & \\ a_{N1} & a_{N2} & .. & a_{NN}\end{bmatrix}$$

Then the system of linear equations, in matrix notation, can be written as

$$A\mathbf{v}=\mathbf{b}$$

Suppose one can define the inverse matrix A^{-1} such that

$$AA^{-1}=A^{-1}A=\mathbb{I}$$

I is the unit matrix that has the property of leaving matrices A (and vectors v) unchanged under multiplication

$$\mathbb{I}A=A;\ \ \mathbb{I}\mathbf{v}=\mathbf{v}$$

I has 1 as its diagonal element with all the off-diagonal elements being zero, and is given in Eq. 3.2.7.

One can formally obtain the solution for the system of linear equations given by

$$A^{-1}A\mathbf{v}=A^{-1}\mathbf{b}\ \Rightarrow\ \mathbf{v}=A^{-1}\mathbf{b}$$

One needs to determine under what conditions A^{-1} exists, which is analyzed in the following Sections.

3.5 Determinant: 2 × 2 Case

The determinant of a matrix is a scalar (a single number). It holds the key to many important properties of matrices, in particular in defining the inverse of a matrix A^{-1}. Most books start by giving a definition of the determinant; a better approach is to first have an intuitive idea of a determinant to understand its key role in the study of matrices, and see how the definition of the determinant arises naturally from its action on vectors.

The most general 2 × 2 matrix is given by

$$A = \begin{bmatrix} a & c \\ b & d \end{bmatrix}$$

The determinant of matrix A, denoted by $\det(A)$ is given by

$$\det(A) = ad - bc$$

If the rules of matrix multiplication look strange, then the rule for obtaining the determinant looks even more counter-intuitive. To gain some understanding of the determinant, consider the diagonal matrix given by

$$A_d = \begin{bmatrix} a & 0 \\ 0 & d \end{bmatrix}$$

The action on the basis vectors, shown in Fig. 3.1, stretches each vector and yields

$$A_d \begin{bmatrix} 1 \\ 0 \end{bmatrix} = \begin{bmatrix} a \\ 0 \end{bmatrix}; \ A_d \begin{bmatrix} 0 \\ 1 \end{bmatrix} = \begin{bmatrix} 0 \\ d \end{bmatrix}$$

The *area* enclosed by the square with unit length—determined by the initial unit vectors—is equal to 1. The matrix A stretches the x-direction by a and the y-direction by d and transforms the square into a rectangle with sides a and d. The area of unit square, which has unit area, is transformed into ad, as shown in Fig. 3.1. Hence, for this special case, the *area* is increased by a factor of ad, which is the determinant.

The effect on the area of a matrix (general linear transformation) acting on E_2 is given in Fig. 3.2. An area of unit size is mapped into a parallelepiped, with the vectors defining the parallelepiped defined by the linear transformation as follows

$$A = \begin{bmatrix} a & c \\ b & d \end{bmatrix} \Rightarrow A \begin{bmatrix} 1 \\ 0 \end{bmatrix} = \begin{bmatrix} a \\ b \end{bmatrix}; \ A \begin{bmatrix} 0 \\ 1 \end{bmatrix} = \begin{bmatrix} c \\ d \end{bmatrix}$$

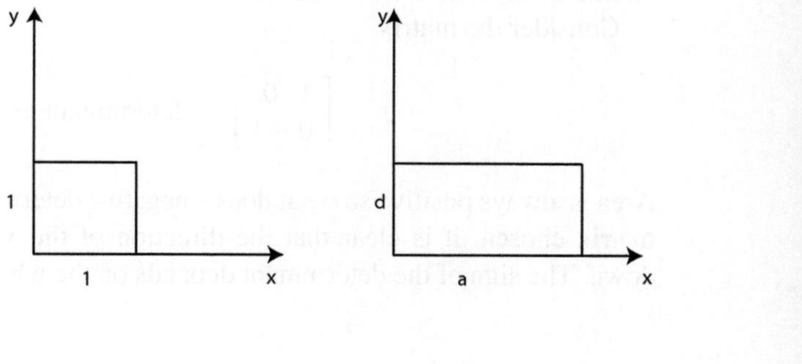

Fig. 3.1 A unit square mapped into a $a \times d$ rectangle. Published with permission of © Belal E. Baaquie 2020. All Rights Reserved

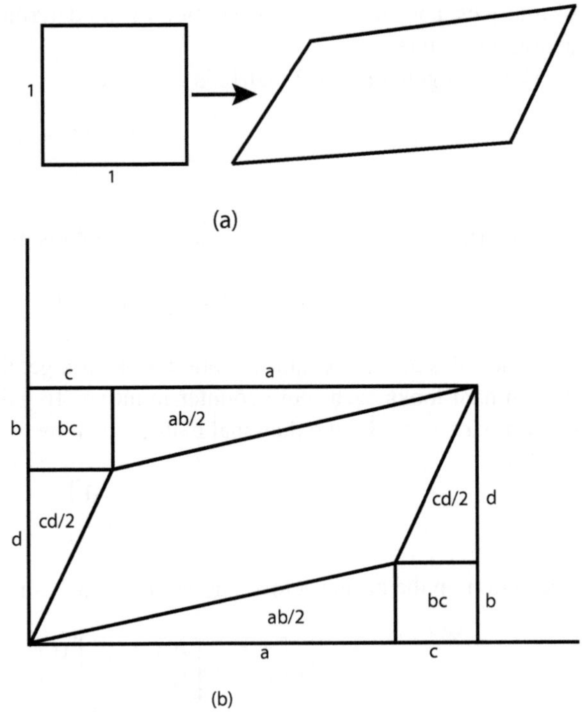

Fig. 3.2 **a** A unit square. **b** Unit square mapped into a parallelepiped. Published with permission of © Belal E. Baaquie 2020. All Rights Reserved

The transformed vectors are shown in Fig. 3.2. The unit area of the original square is mapped to the area of parallelepiped, which is given by the following.

$$\text{Area of parallelepiped} = (a+c)(b+d) - ab - cd - 2bc = ad - bc = \det(A)$$

As expected the determinant of the 2×2 matrix A is given by

$$\det(A) = \det \begin{vmatrix} a & c \\ b & d \end{vmatrix} = ad - bc$$

Hence one has the following important result:
Area S in E_2 after transformation=$\det(A)\times$ Area of S in E_2 before transformation.

Consider the matrix

$$\begin{bmatrix} 1 & 0 \\ 0 & -1 \end{bmatrix} : \text{ determinant} = -1$$

Area is always positive so what does a negative determinant mean? From the special matrix chosen, it is clear that the direction of the y-axis is switched from up to down. The sign of the determinant depends on the whether the linear transformation by matrix A changes the orientation of the vector space E_2, with a net change of orientation resulting in a overall minus sign for the determinant.

Since the determinant changes the area enclosed by vectors of original vector space, one may ask what is the significance of the determinant being zero? To understand this feature of determinants, consider the specific matrix

$$P = \begin{bmatrix} 1 & 1 \\ 1 & 1 \end{bmatrix}: \quad \det(P) = 0$$

The action of P on any vector in E_2 is the following

$$P = \begin{bmatrix} 1 & 1 \\ 1 & 1 \end{bmatrix} \begin{bmatrix} v(1) \\ v(2) \end{bmatrix} = (v(1) + v(2)) \begin{bmatrix} 1 \\ 1 \end{bmatrix} = (v(1) + v(2))\mathbf{u}$$

The vector $\mathbf{u}$ is clearly a very special vector for P. To understand this, let $v(1) + v(2) = \alpha$; then equation above states that for *any* vector $\mathbf{v}$, the action of P on it results in

$$P\mathbf{v} = \alpha \begin{bmatrix} 1 \\ 1 \end{bmatrix} = \alpha\mathbf{u}; \quad \mathbf{v} \in E_2; \quad \alpha \in \Re$$

As one varies the vector $\mathbf{v}$ through all the points in E_2, only the scalar α changes for $P\mathbf{v}$ and *all* the resultant vectors are proportional to the *fixed vector* $\mathbf{u}$. Hence, as shown in Fig. 3.3, *all* vectors of E_2 are mapped into the space E_1 spanned by

$$\mathbf{u} = \begin{bmatrix} 1 \\ 1 \end{bmatrix}$$

where $E_1 = \mathbb{R} = \{\alpha; \quad \alpha \in [-\infty, +\infty]\}$.

In particular, any two vectors (enclosing an area) are both mapped into the same vector $\mathbf{u}$ pointing in the same (or opposite) direction; and hence the mapped vectors enclose *zero area*. The linear span of the vectors mapped by P is one dimensional, and P is said to be projection operator onto the vector $\mathbf{u}$. This remarkable result is the foundation of the theory of eigenvalues and eigenvectors, discussed in Sect. 3.8.

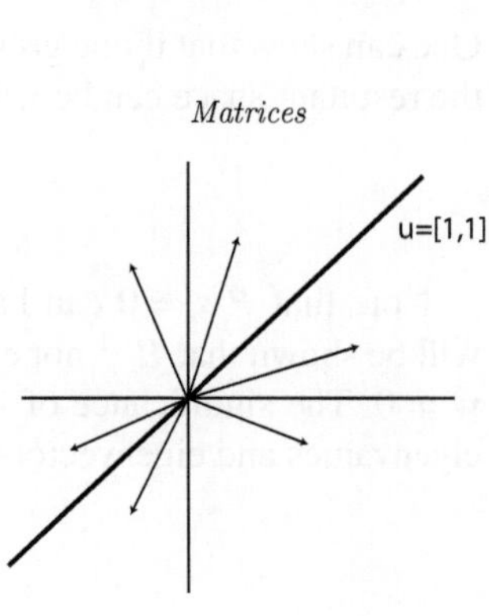

Fig. 3.3 Mapping of all vectors into linear span of vector $u = [1, 1]$. Published with permission of © Belal E. Baaquie 2020. All Rights Reserved

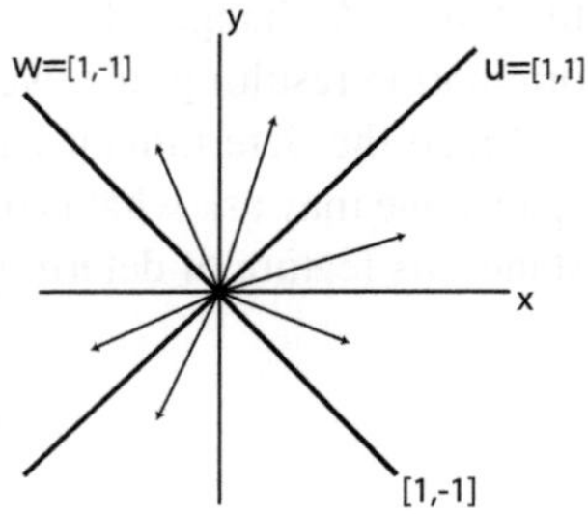

Fig. 3.4 Vector $w = [1, -1]$ orthogonal to vector $u = [1, 1]$. Published with permission of © Belal E. Baaquie 2020. All Rights Reserved

There is a class of vectors that is mapped into the null vector; this is a special vector which is also in the linear span of **u**, with $\alpha = 0$. Consider what happens to vectors proportional to **w** that are orthogonal to **u**? As given below, vectors that are *orthogonal* to the direction of **u** are scalar multiples of **w**, which is given below

$$\mathbf{w} = \begin{bmatrix} 1 \\ -1 \end{bmatrix}; \quad \mathbf{u}^T \cdot \mathbf{w} = 0$$

All vectors orthogonal to **u**, as shown in Figure 3.4, are mapped into the *null vector* since

$$P\mathbf{w} = \begin{bmatrix} 1 & 1 \\ 1 & 1 \end{bmatrix} \begin{bmatrix} 1 \\ -1 \end{bmatrix} = \begin{bmatrix} 0 \\ 0 \end{bmatrix} \equiv 0$$

All vectors mapped into the null vector by the matrix P constitute a space called $\mathcal{K}$, the *kernel* of P. In other words

$$P\mathbf{w} = 0 : \quad \text{for all } \mathbf{w} \in \mathcal{K}$$

where the kernel is the linear span of **w** defined by

$$\mathcal{K} = \{\beta \mathbf{w} : \beta \in \mathbb{R}\}$$

One can show that if one groups all the vectors with the same α into one vector, then the resultant space can be written as

$$E_1 = E_2/\mathcal{K}$$

Note that $P\mathbf{w} = 0$ can have a trivial result of $\mathbf{w} = 0$; but since $\det(P) = 0$, it will be shown that P^{-1} not exist and hence $P\mathbf{w} = 0$ does not *necessarily* imply that $\mathbf{w} = 0$. The significance of vectors **u**, **w** will become clearer when we analyze the eigenvalues and eigenvectors of a matrix A.

In summary all vectors of E_2 are in fact mapped by P into the linear span of $\mathbf{u}$; all vectors $\mathbf{w}$ that are orthogonal to $\mathbf{u}$ are mapped into the null vector, which is special element in the linear span of $\mathbf{u}$.

Matrices that have determinant zero don't have an inverse and lead to the following properties different from numbers. Consider the two matrices

$$A = \begin{bmatrix} 3 & 4 \\ 9/4 & 3 \end{bmatrix}; \; B = \begin{bmatrix} -3 & 4 \\ 9/4 & -3 \end{bmatrix}; \quad \det(A) = 0 = \det(B)$$

Then

$$AB = \begin{bmatrix} 0 & 0 \\ 0 & 0 \end{bmatrix}; \; A \neq 0; \; B \neq 0$$

Another feature is the following. Let

$$A = \begin{bmatrix} 2 & 3 \\ 6 & 9 \end{bmatrix}; \; B = \begin{bmatrix} 1 & 1 \\ 1 & 2 \end{bmatrix}; \; C = \begin{bmatrix} -2 & 1 \\ 3 & 2 \end{bmatrix}$$

with

$$\det(A) = 0; \quad \det(B) \neq 0; \quad \det(C) \neq 0$$

Then

$$AB = \begin{bmatrix} 5 & 8 \\ 15 & 24 \end{bmatrix} = AC; \; B \neq C$$

The action of these singular matrices on vector also have similar results. Consider

$$A\mathbf{v} = \begin{bmatrix} 2 & 3 \\ 6 & 9 \end{bmatrix} \mathbf{v} = 0; \; \mathbf{v} = \begin{bmatrix} -3/2 \\ 1 \end{bmatrix}; \quad \det(A) = 0$$

The significance of $\mathbf{v} \neq 0$ will become clear in the study of eigenvectors.

The result for the two dimensional case generalizes to the volume enclosed by vectors in N-dimensions. For any matrix that has two columns that are identical, the determinant is zero. This is because two basis vectors have been mapped (up to a scaling) into the same vector—and hence leading to the N-dimensional volume enclosed by the transformed vectors to go to zero.

The *rank* of a matrix is the number of linearly independent column (or equivalently row) vectors. For the matrix P, its rank is 1. The rank of a matrix determines the dimension of the subspace on which the matrix projects all vectors. For an $N \times N$ matrix A, if its rank is $n < N$, it implies that $A : E_N \to E_n$, and the linear span into which all vectors are projected is the space E_n. $n < N$ implies that the determinant of the matrix is zero and that A collapses a N-dimensional volume to a n-dimensional volume.

For the inverse of a $N \times N$ matrix to exist, its rank must be N; this in turn implies that its determinant is not zero.

3.6 Inverse of a 2 × 2 Matrix

Consider a simultaneous linear equation in two unknowns x, y given by

$$\begin{aligned} ax + cy &= f_1 \\ bx + dy &= f_2 \end{aligned} \quad \Rightarrow \quad \begin{bmatrix} a & c \\ b & d \end{bmatrix} \begin{bmatrix} x \\ y \end{bmatrix} \equiv A \begin{bmatrix} x \\ y \end{bmatrix} = \begin{bmatrix} f_1 \\ f_2 \end{bmatrix}$$

The solution can be obtained directly and is given by

$$\begin{bmatrix} x \\ y \end{bmatrix} = \frac{1}{ad - bc} \begin{bmatrix} d & -c \\ -b & a \end{bmatrix} \begin{bmatrix} f_1 \\ f_2 \end{bmatrix} = A^{-1} \begin{bmatrix} f_1 \\ f_2 \end{bmatrix}$$

Hence

$$A^{-1} = \frac{1}{ad - bc} \begin{bmatrix} d & -c \\ -b & a \end{bmatrix} = \frac{1}{\det(A)} \begin{bmatrix} d & -c \\ -b & a \end{bmatrix}$$

The main result we have is that[1]

$$ad - bc = \det(A) \neq 0 \quad \Rightarrow \quad A^{-1} \text{ exists}$$

One can verify by matrix multiplication that

$$AA^{-1} = A^{-1}A = \begin{bmatrix} 1 & 0 \\ 0 & 1 \end{bmatrix} = \mathbb{I}$$

3.7 Tensor (Outer) Product; Transpose

Consider the following set of xy-basis states

$$\mathbf{e}_1 = \begin{bmatrix} 1 \\ 0 \end{bmatrix}; \quad \mathbf{e}_2 = \begin{bmatrix} 0 \\ 1 \end{bmatrix}$$

To express a matrix in terms of the basis states, one needs to evaluate the tensor product of the basis states. Recall the scalar product of two vectors has been defined by $\mathbf{v}^T\mathbf{w}$. One can also define the *tensor product* or *outer product* of two vectors (and of matrices in general). The outer product is defined by the symbol $\mathbf{w} \otimes \mathbf{v}^T$, with the explicit component wise representation given by

[1] Recall for complex and real numbers, the expression $1/0$ is undefined.

$$\mathbf{w}\otimes\mathbf{v}^T = \begin{bmatrix} w_1 \\ w_2 \\ \cdots \\ w_N \end{bmatrix} \otimes \begin{bmatrix} v_1 & v_2 \cdots v_N \end{bmatrix} \equiv \begin{bmatrix} w_1v_1 & w_1v_2 & \cdots & w_1v_N \\ & \cdots & & \\ w_iv_1 & w_iv_2 & \cdots & w_iv_N \\ & \cdots & & \\ w_Nv_1 & w_Nv_2 & \cdots & w_Nv_N \end{bmatrix}$$

In general, for N-component vectors w_i, v_j, the outer product, denoted by A, is given by

$$A = \mathbf{w}\otimes\mathbf{v}^T;\ \ A_{ij} = w_iv_j;\ \ i, j = 1, \ldots N$$

For xy-basis states, one has the following four outer products

$$\mathbf{e}_1\otimes\mathbf{e}_1^T = \begin{bmatrix} 1 \\ 0 \end{bmatrix}\otimes\begin{bmatrix} 1 & 0 \end{bmatrix} = \begin{bmatrix} 1 & 0 \\ 0 & 0 \end{bmatrix};\ \mathbf{e}_1\otimes\mathbf{e}_2^T = \begin{bmatrix} 1 \\ 0 \end{bmatrix}\otimes\begin{bmatrix} 0 & 1 \end{bmatrix} = \begin{bmatrix} 0 & 1 \\ 0 & 0 \end{bmatrix}$$

and

$$\mathbf{e}_2\otimes\mathbf{e}_1^T = \begin{bmatrix} 0 \\ 1 \end{bmatrix}\otimes\begin{bmatrix} 1 & 0 \end{bmatrix} = \begin{bmatrix} 0 & 0 \\ 1 & 0 \end{bmatrix};\ \mathbf{e}_2\otimes\mathbf{e}_2^T = \begin{bmatrix} 0 \\ 1 \end{bmatrix}\otimes\begin{bmatrix} 0 & 1 \end{bmatrix} = \begin{bmatrix} 0 & 0 \\ 0 & 1 \end{bmatrix}$$

The tensor product of the basis states yields a basis for the 2×2 matrix since each position in the matrix corresponds to one matrix with a nonzero entry only at that position. For the N-dimensional case of E_N, define the tensor product of any two basis vectors by

$$\mathbf{e}_i\otimes\mathbf{e}_j^T;\ \ i, j = 1, \ldots, N$$

The fact that the tensor product of the basis vectors is sufficient to represent any vector of E_N requires that the basis states must satisfy the *completeness equation* given by

$$\sum_{i=1}^{N}\mathbf{e}_i\otimes\mathbf{e}_i^T = \mathbb{I}_{N\times N} \tag{3.7.1}$$

For the 2×2 matrices, one has

$$\mathbf{e}_1\otimes\mathbf{e}_1^T + \mathbf{e}_2\otimes\mathbf{e}_2^T = \begin{bmatrix} 1 & 0 \\ 0 & 0 \end{bmatrix} + \begin{bmatrix} 0 & 0 \\ 0 & 1 \end{bmatrix} = \begin{bmatrix} 1 & 0 \\ 0 & 1 \end{bmatrix} = \mathbb{I}$$

Completeness Eq. 3.7.1 yields the following

$$\mathbf{v} = \mathbb{I}\mathbf{v} = \sum_{i=1}^{N}\mathbf{e}_i\otimes(\mathbf{e}_i^T\cdot\mathbf{v}) = \sum_{i=1}^{N}v_i\mathbf{e}_i \tag{3.7.2}$$

where

$$v_i = \mathbf{e}_i^T \cdot \mathbf{v}: \quad \text{components of } \mathbf{v}$$

Due to the completeness equation, for an arbitrary matrix[2]

$$A = \sum_{ij} A_{ij}\mathbf{e}_i \otimes \mathbf{e}_j^T = \sum_{ij} A_{ij}\mathbf{e}_i\mathbf{e}_j^T \tag{3.7.3}$$

Note that

$$\mathbf{e}_I^T A \mathbf{e}_J = \sum_{ij} A_{ij}(\mathbf{e}_I^T \cdot \mathbf{e}_i) \otimes (\mathbf{e}_j^T \cdot \mathbf{e}_J) = \sum_{ij} A_{ij}\delta_{I-i}\delta_{J-j} = A_{IJ}$$

In the notation of tensor product, the action of A on a general vector is given by

$$A\begin{bmatrix} v_1 \\ v_2 \end{bmatrix} = \sum_{ij} A_{ij}\mathbf{e}_i \otimes (\mathbf{e}_j^T \cdot \sum_k v(k)\mathbf{e}_k) = \sum_{ijk} A_{ij}\mathbf{e}_i v_k \delta_{j-k}$$
$$= \sum_{ij} A_{ij}v_j\mathbf{e}_i = \sum_i w_i\mathbf{e}_i \tag{3.7.4}$$

where

$$\mathbf{w} = \sum_i w_i\mathbf{e}_i = \begin{bmatrix} w_1 \\ w_2 \end{bmatrix} = A\mathbf{v} = \begin{bmatrix} a_{11} & a_{12} \\ a_{21} & a_{22} \end{bmatrix}\begin{bmatrix} v_1 \\ v_2 \end{bmatrix} \tag{3.7.5}$$

Then, from Eq. 3.7.4

$$w_I = \mathbf{e}_I^T \cdot \mathbf{w} = \sum_{ij} A_{ij}v_j(\mathbf{e}_I^T \cdot \mathbf{e}_i) = \sum_{ij} A_{ij}v_j\delta_{i-I}$$
$$\Rightarrow w_I = \sum_j A_{Ij}v_j \tag{3.7.6}$$

3.7.1 *Transpose*

The transpose of a vector is a special case of the transposition of a matrix. Similar to the case of a vector, where transposition entails mapping a column vector into a row

[2]Where the symbol $\otimes$ is sometimes dropped if the expression has no ambiguity.

vector, in case of a matrix, all the columns and rows are interchanged. In symbols, the transpose of a matrix A and vector $\mathbf{v}$ is defined by

$$(A^T)_{ij} = (A)_{ji} : \text{ Transpose} \tag{3.7.7}$$

$$\begin{aligned}(A+B)^T &= A^T + B^T \\ (ABC)^T &= C^T B^T A^T \\ ((A^T)^T) &= A \\ (\mathbf{v})^T &= \mathbf{v}^T;\ ((\mathbf{v})^T)^T = \mathbf{v}\end{aligned} \tag{3.7.8}$$

Hence

$$A^T = \left(\sum_{ij} A_{ij}\mathbf{e}_i \otimes \mathbf{e}_j^T\right)^T = \sum_{ij} A_{ij}(\mathbf{e}_i \otimes \mathbf{e}_j^T)^T$$

$$= \sum_{ij} A_{ij}\mathbf{e}_j \otimes \mathbf{e}_i^T = \sum_{ij} A_{ji}^T\mathbf{e}_j \otimes \mathbf{e}_i^T \tag{3.7.9}$$

The discussion on determinant has focused on transforming the basis states; an equivalent conclusion is reached by analyzing the effect of a linear transformation on the dual basis states $\mathbf{e}^T$. The change of volume is identical since both the vector and its dual provide an equivalent description of the linear transformation. The dual basis vectors are transformed by the transpose of the linear transformation given by

$$A^T : E_2 \to E_2$$

Hence the determinant has the following property

$$\det(A) = \det(A^T) \tag{3.7.10}$$

Taking the transpose of Eq. 3.7.5 following the rules given in Eq. 3.7.8 yields

$$\mathbf{w}^T = \mathbf{v}^T A^T$$

To verify this definition is consistent, consider

$$w_I = \mathbf{v}^T A^T \mathbf{e}_I = \sum_k v_k \sum_{ij} A_{ji}^T (\mathbf{e}_k^T \cdot \mathbf{e}_j) \otimes (\mathbf{e}_i^T \cdot \mathbf{e}_I) = \sum_{ijk} v_k A_{ij} \delta_{k-j} \delta_{i-I}$$

The result is consistent with Eq. 3.7.6 and given by

$$w_I = \sum_i A_{Ij} v_j$$

Hence, from above vector **v** and matrix A obey the following rules

$$\mathbf{w} = A\mathbf{v} \iff \mathbf{w}^T = \mathbf{v}^T A^T$$

3.8 Eigenvalues and Eigenvectors

The study of determinants have revealed that there are vectors, in the previous Section vectors **u**, **w** that have a special significance. Understanding the properties of determinants allows one to address this question.

Consider the following matrix equation, where A is a $N \times N$ matrix, **v** is a N-dimensional vector and λ is a scalar

$$A\mathbf{v} = \lambda\mathbf{v}; \quad \mathbf{v}^T \cdot \mathbf{v} = 1 \tag{3.8.1}$$

The norm of the eigenvector is taken to be 1 since otherwise there is an arbitrary scale, since any scalar multiple of an eigenvector is also an eigenvector.

The vector **v** is very special, since its direction is not changed by the action of A but it is rescaled by a scalar λ. The vector **v** and scalar λ are called eigenvector and eigenvalue respectively. The eigenvalue equation can be re-written as

$$(A - \lambda\mathbb{I})\mathbf{v} = \mathbf{0}$$

If $M = A - \lambda\mathbb{I}$ has an inverse, then the solution if trivial since $\mathbf{v} = M^{-1}(\mathbf{0}) = \mathbf{0}$. Section 3.6 shows that an inverse of a matrix A does not exist if its determinant is zero. Hence, to have nontrivial eigenvectors the eigenvalues λ must be *finely tuned* to satisfy following condition

$$\det(A - \lambda\mathbb{I}) = 0$$

For a $N \times N$ matrix, the eigenvalue above is a Nth order polynomial for λ and hence has N solutions that yield the eigenvalues given by $\lambda_1, \cdots, \lambda_N$; not all the eigenvalues are necessarily distinct or real.

Once the eigenvalues λ_i have been fixed, for each eigenvalue λ_i the eigenvalue equation given in Eq. 3.8.1 becomes equivalent to a system of simultaneous linear equations.

Example
Consider the following matrix

$$A = \begin{bmatrix} 0 & 1 \\ 1 & 0 \end{bmatrix}$$

The eigenvalue equation is given by

$$\det(A - \lambda\mathbb{I})] = \det\begin{vmatrix} -\lambda & 1 \\ 1 & -\lambda \end{vmatrix} = \lambda^2 - 1 = 0 \Rightarrow \lambda_\pm = \pm 1$$

The two eigenvectors $\mathbf{v}_\pm$ corresponding to $\lambda_\pm$ are orthonormal (orthogonal and with unit length (normal))

$$\mathbf{v}_-^T \cdot \mathbf{v}_+ = 0; \;\; \mathbf{v}_-^T \cdot \mathbf{v}_- = 1; \;\; \mathbf{v}_+^T \cdot \mathbf{v}_+ = 1$$

Case I: $\lambda_+ = 1$

$$(A - \lambda\mathbb{I})\mathbf{v}_+ = \begin{bmatrix} -1 & 1 \\ 1 & -1 \end{bmatrix}\begin{bmatrix} v_+(1) \\ v_+(2) \end{bmatrix} = 0 \Rightarrow \mathbf{v}_+ = \frac{1}{\sqrt{2}}\mathbf{u} = \frac{1}{\sqrt{2}}\begin{bmatrix} 1 \\ 1 \end{bmatrix}$$

For $\lambda_+ = 1$ the eigenvalue equation Eq. 3.8.1 yields the following simultaneous linear equations

$$A\mathbf{v}_+ - = \begin{bmatrix} 0 & 1 \\ 1 & 0 \end{bmatrix}\begin{bmatrix} v_+(1) \\ v_+(2) \end{bmatrix} = +1\begin{bmatrix} v_+(1) \\ v_+(2) \end{bmatrix} \;\; \Rightarrow v_+(1) = v_+(2) = v$$

$$\mathbf{v}_+ = \mathcal{N}\begin{bmatrix} v \\ v \end{bmatrix}; \;\; |\mathbf{v}|^2 = \mathbf{v}_+^T\mathbf{v}_+ = 2v^2 = \frac{1}{\mathcal{N}}$$

Normalizing the eigenvector yields

$$\mathbf{v}_+ = \frac{1}{v\sqrt{2}}\begin{bmatrix} v \\ v \end{bmatrix} \;\; \Rightarrow \mathbf{v}_+ = \frac{1}{\sqrt{2}}\mathbf{u} = \frac{1}{\sqrt{2}}\begin{bmatrix} 1 \\ 1 \end{bmatrix}$$

Case II: $\lambda_- = -1$

$$(A - \lambda\mathbb{I})\mathbf{v}_- = \begin{bmatrix} 1 & 1 \\ 1 & 1 \end{bmatrix}\begin{bmatrix} v_-(1) \\ v_-(2) \end{bmatrix} = 0 \Rightarrow \mathbf{v}_- = \frac{1}{\sqrt{2}}\mathbf{w} = \frac{1}{\sqrt{2}}\begin{bmatrix} 1 \\ -1 \end{bmatrix}$$

One can directly verify that $\mathbf{v}_\pm$ satisfy the eigenvector equations

$$\begin{bmatrix} 0 & 1 \\ 1 & 0 \end{bmatrix}\begin{bmatrix} 1 \\ -1 \end{bmatrix} = -\begin{bmatrix} 1 \\ -1 \end{bmatrix}; \;\; \begin{bmatrix} 0 & 1 \\ 1 & 0 \end{bmatrix}\begin{bmatrix} 1 \\ 1 \end{bmatrix} = \begin{bmatrix} 1 \\ 1 \end{bmatrix}$$

The eigenvectors of A encode the information that $\det(A - \lambda\mathbb{I}) = 0$. As one can see from its definition, the eigenvectors are unchanged, up to a rescaling, under the action of A on the space E_2. Note that the vectors boldface **u, w** discussed earlier in Sect. 3.5 are the eigenvectors **v+** and **v-**, respectively.

3.8.1 Matrices: Spectral Decomposition

For a class of matrices that will be defined later, the eigenvalues and eigenvectors have all the information about the matrix A: the spectral decomposition is an expression of this.

Consider the expression

$$\mathbf{v}_+ \otimes \mathbf{v}_+^T = \frac{1}{2}\begin{bmatrix}1\\1\end{bmatrix} \otimes \begin{bmatrix}1 & 1\end{bmatrix} = \frac{1}{2}\begin{bmatrix}1 & 1\\1 & 1\end{bmatrix}$$

and

$$\mathbf{v}_- \otimes \mathbf{v}_-^T = \frac{1}{2}\begin{bmatrix}1\\-1\end{bmatrix} \otimes \begin{bmatrix}1 & -1\end{bmatrix} = \frac{1}{2}\begin{bmatrix}1 & -1\\-1 & 1\end{bmatrix}$$

It follows that

$$\lambda_+ \mathbf{v}_+ \otimes \mathbf{v}_+^T + \lambda_- \mathbf{v}_- \otimes \mathbf{v}_-^T = \frac{1}{2}\begin{bmatrix}1 & 1\\1 & 1\end{bmatrix} - \frac{1}{2}\begin{bmatrix}1 & -1\\-1 & 1\end{bmatrix} = \begin{bmatrix}0 & 1\\1 & 0\end{bmatrix}$$

and hence

$$A = \lambda_+ \mathbf{v}_+ \otimes \mathbf{v}_+^T + \lambda_- \mathbf{v}_- \otimes \mathbf{v}_-^T : \text{ Spectral decomposition}$$

Consider an $N \times N$ matrix A with N orthogonal eigenvectors $\mathbf{v}_n$ and eigenvalues λ_n, called the spectrum of A. It is shown in Sect. 4.13.1 that a symmetric matrix has the following decomposition

$$A = \sum_{n=1}^{N} \lambda_n \mathbf{v}_n \otimes \mathbf{v}_n^T; \quad \mathbf{v}_n^T \cdot \mathbf{v}_m = \delta_{m-n} \tag{3.8.2}$$

Equation 3.8.2 is called the spectral decomposition of a matrix, and is perhaps one of the most important results in matrix theory.

The eigenvector equations are all explicitly satisfied since

$$A\mathbf{v}_k = \sum_{n=1}^{N} \lambda_n \mathbf{v}_n \otimes (\mathbf{v}_n^T \cdot \mathbf{v}_k) = \sum_{n=1}^{N} \lambda_n \mathbf{v}_n \delta_{k-n} = \lambda_n \mathbf{v}_n$$

What is the following matrix (without the λ_n's)?

$$B = \sum_{n=1}^{N} \mathbf{v}_n \otimes \mathbf{v}_n^T$$

For the special case of eigenvectors of matrix A

$$\mathbf{v}_+ \otimes \mathbf{v}_+^T + \mathbf{v}_- \otimes \mathbf{v}_-^T = \frac{1}{2}\begin{bmatrix} 1 & 1 \\ 1 & 1 \end{bmatrix} + \frac{1}{2}\begin{bmatrix} 1 & -1 \\ -1 & 1 \end{bmatrix} = \begin{bmatrix} 1 & 0 \\ 0 & 1 \end{bmatrix}$$

Hence

$$\mathbf{v}_+ \otimes \mathbf{v}_+^T + \mathbf{v}_- \otimes \mathbf{v}_-^T = \mathbb{I} : \text{ Completeness equation}$$

In general, consider

$$B^2 = \sum_{mn=1}^{N} \mathbf{v}_m \otimes (\mathbf{v}_m^T \cdot \mathbf{v}_n) \otimes \mathbf{v}_n^T = \sum_{mn=1}^{N} \mathbf{v}_m \otimes \mathbf{v}_n^T \delta_{m-n} = \sum_{n=1}^{N} \mathbf{v}_n \otimes \mathbf{v}_n^T = B$$

If A is a *symmetric* matrix that satisfies $A^T = A$, then one can show that for all the eigenvectors $\mathbf{v}_n;\ n = 1, \cdots, N$

$$B\mathbf{v}_n = \mathbf{v}_n;\ n = 1, \ldots, N$$

Hence, one can conclude that B is the unit matrix given by

$$B = \mathbb{I} = \sum_{n=1}^{N} \mathbf{v}_n \otimes \mathbf{v}_n^T : \text{ Completeness equation} \tag{3.8.3}$$

Equation 3.8.3 reflects the fact that there are enough eigenvectors to span the entire linear vector space E_N and hence the set of eigenstates forms a complete set. The completeness equation is also called a *resolution of the identity*.

3.9 Problems

1. Consider matrices A, B, C. Prove the following by showing that matrix elements of both sides are equal.
    - $(AB)^T = B^T A^T$
    - A(BC)=(AB)C
    - Tr(ABC)=Tr(BCA)
2. Consider matrices

$$M = \begin{bmatrix} 2 & 3 & 5 \\ 3 & 1 & 2 \\ 1 & 4 & 3 \end{bmatrix};\ N = \begin{bmatrix} 2 & 3 \\ 3 & 1 \\ 1 & 4 \end{bmatrix}$$

- Evaluate M^T, N^T
- Evaluate $N^T M^T$ and $N^T M$.

3. Consider the vector and matrix

$$\mathbf{u} = \begin{bmatrix} 3 \\ -2 \\ 5 \end{bmatrix}; \quad M = \begin{bmatrix} 2 & 3 & 5 \\ 3 & 1 & 2 \\ 1 & 4 & 3 \end{bmatrix}$$

- Evaluate $N = \mathbf{u} \otimes \mathbf{u}^T$
- Evaluate NM and MN.

4. Consider the following spectral decomposition of a symmetric

$$A = \sum_{n=1}^{N} \lambda_n \mathbf{v}_n \otimes \mathbf{v}_n^T; \quad \mathbf{v}_n^T \cdot \mathbf{v}_m = \delta_{m-n}$$

Find the matrix elements A_{ij}.

5. Consider the matrix A given by

$$A = \begin{bmatrix} -1 & 1 \\ -1 & -1 \end{bmatrix}$$

- What is the linear span generated by applying the matrix A to all the vectors of E_2?
- Find the eigenvectors and eigenvalues of matrix A.
- Write the matrix A in terms of the eigenvectors and eigenvalues (spectral decomposition).
- Show that the eigenvectors of the matrix A satisfy the completeness equation.
- Can the matrix A be diagonalized?

6. Find the eigenvectors and eigenvalues of the following matrix

$$M = \begin{bmatrix} 1 & 2 & 3 \\ 4 & 5 & 6 \\ 7 & 8 & 9 \end{bmatrix}$$

7. Let $\lambda_1 = 3$; $\lambda_2 = 4$ be eigenvalues and with eigenvectors

$$\mathbf{v}_1 = \frac{1}{\sqrt{2}} \begin{bmatrix} 1 \\ 1 \end{bmatrix}; \quad \mathbf{v}_2 = \frac{1}{\sqrt{2}} \begin{bmatrix} 1 \\ -1 \end{bmatrix}$$

Find the matrix M for which

$$M\mathbf{v}_n = \lambda_n \mathbf{v}_n; \quad n = 1, 2$$

Chapter 4
Square Matrices

Abstract Square matrices are a special class of matrices that map a N dimensional linear vector space to itself. Many important properties, such as the definition of an inverse and of a determinant are only possible for square matrices. Determinants are defined for the general $N \times N$ matrices, and the evaluation of a general determinant is discussed; the inverse of a square matrix is defined. The symmetric and Hermitian $N \times N$ matrices are studied in some detail. The Leontief input-output model is briefly discussed.

4.1 Introduction

A number of properties of square matrices are given in Table 4.1. Based on these properties, special square matrices can be defined that occur in many applications in economics and finance.

Let A be a complex valued $N \times N$ square matrix. Suppose A is an invertible matrix. The following are some of the matrices related to A, and given in the Table 4.1.

- Transpose: $A^T_{ij} = A_{ji}$
- Complex conjugate: $A^*_{ij} = (A_{ji})^*$
- Hermitian conjugate: $A^\dagger_{ij} = (A^*_{ij})^T = A^*_{ji}$
- Inverse:

$$A^{-1}_{ij} = (\frac{1}{A})_{ij}$$

The following are some special $N \times N$ square matrices that have far reaching applications.

- Symmetric matrix: $S^T = S \ \Rightarrow \ S^T_{ij} = S_{ji} = S_{ij}$
- Hermitian matrix: $H^\dagger = H \ \Rightarrow \ H^\dagger_{ij} = H^*_{ji} = H_{ij}$
- Orthogonal matrix: $O^T O = \mathbb{I} \Rightarrow \ O^{-1} = O^T$
- Unitary matrix: $U^\dagger U = \mathbb{I} \Rightarrow \ U^{-1} = U^\dagger$

B. E. Baaquie, *Mathematical Methods and Quantum Mathematics for Economics and Finance*, https://doi.org/10.1007/978-981-15-6611-0_4

Table 4.1 Matrix A and Related Matrices

Matrix	Symbol	Components	Example
	A	A_{ij}	$\begin{bmatrix} 1 & 2+i \\ -i & 2 \end{bmatrix}$
Transpose	A^T	A^T_{ij}	$\begin{bmatrix} 1 & -i \\ 2+i & 2 \end{bmatrix}$
Complex conjugate	A^*	A^*_{ij}	$\begin{bmatrix} 1 & 2-i \\ +i & 2 \end{bmatrix}$
Hermitian conjugate	$A^\dagger = (A^*)^T$	$A^\dagger_{ij}$	$\begin{bmatrix} 1 & i \\ 2-i & 2 \end{bmatrix}$
Inverse	A^{-1}	A^{-1}_{ij}	$\frac{1}{1+2i}\begin{bmatrix} 2 & -(2+i) \\ i & 1 \end{bmatrix}$

4.2 Determinant: 3 x 3 Case

Determinants are defined only for square matrices. The determinant of a matrix M is written as $|M|$. A few special cases are analyzed so as to generalize the definition of the determinant for a $N \times N$ matrix.

The determinant for a 2×2 was discussed in Sect. 3.5 and was defined by

$$|M| = \begin{vmatrix} a & b \\ c & d \end{vmatrix} = ad - bc \tag{4.2.1}$$

Consider the 3×3 matrix given by

$$M = \begin{bmatrix} a & b & c \\ d & e & f \\ g & h & i \end{bmatrix}$$

The determinant is given by

$$|M| = \begin{vmatrix} a & b & c \\ d & e & f \\ g & h & f \end{vmatrix} = a\begin{vmatrix} e & f \\ h & i \end{vmatrix} - b\begin{vmatrix} d & e \\ g & h \end{vmatrix} + c\begin{vmatrix} d & e \\ g & h \end{vmatrix} \tag{4.2.2}$$

From Eq. 4.2.1

$$|M| = a(ei - hf) - b(di - fg) + c(dh - eg)$$

The evaluation of the 3×3 matrix is finally reduced to the evaluation of 2×2 matrices. The $N \times N$ can be similarly reduced to the evaluation finally of 2×2 matrices.

Example. Consider the following determinant

$$|M| = \begin{vmatrix} 2 & 3 & 5 \\ 3 & 1 & 2 \\ 1 & 4 & 3 \end{vmatrix} = 2\begin{vmatrix} 1 & 2 \\ 4 & 3 \end{vmatrix} - 3\begin{vmatrix} 3 & 2 \\ 1 & 3 \end{vmatrix} + 5\begin{vmatrix} 3 & 1 \\ 1 & 4 \end{vmatrix}$$

Hence, from Eq. 4.2.1

$$|M| = -10 - 21 + 55 = 24$$

The sign alternates when forming the determinant from matrix, as follows

$$\begin{matrix} + & - & + \\ - & + & - \\ + & - & + \end{matrix}$$

4.3 Properties of Determinants

Interchanging two rows or two columns yields an overall minus sign. For 3 × 3 matrix M, one can verify

$$|M| = \begin{vmatrix} a & b & c \\ d & e & f \\ g & h & f \end{vmatrix} = -\begin{vmatrix} d & e & f \\ a & b & c \\ g & h & f \end{vmatrix} = -\begin{vmatrix} b & a & c \\ e & d & f \\ h & g & f \end{vmatrix} \tag{4.3.1}$$

Due to Eq. 4.3.1, a matrix that has two columns or two rows being identical is zero, since interchanging them leaves the determinant unchanged but it is now minus itself. For example, one can verify explicitly that for the 3 × 3 matrix M

$$M = \begin{bmatrix} a & a & c \\ d & d & f \\ g & g & i \end{bmatrix} = 0 = \begin{vmatrix} a & b & c \\ a & b & c \\ g & h & f \end{vmatrix} \tag{4.3.2}$$

Determinants behave like linear functions on the rows of the matrix. For the 2 × 2 matrix

$$|M| = \begin{vmatrix} ka & kb \\ c & d \end{vmatrix} = k\begin{vmatrix} a & b \\ c & d \end{vmatrix} \tag{4.3.3}$$

$$|M| = \begin{vmatrix} a + ka' & b + kb' \\ c & d \end{vmatrix} = \begin{vmatrix} a & b \\ c & d \end{vmatrix} + k\begin{vmatrix} a' & b' \\ c & d \end{vmatrix} \tag{4.3.4}$$

To verify above equation note the left hand side of Eq. 4.3.3

$$|M| = \begin{vmatrix} a+ka' & b+kb' \\ c & d \end{vmatrix} = (a+ka')d - (b+kb')c = ad - bc + k(a'd - b'c)$$

and hence is equal to the right hand side of Eq. 4.3.3.

Adding a row to the other rows doesn't change the value of the determinant. From Eqs. 4.3.2 and 4.3.3

$$|M| = \begin{vmatrix} a+kc & b+kd \\ c & d \end{vmatrix} = \begin{vmatrix} a & b \\ c & d \end{vmatrix} + k\begin{vmatrix} c & d \\ c & d \end{vmatrix} = \begin{vmatrix} a & b \\ c & d \end{vmatrix}$$

Similarly for adding a column to another column leaves the determinant unchanged

$$|M| = \begin{vmatrix} a+kb & b \\ c+kd & d \end{vmatrix} = \begin{vmatrix} a & b \\ c & d \end{vmatrix} + k\begin{vmatrix} b & b \\ d & d \end{vmatrix} = \begin{vmatrix} a & b \\ c & d \end{vmatrix}$$

The proof for higher dimensional matrices is similar.

Example. Is the determinant of the following matrix zero?

$$|M| = \begin{vmatrix} 1 & 2 & 3 \\ 4 & 5 & 6 \\ 7 & 8 & 9 \end{vmatrix}$$

Take two steps: (i) subtract column 1 from column 2 and then (ii) column 1 from column 3; this yields

$$|M| = \begin{vmatrix} 1 & 2 & 3 \\ 4 & 5 & 6 \\ 7 & 8 & 9 \end{vmatrix} = \begin{vmatrix} 1 & 1 & 3 \\ 4 & 1 & 6 \\ 7 & 1 & 9 \end{vmatrix} = \begin{vmatrix} 1 & 1 & 2 \\ 4 & 1 & 2 \\ 7 & 1 & 2 \end{vmatrix} = 2\begin{vmatrix} 1 & 1 & 1 \\ 4 & 1 & 1 \\ 7 & 1 & 1 \end{vmatrix} = 0$$

The discussion in Sect. 3.5 on determinants was based on the linear transformation acting on the basis states; an equivalent conclusion is reached by analyzing the effect of a linear transformation on the dual basis states $\mathbf{e}^T$. The change of volume is identical since both the vector and its dual provide an equivalent description of the linear transformation. The dual basis vectors are transformed by the transpose of the linear transformation given by

$$A^T : E_N \to E_N$$

Hence the determinant has the following property

$$\det(A) = \det(A^T) \tag{4.3.5}$$

Under a linear transformation, all N-dimensional volumes are scaled by a factor given by $\det(A)$.

The product of two transformations A, B is given by

$$C = AB$$

Two consecutive transformations scale the area (volume) two times, the first followed by the second. Hence, we have the important result that

$$\det(AB) = \det(A)\det(B) \tag{4.3.6}$$

From Eq. 4.3.6 it follows, if $|A|$ has an inverse

$$\det(A^{-1}) = \frac{1}{\det(A)}$$

4.4 *N* x *N* Determinant

The general rule for evaluating the determinant of a matrix M is discussed. For a general linear transformation defined by matrix M

$$M : E_N \to E_N$$

The determinant of a square requires the concept of the *co-factor*, denoted by Δ_{ij}. Consider the determinant for the 3 × 3 matrix given in Eq. 4.2.2; writing the determinant in a more general notation yields

$$|M| = \begin{vmatrix} a & b & c \\ d & e & f \\ g & h & f \end{vmatrix} = a\begin{vmatrix} e & f \\ h & i \end{vmatrix} - b\begin{vmatrix} d & e \\ g & h \end{vmatrix} + c\begin{vmatrix} d & e \\ g & h \end{vmatrix}$$
$$= \begin{vmatrix} a_{11} & a_{12} & a_{13} \\ a_{21} & a_{22} & a_{23} \\ a_{31} & a_{32} & a_{33} \end{vmatrix} = a_{11}|\Delta_{11}| - a_{12}|\Delta_{12}| + a_{13}|\Delta_{13}|$$

where

$$\Delta_{11} = \begin{bmatrix} a_{22} & a_{23} \\ a_{32} & a_{33} \end{bmatrix} \; ; \; \Delta_{12} = \begin{bmatrix} a_{21} & a_{23} \\ a_{31} & a_{33} \end{bmatrix} \; ; \; \Delta_{13} = \begin{bmatrix} a_{21} & a_{22} \\ a_{31} & a_{32} \end{bmatrix}$$

It can be verified that the determinant can also be evaluated using a column expansion to yield the following

$$|M| = a_{11}|\Delta_{11}| - a_{21}|\Delta_{21}| + a_{31}|\Delta_{31}|$$

Consider a $N \times N$ matrix M with $M_{ij} = a_{ij}$. The cofactor for an element a_{1j} is a $(N-1) \times (N-1)$ determinant denoted by $|\Delta_{1j}|$. As can be seen by the example of the 3 × 3 matrix , the cofactor is obtained from M by eliminating the row and column that intersect at the element a_{1j} and obtaining a $(N-1)\times(N-1)$ matrix, denoted by Δ_{1j}; the cofactor is the determinant of the matrix Δ_{1j} is denoted by $|\Delta_{1j}|$.

One choice is to enumerate the determinant, as done in the 3×3 case, by using the first row to define the determinant. One then obtains

$$|M| = \sum_{j=1}^{N}(-1)^{j+1}a_{1j}|\Delta_{1j}| = a_{11}|\Delta_{11}| - a_{12}|\Delta_{12}| + \cdots + (-1)^{1+N}a_{1N}|\Delta_{1N}|$$

In general, one can choose any of the rows or columns for enumerating the determinant; the reason one gets the same answer is because upto a sign the determinant is unchanged by the exchange of two columns. Furthermore, since the determinant of a matrix and its transpose are equal, enumerating the determinant along a row can be transformed into an enumeration using a column. Hence, one has, for any choice of $k = 1, \cdots, N$, the following

$$|M| = \sum_{j=1}^{N}(-1)^{j+k}a_{kj}|\Delta_{kj}| \tag{4.4.1}$$

where the factor of $(-1)^{j+k}$ keeps track of the change in the sign of the determinant in moving from one row to the next row. Since the determinant of a matrix and its transpose are equal, enumerating the determinant along a row can be transformed into an enumeration using a column. Hence, one also has, for any choice of $k = 1, \cdots, N$, the following

$$|M| = \sum_{j=1}^{N}(-1)^{j+k}a_{jk}|\Delta_{jk}| \tag{4.4.2}$$

4.4.1 *Inverse of a N x N Matrix*

The Δ_{ij} matrix is obtained from the matrix M by removing the rows and columns that intersect at the element a_{ij}. Consider the matrix A with matrix elements given by the cofactors of M, together with the relevant alternating sign. More precisely

$$A_{ij} = \frac{1}{|M|}(-1)^{i+j}|\Delta_{ij}| \tag{4.4.3}$$

It can be shown that the inverse of M is given by

$$M^{-1} = A^T \quad \Rightarrow A^T M = \mathbb{I} \tag{4.4.4}$$

Noteworthy 4.1: Example

$$M = \begin{bmatrix} a & c \\ b & d \end{bmatrix} = \begin{bmatrix} a_{11} & a_{12} \\ a_{21} & a_{22} \end{bmatrix}$$

The cofactors are

$$|\Delta_{11}| = d \ ; \ |\Delta_{12}| = b \ ; \ |\Delta_{21}| = c \ ; \ |\Delta_{22}| = a$$

Hence, from Eq. 4.4.2, choosing $k = 2$, yields

$$\begin{aligned} |M| = \det[M] &= \sum_{j=1}^{2} (-1)^{j+2} a_{j2} |\Delta_{j2}| \\ &= (-1)^{1+2} a_{12} |\Delta_{12}| + (-1)^{2+2} a_{22} |\Delta_{22}| = -cb + da \end{aligned}$$

The cofactor and inverse of M, from Eqs. 4.4.3 and 4.4.4, are given by

$$A_{11} = \frac{1}{|M|} d \ ; \ A_{12} = -\frac{1}{|M|} b \ ; \ A_{21} = -\frac{1}{|M|} c \ ; \ A_{22} = \frac{1}{|M|} a$$

$$A = \frac{1}{|M|} \begin{bmatrix} d & -b \\ -c & a \end{bmatrix} \Rightarrow M^{-1} = A^T = \frac{1}{|M|} \begin{bmatrix} d & -c \\ -b & a \end{bmatrix}$$

Consider the diagonal element of Eq. 4.4.4; from Eq. 4.4.1

$$\begin{aligned} (MA^T)_{ii} &= \sum_{k=1}^{N} a_{ik} A^T_{ki} = \frac{1}{|M|} \sum_{k=1}^{N} a_{ik} (-1)^{i+k} |\Delta^T_{ki}| \\ &= \frac{1}{|M|} \sum_{k=1}^{N} a_{ik} (-1)^{i+k} |\Delta_{ik}| = 1 \end{aligned} \tag{4.4.5}$$

To prove Eq. 4.4.4, it has to be further shown that *all* the off-diagonal elements are zero; hence, one needs to show that

$$i \neq j \ \Rightarrow \ (MA^T)_{ij} = \sum_{k=1}^{N} a_{ik} A^T_{kj} = \frac{1}{|M|} \sum_{k=1}^{N} a_{ik} (-1)^{i+k} |\Delta_{jk}| = 0 \tag{4.4.6}$$

Hence, from Eqs. 4.4.5 and 4.4.6

$$(MA^T)_{ij} = \delta_{i-j} \ \Rightarrow \ M^{-1} = A^T$$

The proof of Eq. 4.4.6 for the $N \times N$ matrix requires a lot of notation and concepts. Instead, to illustrate the general case, a proof is given for $N = 3$. Consider the following 3×3 matrix

$$M = \begin{bmatrix} a_{11} & a_{12} & a_{13} \\ a_{21} & a_{22} & a_{23} \\ a_{31} & a_{32} & a_{33} \end{bmatrix}$$

From Eq. 4.4.6 one needs to show that all off-diagonal elements are zero; consider, for example, the specific case of $(MA^T)_{13}$. From Eq. 4.4.6 one needs to show that

$$(MA^T)_{13} = 0 \;\Rightarrow\; 0 = a_{11}|\Delta_{31}| - a_{12}|\Delta_{32}| + a_{13}|\Delta_{33}| \tag{4.4.7}$$

Note in the equation above, all the cofactors are of the form Δ_{3i} ; $i = 1, 2, 3$ and do not refer to any element a_{3i} ; $i = 1, 2, 3$ of the third row. To prove Eq. 4.4.7, consider the associated matrix M'

$$M' = \begin{bmatrix} a_{11} & a_{12} & a_{13} \\ a_{21} & a_{22} & a_{23} \\ a'_{31} & a'_{32} & a'_{33} \end{bmatrix}$$

Enumerating the determinant $|M'|$ using Eq. 4.4.1 and with $k = 3$ yields

$$|M'| = a'_{31}|\Delta_{31}| - a'_{32}|\Delta_{32}| + a'_{33}|\Delta_{33}| \tag{4.4.8}$$

Note the cofactors appearing in Eq. 4.4.8 are the same that appear in Eq. 4.4.7.

Now make the following special choice for the third row of M' to yield Γ given by

$$M' = \begin{bmatrix} a_{11} & a_{12} & a_{13} \\ a_{21} & a_{22} & a_{23} \\ a'_{31} & a'_{32} & a'_{33} \end{bmatrix} \rightarrow \Gamma = \begin{bmatrix} a_{11} & a_{12} & a_{13} \\ a_{21} & a_{22} & a_{23} \\ a_{11} & a_{12} & a_{13} \end{bmatrix}$$

Γ has two identical rows and hence its determinant is zero

$$|\Gamma| = 0$$

Enumerating $|\Gamma|$ using the last row as in Eq. 4.4.8 yields

$$0 = |\Gamma| = a_{11}|\Delta_{31}| - a_{12}|\Delta_{32}| + a_{13}|\Delta_{33}| = (MA^T)_{13}$$

Hence, result above proves Eq. 4.4.7.

Using a similar technique for the $N \times N$ matrix, one can prove Eq. 4.4.6.

4.5 Leontief Input-Output Model

Consider an economy producing $x_i,\ i = 1 \cdots N$ worth of commodities. All commodities are expressed in terms of the dollar value. For example, a company produces steel has an output of \$3 m per year. All commodities need other commodities, including itself, to be able to produce x_i worth of commodities; let a_{ij} be the $\$j$ amount of commodity that is required to produce \$1 worth of ith commodity. For example, $a_{32} = \$0.43$ is the dollar value of commodity 2 required to produce \$1 amount of commodity 3.

$A = a_{ij}$ is the input-output matrix. For the *open model* the economy is partitioned into two sectors, the production of commodities, that yields the amount x_i and the demand d_i outside the sphere of production for these commodities—for the purpose of consumption or for exporting. The input-output model is not an equilibrium model, but instead is a model that determines the correct quantity of production of x_i consistent with the inputs and demand. The input-output model yields, similar to Eq. 2.7.4, the following system of linear equations

$$\mathbf{x} = A\mathbf{x} + \mathbf{d}$$

In matrix notation

$$\begin{bmatrix} x_1 \\ \vdots \\ x_i \\ \vdots \\ x_N \end{bmatrix} = \begin{bmatrix} a_{11} & a_{12} & .. & a_{1N} \\ a_{21} & a_{22} & .. & a_{2N} \\ & .... & & \\ a_{i1} & a_{i2} & .. & a_{iN} \\ & .... & & \\ a_{N1} & a_{N2} & .. & a_{NN} \end{bmatrix} \begin{bmatrix} x_1 \\ \vdots \\ x_i \\ \vdots \\ x_N \end{bmatrix} + \begin{bmatrix} d_1 \\ \vdots \\ d_i \\ \vdots \\ d_N \end{bmatrix} \tag{4.5.1}$$

In components, the Leontief equation is written as

$$x_i = \sum_{j=1}^{N} a_{ij} x_j + d_i \ ; \ i = 1 \cdots N$$

There are constraints on the coefficients a_{ij}. To produce \$1 worth of commodity x_I, the inputs must be less than \$1; hence

$$\sum_{j=1}^{N} a_{Ij} < 1 \ ; \ I = 1, \cdots N$$

The solution for the correct amount of output $\mathbf{x}$ is given by

$$(\mathbb{I} - A)\mathbf{x} = \mathbf{d} \ \Rightarrow \ \mathbf{x} = (\mathbb{I} - A)^{-1}\mathbf{d}$$

Example. Consider a two-commodity economy with input-output matrix A and demand **d** given by

$$A = \begin{bmatrix} 0.5 \; 0.2 \\ 0.4 \; 0.1 \end{bmatrix} \; ; \; d = \begin{bmatrix} 7 \\ 4 \end{bmatrix} \Rightarrow x = (\mathbb{I} - A)^{-1} d = \begin{bmatrix} 19.19 \\ 12.97 \end{bmatrix} \tag{4.5.2}$$

4.5.1 Hawkins-Simon Condition

A major constraint on the Leontief equation is that all the quantities x_i must be **positive**. The demand vector **d** is positive. The Hawkins-Simon conditions needs to specified on the matrix $\mathbb{I} - A$ for a positive solution for x_i. Let $B = \mathbb{I} - A$ with entries

$$B = \mathbb{I} - A = \begin{bmatrix} b_{11} & b_{12} & .. & b_{1N} \\ b_{21} & b_{22} & .. & b_{2N} \\ .... \\ b_{i1} & b_{i2} & .. & b_{iN} \\ .... \\ b_{N1} & b_{N2} & .. & b_{NN} \end{bmatrix}$$

The Hawkins-Simon conditions states that *all the co-factors* of B, denoted by Δ_{ij}, must be positive; for example the following series of determinants must be positive

$$b_{11} > 0 \; ; \; \begin{vmatrix} b_{11} & b_{12} \\ b_{21} & b_{22} \end{vmatrix} > 0 \; ; \; \cdots \; ; \; \begin{vmatrix} b_{11} & b_{12} & .. & b_{1N} \\ b_{21} & b_{22} & .. & b_{2N} \\ .... \\ b_{i1} & b_{i2} & .. & b_{iN} \\ .... \\ b_{N1} & b_{N2} & .. & b_{NN} \end{vmatrix} > 0$$

For the example

$$B = (1 - A) = \begin{bmatrix} 0.5 & -0.2 \\ -0.4 & 0.9 \end{bmatrix} \; ; \; b_{11} = 0.5 > 0 \; ; \; |B| = 0.45 - 0.08 = 0.37 > 0$$

4.5.2 Two Commodity Case

The Leontief matrix for two commodities is given by

$$A = \begin{bmatrix} a_{11} \; a_{12} \\ a_{21} \; a_{22} \end{bmatrix}$$

Note that, as required by the Leontief matrix, the coefficients obey the following inequalities

$$a_{ij} > 0 \ : \ a_{11} + a_{12} < 1 \ ; \ a_{21} + a_{22} < 1$$

The Hawkins-Simon condition requires the matrix

$$B = \mathbb{I} - A = \begin{bmatrix} 1 - a_{11} & -a_{12} \\ -a_{21} & 1 - a_{22} \end{bmatrix}$$

The first condition is

$$b_{11} > 0 \ \Rightarrow \ 1 - a_{11} > 0 \ \Rightarrow \ a_{11} < 1$$

Above condition requires that the amount of first commodity required to produce a unit of the first commodity needs to be less than one unit. This is self-evident, since if one produces a commodity that requires an input of the same commodity greater than is produced, then this commodity will not be worth producing.

The second condition requires

$$\begin{vmatrix} b_{11} & b_{12} \\ b_{21} & b_{22} \end{vmatrix} > 0 \ \Rightarrow \ (1 - a_{11})(1 - a_{22}) - a_{12}a_{21} > 0$$

Since $a_{11} < 1$ implies $(1 - a_{11})a_{22} > 0$, the above condition reduces to

$$a_{11} + a_{12}a_{21} < 1$$

Recall a_{12} is the input of the second commodity required to produce one unit of the first commodity, and a_{21} is the input of the second commodity required to produce a unit of the first commodity. The term $a_{11} + a_{12}a_{21}$ is the amount of the first commodity used in the production of the second commodity; this has to be less than unity, since otherwise it would not be viable to produce the second commodity. Hence, the second Hawkins-Simon inequality is required to produce nonnegative amounts of the two commodities.

4.6 Symmetric Matrices

Symmetric matrices arise in many applications. The Hessian matrix is always a symmetric matrix. Gaussian integration for N variables is based on the properties of symmetric matrices (Table 4.2).

Consider the exchange rates between MYR, SGD and Rupiah.

Table 4.2 Exchange Rates (11 October 2016.)

	MYR	SGD	Rupiah
MYR	1	0.33	3106.5
SGD	0.330	1	9423.70
Rupiah	3106.5	9423.70	1

The representation of the FX is given by the symmetric matrix[1]

$$S = \begin{bmatrix} 1 & 0.33 & 3106.5 \\ 0.33 & 1 & 9423.70 \\ 13106.5 & 9423.70 & 1 \end{bmatrix} = S^T \tag{4.6.1}$$

Matrix algebra is a useful tool for writing the payoff of FX options and the condition of no arbitrage for multiple currencies.

4.7 2 × 2 Symmetric Matrix

To illustrate some important features of symmetric matrices, consider the following 2 × 2 real symmetric matrix

$$S = \begin{bmatrix} \alpha & \beta \\ \beta & \gamma \end{bmatrix} = S^T \; ; \; \alpha, \beta, \gamma : \text{real} \tag{4.7.1}$$

The eigenvalue equation is given by

$$\det(S - \lambda \mathbb{I}) = \det \begin{vmatrix} \alpha - \lambda & \beta \\ \beta & \gamma - \lambda \end{vmatrix} = 0 \tag{4.7.2}$$

Solving the eigenvalue equation yields

$$\lambda_{\pm} = \frac{1}{2}(\alpha + \gamma) \pm \sqrt{\frac{1}{4}(\alpha - \gamma)^2 + \beta^2} : \text{real} \tag{4.7.3}$$

Note the eigenvalues are *always real*. Once the eigenvalues have been ascertained, the eigenvectors require solving a system of simultaneous linear equations.

The eigenvector for λ_+, denoted by η_+, is given by the following eigenvector equation

[1]Note that $MYR \to SGD \to$ Rupiah $\approx MYR \to$ Rupiah. Transaction costs explains the lack of exact equality.

$$\eta_+ = \begin{bmatrix} a \\ b \end{bmatrix} \Rightarrow \begin{bmatrix} \alpha & \beta \\ \beta & \gamma \end{bmatrix}\begin{bmatrix} a \\ b \end{bmatrix} = \lambda_+ \begin{bmatrix} a \\ b \end{bmatrix}$$

One obtains the following simultaneous linear equations

$$\begin{aligned} (-\lambda_+ + \alpha)a + \beta b &= 0 \\ \beta a + (-\lambda_+ + \gamma)b &= 0 \end{aligned} \tag{4.7.4}$$

Since the eigenvector is normalized, take $a = 1$; this yields from Eq. 4.7.4

$$b = \frac{-\gamma + \lambda_+}{\beta}$$

and a similar equation for the eigenvector of λ_-, denoted by η_-. The (normalized) eigenvectors of the symmetric matrix S are given by

$$\eta_+ = N_+ \begin{bmatrix} 1 \\ \frac{-\alpha+\lambda_+}{\beta} \end{bmatrix} ; \ \eta_- = N_- \begin{bmatrix} \frac{-\gamma+\lambda_-}{\beta} \\ 1 \end{bmatrix} \tag{4.7.5}$$

The two eigenvectors are orthogonal since, from Eq. 4.7.5

$$\begin{aligned} \eta_+^T \cdot \eta_- &= N_+ N_- \Big(\frac{-\gamma + \lambda_-}{\beta} + \frac{-\alpha + \lambda_+}{\beta} \Big) \\ &= \frac{N_+ N_-}{\beta} \Big(-\gamma - \alpha + \lambda_+ + \lambda_- \Big) = 0 \end{aligned}$$

The normalization is fixed by

$$1 = \eta_+^T \cdot \eta_+ = N_+^2 \Big(1 + (\frac{-\alpha + \lambda_+}{\beta})^2 \Big) \Rightarrow N_+ = \sqrt{\frac{\beta^2}{(-\alpha + \lambda_+)^2 + \beta^2}}$$

and similarly for N_-. Hence

$$\eta_\pm^T \cdot \eta_\pm = 1 \ ; \ \eta_+^T \cdot \eta_- = 0$$

$$N_+ = \sqrt{\frac{\beta^2}{(-\alpha + \lambda_+)^2 + \beta^2}} \ ; \ N_- = \sqrt{\frac{\beta^2}{(-\gamma + \lambda_-)^2 + \beta^2}}$$

The spectral decomposition and completeness equation are given by

$$S = \lambda_+ \eta_+ \otimes \eta_+^T + \lambda_- \eta_- \otimes \eta_-^T = \begin{bmatrix} \alpha & \beta \\ \beta & \gamma \end{bmatrix} \tag{4.7.6}$$

$$\eta_+ \otimes \eta_+^T + \eta_- \otimes \eta_-^T = \begin{bmatrix} 1 & 0 \\ 0 & 1 \end{bmatrix} = \mathbb{I} \tag{4.7.7}$$

In summary, the eigenvalues of a symmetric matrix are real and the eigenvectors corresponding to the two distinct eigenvalues are orthogonal.

4.8 $N \times N$ Symmetric Matrix

Consider a $N \times N$ symmetric matrix S with components

$$S = S^T \quad \Rightarrow \quad S_{ij} = S_{ji} \;\; i, j = 1, 2 \cdots, N$$

The eigenvalue equation yields

$$S\mathbf{v}_I = \lambda_I \mathbf{v}_I \;\; ; \;\; I = 1, 2, \cdots, N \tag{4.8.1}$$

where the N eigenvalues are given by the secular equation

$$\det(S - \lambda\mathbb{I}) = 0$$

Symmetric matrices have two important features, which we have seen holds for the special 2×2 case.

- All the eigenvalues λ_I are real, which are assumed to be all distinct[2]
- The N eigenvectors $\mathbf{v}_I$ are linearly independent

To generalize the results of E_2 case to E_N, the notation needs to be enhanced. A N-component vector $\mathbf{v}_I$ has components $v_I(j)$ given by[3]

$$\mathbf{e}_j^T \cdot \mathbf{v}_I = v_I(j) \equiv v_{jI} \;\; ; \;\; \mathbf{v}_I = \sum_j v_{jI} \mathbf{e}_j$$

Note that we have the convention

$$v_I(j) \equiv v_{jI}$$

[2]The case of eigenvalues having the same value can be obtained from the unequal case using the so called Gram-Schmidt procedure.

[3] In particular, the components of basis vector $\mathbf{e}_I$ are given by

$$\mathbf{e}_j^T \cdot \mathbf{e}_I = e_I(j) = \delta_{I-j}$$

.

Hence

$$\mathbf{v}_J^T \cdot \mathbf{v}_I = \sum_j v_{jJ} v_{jI} \ ; \ \mathbf{v}_I^T \cdot \mathbf{v}_I = \sum_j v_{jI}^2 = 1$$

The notation is adopted for reasons that will become clear further down. Anticipating later results, it will be shown that the symmetric matrix S has N-eigenvectors $\mathbf{v}_I$, which are all mutually orthogonal. In the notation of Eq. 3.7.3, S can be written in terms of its component S_{ij} as follows

$$S = \sum_{ij} S_{ij} \mathbf{e}_i \otimes \mathbf{e}_j^T \ ; \ \mathbf{v}_I = \sum_j v_{jI} \mathbf{e}_j \tag{4.8.2}$$

Therefore

$$S\mathbf{v}_I = \sum_{ijk} S_{ij} v_{kI} \mathbf{e}_i \otimes (\mathbf{e}_j^T \cdot \mathbf{e}_k) = \sum_{ij} S_{ij} v_{jI} \mathbf{e}_i = \lambda_I \sum_i v_{iI} \mathbf{e}_i$$

The linear independence of the basis vectors yields, in terms of the components of the S matrix and eigenvector $\mathbf{v}$, the eigenvalue equation

$$\sum_j S_{ij} v_{jI} = \lambda_I v_{iI} \tag{4.8.3}$$

The N-component vectors can be arranged into a column vector and a row vector

$$\mathbf{v}_I = \begin{bmatrix} v_{1I} \\ \vdots \\ v_{iI} \\ \vdots \\ v_{NI} \end{bmatrix} \ ; \ \mathbf{v}_I^T = \begin{bmatrix} v_{1I} & \cdots & v_{iI} & \cdots & v_{NI} \end{bmatrix} \tag{4.8.4}$$

The eigenvectors corresponding to different eigenvalues are orthogonal. To prove this, consider the scalar product

$$\mathbf{v}_I^T \cdot \mathbf{v}_J = \mathbf{v}_J^T \cdot \mathbf{v}_I$$

For eigenvalues λ_I, λ_J the eigenvector equation yields

$$\mathbf{v}_I^T S \mathbf{v}_J = \lambda_J \mathbf{v}_I^T \cdot \mathbf{v}_J \ ; \ \mathbf{v}_I^T S \mathbf{v}_J = (\mathbf{v}_J^T S \mathbf{v}_I)^T = \lambda_I \mathbf{v}_I^T \cdot \mathbf{v}_J$$

Subtracting the above results, and using $\lambda_I \neq \lambda_J$, yields

$$0 = (\lambda_I - \lambda_J) \mathbf{v}_I^T \cdot \mathbf{v}_J \ \Rightarrow \ \mathbf{v}_I^T \cdot \mathbf{v}_J = 0 \tag{4.8.5}$$

Orthogonality of the eigenvector proves that all the eigenvectors are linearly independent. Since $\mathbf{v}_I$ are N orthogonal vectors, they form a complete basis of the space E_N. Hence, the orthonormality of the eigenvectors yields

$$\mathbb{I} = \sum_I \mathbf{v}_I \otimes \mathbf{v}_I^T$$
$$\Rightarrow \delta_{i-j} = \sum_I v_I(i)\, v_I(j) \ : \ \text{completeness equation} \tag{4.8.6}$$

The completeness equation, or the resolution of the identity, given in Eq. 4.8.6 shows that the eigenvectors are a possible basis states for E_N. This is discussed in some detail in Sect. 4.13.

Using the completeness equation given in Eq. 4.8.6 yields the spectral decomposition for S as follows

$$S = S\mathbb{I} = \sum_I S\mathbf{v}_I \otimes \mathbf{v}_I^T = \sum_I \lambda_I \mathbf{v}_I \otimes \mathbf{v}_I^T \tag{4.8.7}$$

The spectral representation allows one to form any function of S. For starters, note that

$$S^2 = \sum_{I,J=1}^{N} \lambda_I \lambda_J \mathbf{v}_I \otimes (\mathbf{v}_I^T \cdot \mathbf{v}_J) \otimes \mathbf{v}_J^T = \sum_{I=1}^{N} \lambda_I^2 \mathbf{v}_I \otimes \mathbf{v}_I^T$$

Hence

$$S^n = \sum_{I=1}^{N} \lambda_I^n \mathbf{v}_I \otimes \mathbf{v}_I^T \tag{4.8.8}$$

In general, for any function $F(S)$, one has

$$F(S) = \sum_{I=1}^{N} F(\lambda_I) \mathbf{v}_I \otimes \mathbf{v}_I^T \tag{4.8.9}$$

In particular

$$\exp(S) = \sum_{I=1}^{N} \exp(\lambda_I) \mathbf{v}_I \otimes \mathbf{v}_I^T \tag{4.8.10}$$

Equations 4.8.9 and 4.8.10 are the results, expressed in terms of the spectral resolution. The same results are given later, in Eqs. 4.10.10 and 4.10.9, in matrix form.

4.9 Orthogonal Matrices

Consider N orthonormal vectors that form an orthonormal basis $\mathbf{f}_I$, $I = 1, 2, \cdots, N$, with each vector being N-components. Hence

$$\mathbf{f}_I^T \cdot \mathbf{f}_J = \delta_{I-J} \;\Rightarrow\; \mathbf{f}_I^T \cdot \mathbf{f}_I = 1 \tag{4.9.1}$$

The basis states can be arranged into the following $N \times N$ matrix

$$\mathcal{O} = \left[\mathbf{f}_1 \cdots \mathbf{f}_I \,..\, \mathbf{f}_N\right] = \begin{bmatrix} f_{11} & \cdots f_{1I} & \cdots f_{1N} \\ f_{21} & \cdots f_{2I} & \cdots f_{2N} \\ \vdots & & \\ f_{N1} & \cdots f_{NI} & \cdots f_{NN} \end{bmatrix} \;;\; \mathcal{O}^T = \begin{bmatrix} \mathbf{f}_1^T \\ \vdots \\ \mathbf{f}_I^T \\ \vdots \\ \mathbf{f}_N^T \end{bmatrix}$$

$\mathcal{O}$ is an orthogonal matrix. To see this consider consider, in components, the matrix multiplication of $\mathcal{O}^T$ with $\mathcal{O}$, given by

$$\mathcal{O}^T\mathcal{O} = \begin{bmatrix} f_{11} & \cdots f_{I1} & \cdots f_{N1} \\ \vdots & \vdots & \vdots \\ f_{1J} & \cdots f_{IJ} & \cdots f_{NJ} \\ \vdots & & \\ f_{1N} & \cdots f_{IN} & \cdots f_{NN} \end{bmatrix} \begin{bmatrix} f_{11} & \cdots f_{1J} & \cdots f_{1N} \\ \vdots & \vdots & \vdots \\ f_{I1} & \cdots f_{IJ} & \cdots f_{IN} \\ \vdots & & \\ f_{N1} & \cdots f_{NJ} & \cdots f_{NN} \end{bmatrix}$$

Matrix multiplication given above can be written compactly using the following notation

$$\mathcal{O}^T\mathcal{O} = \begin{bmatrix} \mathbf{f}_1^T \cdot \mathbf{f}_1 & \cdots \mathbf{f}_1^T \cdot \mathbf{f}_I & \cdots \mathbf{f}_1^T \cdot \mathbf{f}_N \\ \vdots & \vdots & \vdots \\ \mathbf{f}_I^T \cdot \mathbf{f}_1 & \cdots \mathbf{f}_I^T \cdot \mathbf{f}_I & \cdots \mathbf{f}_1^T \cdot \mathbf{f}_N \\ \vdots & & \\ \mathbf{f}_N^T \cdot \mathbf{f}_1 & \cdots \mathbf{f}_N^T \cdot \mathbf{f}_I & \cdots \mathbf{f}_N^T \cdot \mathbf{f}_N \end{bmatrix} = \begin{bmatrix} \mathbf{f}_1^T \\ \vdots \\ \mathbf{f}_I^T \\ \vdots \\ \mathbf{f}_N^T \end{bmatrix} \otimes \left[\mathbf{f}_1 \cdots \mathbf{f}_I \,..\mathbf{f}_N\right] \tag{4.9.2}$$

The notation in Eq. 4.9.2 shows that matrix multiplication of $\mathcal{O}^T$ into $\mathcal{O}$ can be viewed as the tensor product of a column vector (where each element is a (dual) *row* vector given by $\mathbf{f}_I^T$) with a row vector (where each element is a *column* vector given by $\mathbf{f}_I$).

Hence, from above and Eqs. 4.9.2 and 4.9.1

$$\mathcal{O}^T\mathcal{O} = \mathcal{O}\mathcal{O}^T = \begin{bmatrix} 1 & 0 & \cdots 0 \\ \vdots & \vdots & \vdots \\ 0 \cdots 1 \cdots 0 \\ \vdots \\ 0 \cdots 0 \cdots 1 \end{bmatrix} = \mathbb{I} \tag{4.9.3}$$

4.10 Symmetric Matrix: Diagonalization

Suppose all the eigenvalues and eigenvectors of a symmetric matrix have been determined. We discuss a procedure that uses the eigenvalues and eigenvectors to recursively diagonalize a symmetric matrix. The result is an expression of the spectral decomposition and of the completeness equation.

Suppose that $\mathbf{v}_1$ is an eigenvector of S. Choose orthonormal vectors $\mathbf{f}_I$, $I = 2, 3, \cdots, N$ such that

$$\mathbf{v}_1^T \cdot \mathbf{v}_1 = 1 \; ; \; \mathbf{v}_1^T \cdot \mathbf{f}_I = 0 = \mathbf{f}_I^T \cdot \mathbf{v}_1 \; ; \; \mathbf{f}_I^T \cdot \mathbf{f}_J = \delta_{I-J} \tag{4.10.1}$$

Define the orthogonal matrix

$$\mathcal{P} = \left[\mathbf{v}_1 \, \mathbf{f}_2 \cdots \mathbf{f}_I \, .. \, \mathbf{f}_N\right] \; ; \; \mathcal{P}^T\mathcal{P} = \mathbb{I}$$

Consider the matrix

$$\mathcal{D} = \mathcal{P}^T S \mathcal{P} \tag{4.10.2}$$

Note that $\mathcal{D}$ is symmetric since

$$\mathcal{D}^T = \mathcal{P}^T S^T \mathcal{P} = \mathcal{P}^T S \mathcal{P} = \mathcal{D}$$

Writing out Eq. 4.10.2 yields

$$\mathcal{P}^T S \mathcal{P} = \begin{bmatrix} \mathbf{v}_1^T \\ \vdots \\ \mathbf{f}_I^T \\ \vdots \\ \mathbf{f}_N^T \end{bmatrix} S \left[\mathbf{v}_1 \, \mathbf{f}_2 \cdots \mathbf{f}_I \, .. \mathbf{f}_N\right] = \begin{bmatrix} \mathbf{v}_1^T \\ \vdots \\ \mathbf{f}_I^T \\ \vdots \\ \mathbf{f}_N^T \end{bmatrix} \otimes \left[\lambda_1 \mathbf{v}_1 \cdots S\mathbf{f}_I \, .. S\mathbf{f}_N\right]$$

From Eqs. 4.9.1 and 4.9.2

$$\mathcal{P}^T S\mathcal{P} = \begin{bmatrix} \lambda_1 & \cdots \mathbf{v}_1^T S\mathbf{f}_I & \cdots \mathbf{v}_1^T S\mathbf{f}_N \\ \vdots & \vdots & \vdots \\ \mathbf{f}_1^T \cdot \mathbf{v}_1 & \cdots * & * \\ \vdots & & \\ \mathbf{f}_N^T \cdot \mathbf{v}_1 & \cdots * & \cdots * \end{bmatrix} = \begin{bmatrix} \lambda_1 \cdots 0 \cdots 0 \\ \vdots \quad \vdots \quad \vdots \\ 0 \cdots * \quad * \\ \vdots \\ 0 \cdots * \cdots * \end{bmatrix} \tag{4.10.3}$$

where entries denoted by $*$ are nonzero. The first row (except the diagonal term) is zero due to the following: since S is symmetric

$$S\mathbf{v}_1 = \lambda_1 \mathbf{v}_1 \Rightarrow \mathbf{v}_1^T S = \lambda_1 \mathbf{v}_1^T \ ; \ \mathbf{v}_1^T S\mathbf{f}_I = \lambda_1 \mathbf{v}_1^T \cdot \mathbf{f}_I = 0$$

where the last equation above follows from Eq. 4.10.1. Hence

$$\mathcal{P}^T S\mathcal{P} = \begin{bmatrix} \lambda_1 \cdots 0 \cdots 0 \\ \vdots \quad \vdots \quad \vdots \\ 0 \quad S' \\ \vdots \\ 0 \quad \cdots \quad \cdots \end{bmatrix} \tag{4.10.4}$$

$\mathcal{P}^T S\mathcal{P}$ is symmetric since S is symmetric; hence, the matrix S' is another $(N-1) \times (N-1)$ *symmetric matrix*

$$(S')^T = S'$$

As shown in Eq. 4.8.5, the eigenvectors corresponding to different eigenvalues are orthogonal.

To complete the procedure for diagonalizing S, consider the eigenvector $\mathbf{v}_2$ with (distinct) eigenvalue λ_2. The recursive process of diagonalization can be carried another step by including the eigenvector $\mathbf{v}_2$, which is orthogonal to $\mathbf{v}_1$, as one of the basis state—as well as another $N-2$ orthonormal basis vectors (which in general are different from what we started) and denoted by $\mathbf{h}_I$. Define the orthogonal matrix

$$\mathcal{P}' = \begin{bmatrix} \mathbf{v}_1 \ \mathbf{v}_2 \cdots \mathbf{h}_I \cdots \mathbf{h}_N \end{bmatrix} \ ; \ (P')^T \mathcal{P}' = \mathbb{I}$$

The symmetric matrix S' is then partially diagonalized.

This procedure is carried out N-times to achieve the complete diagonalization of the symmetric matrix S. We obtain the final result

$$\mathcal{V}^T S \mathcal{V} = \begin{bmatrix} \lambda_1 & 0 & .. & 0 \\ 0 & \lambda_2 & .. & 0 \\ \vdots & & & \\ 0 & 0 & .. & \lambda_N \end{bmatrix} = \mathcal{D} \tag{4.10.5}$$

The orthogonal matrix $\mathcal{V}$ is given by

$$\mathcal{V} = \begin{bmatrix} \mathbf{v}_1 \cdots \mathbf{v}_I .. \mathbf{v}_N \end{bmatrix} \; ; \; \mathcal{V}^T \mathcal{V} = \mathbb{I} \tag{4.10.6}$$

$\mathcal{V}$ is orthogonal and, from Eq. 4.3.5, $\det(\mathcal{V}) = \det(\mathcal{V}^T)$; hence

$$\det(\mathcal{V}^T)\det(\mathcal{V}) = \det(\mathcal{V})^2 = \det(\mathbb{I}) = 1$$
$$\Rightarrow \det(\mathcal{V}) = \pm 1 \; ; \; \mathcal{V}^T \mathcal{V} = \mathbb{I} \tag{4.10.7}$$

The proof of diagonalization has utilized the fact that, as given in Eq. 4.8.5, the N eigenvectors of S are orthonormal

$$\mathbf{v}_I^T \cdot \mathbf{v}_J = \delta_{I-J}$$

Noteworthy 4.2: Example

Consider matrix A given by

$$A = \begin{bmatrix} 0 & 1 \\ 1 & 0 \end{bmatrix}$$

The transformation diagonalizing A is given by

$$R = \begin{bmatrix} \mathbf{v}_+ & \mathbf{v}_- \end{bmatrix} = \frac{1}{\sqrt{2}} \begin{bmatrix} 1 & 1 \\ 1 & -1 \end{bmatrix} \; ; \; R^T = \frac{1}{\sqrt{2}} \begin{bmatrix} 1 & 1 \\ 1 & -1 \end{bmatrix} \tag{4.10.8}$$

The matrix A is diagonalized by matrix R since

$$R^T A R = \frac{1}{2} \begin{bmatrix} 1 & 1 \\ 1 & -1 \end{bmatrix} \begin{bmatrix} 0 & 1 \\ 1 & 0 \end{bmatrix} \begin{bmatrix} 1 & 1 \\ 1 & -1 \end{bmatrix} = \begin{bmatrix} 1 & 0 \\ 0 & -1 \end{bmatrix} = \begin{bmatrix} \lambda_+ & 0 \\ 0 & \lambda_- \end{bmatrix}$$

4.10.1 Functions of a Symmetric Matrix

The diagonalization leads to the following expression for S

$$S = \mathcal{V} \mathcal{D} \mathcal{V}^T \; ; \; \mathcal{V} \mathcal{D} \mathcal{V}^T = \mathbb{I}$$

The diagonalization of the (symmetric) matrix shows that S is the collection of its eigenvalues and eigenvectors.

We study two important functions of a matrix, which is its determinant and its inverse. The determinant, from Eqs. 4.3.6 and 4.10.7, is given by

$$\det(S) = \det(\mathcal{V})\det(\mathcal{D})\det(\mathcal{V}^T) \Rightarrow \det(S) = \det(\mathcal{D}) = \prod_{i=1}^{N} \lambda_i$$

Note that the determinant is zero if *any* eigenvalue is zero. As can be verified by direct substitution, the inverse of matrix S is given by

$$S = \mathcal{V}\mathcal{D}^{-1}\mathcal{V}^T = \mathcal{V}\begin{bmatrix} \frac{1}{\lambda_1} & 0 & .. & 0 \\ 0 & \frac{1}{\lambda_2} & .. & 0 \\ \vdots & & & \\ 0 & 0 & .. & \frac{1}{\lambda_N} \end{bmatrix}\mathcal{V}^T$$

One can now see why the determinant needs to be nonzero for the inverse to exist; this is because every eigenvalue is inverted to obtain the inverse, and any eigenvalue being zero would cause the inverse to be undefined.

Any arbitrary power S^n is given by the following

$$S^n = \underbrace{\mathcal{V}\mathcal{D}\mathcal{V}^T\mathcal{V}\mathcal{D}\mathcal{V}^T \cdots \mathcal{V}\mathcal{D}\mathcal{V}^T}_{n\text{-times}} = \mathcal{V}\mathcal{D}^n\mathcal{V}^T = \mathcal{V}\begin{bmatrix} \lambda_1^n & 0 & .. & 0 \\ 0 & \lambda_2^n & .. & 0 \\ \vdots & & & \\ 0 & 0 & .. & \lambda_N^n \end{bmatrix}\mathcal{V}^T$$

Hence, using Taylor expansion, the exponential of a matrix is given by

$$\exp(S) = \mathcal{V}\begin{bmatrix} e^{\lambda_1} & 0 & .. & 0 \\ 0 & e^{\lambda_2} & .. & 0 \\ \vdots & & & \\ 0 & 0 & .. & e^{\lambda_N} \end{bmatrix}\mathcal{V}^T \qquad (4.10.9)$$

In general, any function F of the matrix S is given by

$$F(S) = \mathcal{V}\begin{bmatrix} F(\lambda_1) & 0 & .. & 0 \\ 0 & F(\lambda_2) & .. & 0 \\ \vdots & & & \\ 0 & 0 & .. & F(\lambda_N) \end{bmatrix}\mathcal{V}^T \qquad (4.10.10)$$

The transformation encoded in the matrix $\mathcal{V}$ needs a coordinate system to be written out explicitly. The eigenvalues are independent of the coordinate system, and the trace function of matrix is defined to pick out the eigenvalues. Define the trace of a matrix A by summing over all the diagonal elements

$$\text{Trace}(A) = \sum_i A_{ii}$$

It follows that the Trace is a cyclic function of the matrices in the sense that

$$\text{Trace}(ABC) = \text{Trace}(CAB) = \text{Trace}(BCA)$$

For symmetric matrix S, trace yields the following

$$\text{ch}_n = \text{Trace}(S^n) = \text{Trace}(\mathcal{V}\mathcal{D}^n\mathcal{V}^T) = \text{Trace}(\mathcal{D}^n) = \sum_i \lambda_i^n$$

Hence, on evaluating all the trace functions $\text{ch}_1, \text{ch}_2, \cdots, \text{ch}_N$ one can calculate all the N-eigenvalues $\lambda_i,\ i = 1, 2, \cdots, N$.

4.11 Hermitian Matrices

All the results for symmetric matrices can also be derived for Hermitian matrices: all transpositions are replaced by Hermitian conjugation, which recall entails transposition as well as the complex conjugation of all the elements. In addition, one needs to keep track of the fact that the elements of a Hermitian matrix, in general, are complex. In particular, the dual of a vector is defined by transposition and complex conjugation and given by

$$\mathbf{e}_i^\dagger = (\mathbf{e}_i^*)^T$$

Consider a general 2×2 Hermitian matrix

$$H = \begin{bmatrix} \alpha & \beta \\ \beta^* & \gamma \end{bmatrix} = H^\dagger \ ; \ \alpha, \gamma : \text{real} \ ; \ \beta : \text{complex} \tag{4.11.1}$$

Solving the eigenvalue equation yields real eigenvalues given by

$$\lambda_\pm = \frac{1}{2}(\alpha + \gamma) \pm \sqrt{\frac{1}{4}(\alpha - \gamma)^2 + |\beta|^2} : \text{real}$$

The (normalized) eigenvectors of the Hermitian matrix are given by a procedure similar to the one carried out in detail in Sect. 4.7 for a 2×2 symmetric matrix. The eigenvectors are given by the following

$$\mathbf{e}_+ = N_+ \begin{bmatrix} 1 \\ \frac{\gamma-\lambda_+}{\beta^*} \end{bmatrix} \; ; \; \mathbf{e}_- = N_- \begin{bmatrix} \frac{\gamma-\lambda_-}{\beta^*} \\ 1 \end{bmatrix}$$

$$N_\pm = \sqrt{\frac{|\beta|^2}{(\gamma-\lambda_\pm)^2 + |\beta|^2}}$$

$$\mathbf{e}_\pm^\dagger \cdot \mathbf{e}_\pm = 1 \; ; \; \mathbf{e}_+^\dagger \cdot \mathbf{e}_- = 0$$

The completeness equation is given by

$$\mathbf{e}_+^\dagger \cdot \mathbf{e}_+ + \mathbf{e}_-^\dagger \cdot \mathbf{e}_- = \begin{bmatrix} 1 & 0 \\ 0 & 1 \end{bmatrix} = \mathbb{I} \tag{4.11.2}$$

The eigenvalue equation for a $N \times N$ Hermitian matrix H is given by

$$H\mathbf{u}_I = \lambda_I \mathbf{u}_I \; ; \; \mathbf{u}_I^\dagger \cdot \mathbf{u}_J = \delta_{I-J}$$

where $\mathbf{u}_I$ are the complex valued orthonormal eigenvectors.

Similar to a symmetric matrix, all the eigenvalues of a Hermitian matrix are real and H can be diagonalized by unitary matrices. In other words, the $N \times N$ Hermitian matrix H is diagonalized by a $N \times N$ unitary matrix U that is given by

$$\mathcal{U} = \begin{bmatrix} \mathbf{u}_1 \cdots \mathbf{u}_I \cdots \mathbf{u}_N \end{bmatrix} \; ; \; \mathcal{U}^\dagger = \begin{bmatrix} \mathbf{u}_1^\dagger \\ \vdots \\ \mathbf{u}_I^\dagger \\ \vdots \\ \mathbf{u}_N^\dagger \end{bmatrix} \; ; \; \mathcal{U}^\dagger \mathcal{U} = \mathbb{I}$$

Similar to Eq. 4.10.5, one has

$$\mathcal{U}^\dagger H \mathcal{U} = \begin{bmatrix} \lambda_1 & 0 & .. & 0 \\ 0 & \lambda_2 & .. & 0 \\ \vdots & & & \\ 0 & 0 & .. & \lambda_N \end{bmatrix} = \mathcal{D} \tag{4.11.3}$$

The completeness equation is given by

$$\mathbb{I} = \sum_I \mathbf{u}_I^\dagger \mathbf{u}_I$$

4.12 Diagonalizable Matrices

What is the criterion for a complex square matrix M to be diagonalizable?

Some books make the statement that the matrix M must have a representation given by $M = S^\dagger DS$, where D is diagonal. This statement is not a proof since it *assumes* the answer; instead, we need to first directly analyze M and then decide if a diagonal representation of M, in fact, at all exists.

There are many ways of characterizing matrices that are diagonalizable. A criterion that always holds has been mentioned earlier in Sect. 3.6 in the context of the inverse. 'For the inverse of a $N \times N$ matrix to exist, its rank must be N; this in turn implies that its determinant is not zero.' The rank being N also implies that the matrix is diagonalizable.

Another criterion for a matrix to be diagonalizable is now discussed. Consider two real $\mathcal{D}_A, \mathcal{D}_B$ *diagonal matrices*. Suppose A, B are Hermitian matrices; then, from Sect. 4.11, both of them can be separately diagonalized by a unitary matrix. Now assume further that both Hermitian matrices A, B can be diagonalized by the *same* unitary matrix $\mathcal{U}$, and are given by

$$A = \mathcal{U}\mathcal{D}_A\mathcal{U}^\dagger \ ; \ B = \mathcal{U}\mathcal{D}_B\mathcal{U}^\dagger$$

where $\mathcal{D}_A, \mathcal{D}_B$ are the diagonal matrices. This yields

$$AB = \mathcal{U}\mathcal{D}_A\mathcal{D}_B\mathcal{U}^\dagger = \mathcal{U}\mathcal{D}_B\mathcal{D}_A\mathcal{U}^\dagger = BA \tag{4.12.1}$$

In other words, two matrices A, B can be simultaneously diagonalized by the same matrix $\mathcal{U}$, only if the matrices A, B commute.

Write the matrix M as follows

$$M = \frac{1}{2}(M + M^\dagger) + i \cdot \frac{1}{2i}(M - M^\dagger) \equiv A + iB$$

where

$$A = \frac{1}{2}(M + M^\dagger) = A^\dagger \ ; \ B = \frac{1}{2i}(M - M^\dagger) = B^\dagger$$

Hence, for arbitrary square matrix M, both A, B are Hermitian matrices. If A, B can be simultaneously diagonalized, then M can be diagonalized.

For A, B to be simultaneously diagonalizable, from Eq. 4.12.1 they must commute, that is $AB - BA \equiv [A, B] = 0$. Consider the following commutator

$$[M, M^\dagger] = [A + iB, A - iB] = \frac{1}{2i}[A, B] \tag{4.12.2}$$

If M commutes with $M^\dagger$, then

$$[M, M^{\dagger}] = 0 \tag{4.12.3}$$

Eq. 4.12.3 implies, from Eq. 4.12.2, that

$$[M, M^{\dagger}] = 0 \;\Rightarrow\; [A, B] = 0$$

Since $[A, B] = 0$, from Eq. 4.12.1

$$A = \mathcal{U}\mathcal{D}_A\mathcal{U}^{\dagger} \;;\;\; B = \mathcal{U}\mathcal{D}_B\mathcal{U}^{\dagger}$$

Hence M can be diagonalized and is given by

$$M = A + iB = \mathcal{U}(D_A + iD_B)\mathcal{U}^{\dagger} = \mathcal{U}\mathcal{D}_M\mathcal{U}^{\dagger}$$

where $\mathcal{D}_M$ is a *diagonal matrix* given by

$$\mathcal{D}_M = \mathcal{D}_A + i\mathcal{D}_B$$

We hence conclude that

$$M \text{ is diagonlizable if } [M, M^{\dagger}] = 0$$

In the matrix R is real, then the condition reduces to the following

$$R \text{ is diagonlizable if } [R, R^T] = 0$$

Note a symmetric, unitary and Hermitian matrix each fulfill the criterion of $[M, M^{\dagger}] = 0$.

4.12.1 Non-symmetric Matrix

To illustrate that linear independence also leads to diagonalizability, consider the matrix

$$M = \begin{bmatrix} a & b-a \\ 0 & b \end{bmatrix} \neq M^T$$

It can easily verified that the columns yield two vectors that are linearly independent. Note also that $[M, M^T] = 0$ since

$$MM^T = \begin{bmatrix} a^2 & b(b-a) \\ b(b-a) & b^2 \end{bmatrix} = M^T M$$

Hence we conclude that M can be diagonalized.

There are general features that arise from an explicit diagonalization that we now consider. The eigenvalue equation yields

$$\det \begin{vmatrix} a-\lambda & b-a \\ 0 & b-\lambda \end{vmatrix} = 0 \ : \ \lambda_a = a \ ; \ \lambda_b = b$$

Since M is not symmetric, for a given eigenvalue both the left and right eigenvectors (which are not equal) need to be determined.
Case I: $\lambda_a = a$

$$M\mathbf{v}_a = a\mathbf{v}_a \ \Rightarrow \ \mathbf{v}_a = \begin{bmatrix} 1 \\ 0 \end{bmatrix} \ ; \ \mathbf{w}_a^T M = a\mathbf{w}_a^T \ \Rightarrow \ \mathbf{w}_a^T = \begin{bmatrix} 1 & -1 \end{bmatrix}$$

Case II: $\lambda_b = b$

$$M\mathbf{v}_b = b\mathbf{v}_b \ \Rightarrow \ \mathbf{v}_b = \begin{bmatrix} 1 \\ 1 \end{bmatrix} \ ; \ \mathbf{w}_b^T M = b\mathbf{w}_b^T \ \Rightarrow \ \mathbf{w}_b^T = \begin{bmatrix} 0 & 1 \end{bmatrix}$$

The orthonormality condition for a non-symmetric matrix is given by

$$\mathbf{w}_a^T \cdot \mathbf{v}_a = 1 = \mathbf{w}_b^T \cdot \mathbf{v}_b \ ; \ \mathbf{w}_a^T \cdot \mathbf{v}_b = 0 = \mathbf{w}_b^T \cdot \mathbf{v}_a$$

The spectral decomposition of the matrix is given by

$$\begin{aligned} \lambda_a \mathbf{v}_a \otimes \mathbf{w}_a^T + \lambda_b \mathbf{v}_b \otimes \mathbf{w}_b^T &= a \begin{bmatrix} 1 \\ 0 \end{bmatrix} \otimes \begin{bmatrix} 1 & -1 \end{bmatrix} + b \begin{bmatrix} 1 \\ 1 \end{bmatrix} \otimes \begin{bmatrix} 0 & 1 \end{bmatrix} \\ &= a \begin{bmatrix} 1 & -1 \\ 0 & 0 \end{bmatrix} + b \begin{bmatrix} 0 & 1 \\ 0 & 1 \end{bmatrix} = \begin{bmatrix} a & b-a \\ 0 & b \end{bmatrix} = M \end{aligned} \tag{4.12.4}$$

To diagonalize the matrix define

$$\mathcal{P} = \begin{bmatrix} \mathbf{v}_a & \mathbf{v}_b \end{bmatrix} = \begin{bmatrix} 1 & 1 \\ 0 & 1 \end{bmatrix} \ ; \ \mathcal{P}^{-1} = \begin{bmatrix} 1 & -1 \\ 0 & 1 \end{bmatrix} = \begin{bmatrix} \mathbf{w}_a^T \\ \mathbf{w}_b^T \end{bmatrix} \neq \mathcal{P}^T$$

Then

$$\mathcal{P}^{-1} M \mathcal{P} = \begin{bmatrix} 1 & -1 \\ 0 & 1 \end{bmatrix} \begin{bmatrix} a & b-a \\ 0 & b \end{bmatrix} \begin{bmatrix} 1 & 1 \\ 0 & 1 \end{bmatrix} = \begin{bmatrix} a & 0 \\ 0 & b \end{bmatrix}$$

In general for a $N \times N$ non-symmetric matrix M, for eigenvalue λ_I, let the eigenvector be $\mathbf{v}_I$ and the dual eigenvector $\mathbf{w}_I$; they satisfy the left and right eigenvalue equation as well as the orthonormality condition

$$\mathbf{w}_I^T M = \lambda_I \mathbf{w}_I^T \ ; \ M\mathbf{v}_I = \lambda_I \mathbf{v}_I \ ; \ \mathbf{w}_I^T \cdot \mathbf{v}_J = \delta_{I-J} \ ; \ I, J = 1, 2, \cdots, N$$

The spectral decomposition of the non-symmetric diagonalizable matrix M is given by

$$M = \sum_I \lambda_I \mathbf{v}_I \otimes \mathbf{w}_I^T$$

The matrix diagonalizing M is given by

$$P = \begin{bmatrix} \mathbf{v}_1 \cdots \mathbf{v}_I \cdots \mathbf{v}_N \end{bmatrix} \; ; \; P^{-1} = \begin{bmatrix} \mathbf{w}_1^T \\ \vdots \\ \mathbf{w}_I^T \\ \vdots \\ \mathbf{w}_N^T \end{bmatrix} \quad \Rightarrow \mathcal{P}^{-1}\mathcal{P} = \mathbb{I}$$

and yields

$$\mathcal{P}^{-1}M\mathcal{P} = \begin{bmatrix} \lambda_1 & 0 & .. & 0 \\ 0 & \lambda_2 & .. & 0 \\ \vdots & & & \\ 0 & 0 & .. & \lambda_N \end{bmatrix} \quad \Rightarrow \mathcal{M} = P \begin{bmatrix} \lambda_1 & 0 & .. & 0 \\ 0 & \lambda_2 & .. & 0 \\ \vdots & & & \\ 0 & 0 & .. & \lambda_N \end{bmatrix} P^{-1}$$

Note for the symmetric case $M^T = M$ the left and right eigenvectors are equal $\mathbf{w}_I = \mathbf{v}_I$ and one recovers the result given earlier for symmetric matrices.

Noteworthy 4.3: Matrix Diagonalization

In summary, to diagonalize a diagonalizable $N \times N$ matrix M one needs to take the following steps.

- Determine all the eigenvalues λ_n, $n = 1, \cdots, N$ by solving for the roots of an Nth order polynomial.
- Solve the simultaneous $N \times N$ linear equation to determine all the (left and right) eigenvectors.
- Construct the matrix S, S^{-1} from the (normalized) eigenvectors.
- The matrix M is given by $M = S^{-1}\mathcal{D}S$, where $\mathcal{D}$ is the diagonal matrix with the eigenvalues as the diagonal elements.

4.13 Change of Basis States

The diagonalization of a symmetric matrix S leads to a representation of S given entirely in terms of its eigenvectors and eigenvalues. There is another representation of S given in terms of the canonical basis vectors $\mathbf{e}_i$. The connection between these two representations is analyzed.

The linear vector space E_2 has no intrinsic basis and there is nothing special about any basis states chosen for E_2. In fact, any two linearly independent vectors can serve as a basis and one needs to know how one can relate two different sets of basis states.

The matrix $\mathcal{O}$, from Eq. 4.8.2, is given by

$$\mathcal{O} = \sum_{ij} O_{ij}\mathbf{e}_i \otimes \mathbf{e}_j^T \ ; \ \mathcal{O}^T = \sum_{ij} O_{ij}\mathbf{e}_j \otimes \mathbf{e}_i^T = \sum_{ij} O_{ij}^T\mathbf{e}_i \otimes \mathbf{e}_j^T \tag{4.13.1}$$

Consider the transformation (rotation) of a basis in N-dimensional space E_N. In matrix notation, the coordinate transformation is written as

$$\mathbf{e}_I = \mathcal{O}^T\mathbf{b}_I \ ; \ \mathbf{e}_I^T = \mathbf{b}_I^T\mathcal{O} \tag{4.13.2}$$

Since $\mathcal{O}\mathcal{O}^T = \mathbb{I}$, Eq. 4.13.2 yields

$$\mathbf{b}_i = \mathcal{O}\mathbf{e}_i \ \Rightarrow \ \mathbf{b}_i^T = \mathbf{e}_i^T\mathcal{O}^T$$

The transformation is orthogonal since $\mathcal{O}$ preserves both the angle between any two vectors as well as the keeping the norm of the vector. To see this note that

$$\mathbf{e}_I^T.\mathbf{e}_J = \mathbf{b}_I^T\mathcal{O}\mathcal{O}^T\mathbf{b}_J = \mathbf{b}_I^T \cdot \mathbf{b}_J \ : \ \text{angle} - \text{preserving}$$

and

$$\mathbf{e}_I^T.\mathbf{e}_I = \mathbf{b}_I^T\mathcal{O}\mathcal{O}^T\mathbf{b}_I = \mathbf{b}_I^T \cdot \mathbf{b}_I = 1 \ : \ \text{norm} - \text{preserving}$$

The new basis $\mathbf{e}_i$ is a linear combination of the old basis states $\mathbf{b}_i$.

Matrix A is given by

$$A = \sum_{ij} A_{ij}\mathbf{b}_i \otimes \mathbf{b}_j^T$$

Transform matrix A to A' such that

$$A' = \mathcal{O}^T A\mathcal{O}$$

Consider transforming the basis states of the matrix A from $\mathbf{b}_i$ to $\mathbf{e}_i$. One has the following representation

$$A' = \mathcal{O}^T A\mathcal{O} = \sum_{ij} A_{ij}\mathcal{O}^T\mathbf{b}_i \otimes \mathbf{b}_j^T\mathcal{O} = \sum_{ij} A_{ij}\mathbf{e}_i \otimes \mathbf{e}_j^T \tag{4.13.3}$$

In other words, to obtain matrix A' from A, the basis states of A are rotated by orthogonal transformation $\mathcal{O}$.

4.13.1 Symmetric Matrix: Change of Basis

Transforming to a new basis is useful in diagonalizing a matrix, and is discussed in this Section. From Eq. 4.8.2, S_{ij}—the components of symmetric matrix S in the canonical basis—are given by

$$S = \sum_{ij} S_{ij} \mathbf{e}_i \otimes \mathbf{e}_j^T$$

The basis states $\mathbf{e}_i$ are transformed to obtain the spectral representation of the matrix S. For vector $\mathbf{v}_I$ the following notation is used

$$\mathbf{e}_j^T \cdot \mathbf{v}_I = v_I(j) \equiv v_{jI}$$

The completeness equation given in Eq. 4.8.6 yields, for eigenvectors $\mathbf{v}_I$ of S, the following

$$\delta_{j-k} = \sum_I v_{jI} v_{kI} \ : \ \text{completeness equation} \tag{4.13.4}$$

Eq. 4.8.6 is used to re-write S_{ij} as follows

$$S_{ij} = \sum_k S_{ik} \delta_{j-k} = \sum_{kI} S_{ik} v_{jI} v_{kI} \tag{4.13.5}$$

The eigenvalue equation given in Eq. 4.8.3

$$\sum_j S_{ij} v_{jI} = \lambda_I v_{iI}$$

yields

$$S_{ij} = \sum_{kI} S_{ik} v_{jI} v_{kI} = \sum_I \lambda_I v_{iI} v_{jI} \tag{4.13.6}$$

Hence, from Eq. 4.8.2, we have the following

$$S = \sum_{ij} S_{ij} \mathbf{e}_i \otimes \mathbf{e}_j^T = \sum_{ij} \sum_I \lambda_I v_{iI} v_{jI} \mathbf{e}_i \otimes \mathbf{e}_j^T \tag{4.13.7}$$

The eigenvectors are given as follows by the basis vectors, which can also be viewed as a change of basis states

$$\mathbf{v}_I = \sum_j v_{jI} \mathbf{e}_j \ ; \ \mathbf{v}_I^T = \sum_j v_{jI} \mathbf{e}_j^T$$

Hence, from Eq. 4.13.7 and the new basis vectors, the spectral decomposition of S is given by

$$S = \sum_{I=1}^{N} \lambda_I \mathbf{v}_I \otimes \mathbf{v}_I^T \tag{4.13.8}$$

and we have recovered Eq. 4.8.7. Note that Eq. 4.13.6 is Eq. 4.13.8 written in terms of the components of the eigenvalues of the matrix S and its eigenvectors. The spectral representation of a matrix was given in Eq. 3.8.2. Matrix S was diagonalized by transforming from the canonical basis $\mathbf{e}_i$ to an orthogonal basis given by the eigenvectors $\mathbf{v}_i$.

The diagonalization of a symmetric matrix S discussed in Sect. 4.10 is an instance of the change of basis vectors. The transformation matrix is now constructed out of the eigenvectors of S and provides another derivation of Eq. 4.13.8.

From Eq. 4.10.6, the orthogonal matrix $\mathcal{V}$ is given by

$$\mathcal{O} \equiv \mathcal{V} = \begin{bmatrix} \mathbf{v}_1 \cdots \mathbf{v}_I \cdots \mathbf{v}_N \end{bmatrix} \ ; \ \mathcal{V}^T\mathcal{V} = \mathbb{I}$$

Consider the expression given in Eq. 4.10.5 applied to the spectral representation of S given in

$$\mathcal{V}^T S \mathcal{V} = \sum_{I=1}^{N} \lambda_I \mathcal{V}^T \mathbf{v}_I \otimes \mathbf{v}_I^T \mathcal{V}$$

The dual vector $\mathbf{v}_I^T$ is a $1 \times N$ matrix and is multiplied into the $N \times N$ matrix $\mathcal{V}$. Hence, the result is a $1 \times N$ matrix, which is also a N dimensional dual (row) vector. In particular, we have that

$$\begin{aligned} \mathbf{v}_I^T \mathcal{V} &= \mathbf{v}_I^T \cdot \begin{bmatrix} \mathbf{v}_1 \cdots \mathbf{v}_I \cdots \mathbf{v}_N \end{bmatrix} \\ &= \underbrace{[0, 0, \cdots, 1, \cdots, 0]}_{\text{entry at I th position}} = \mathbf{e}_I^T \end{aligned} \tag{4.13.9}$$

Similarly

$$\mathcal{V}^T \mathbf{v}_I = \mathbf{e}_I \tag{4.13.10}$$

Hence, we have the expected result derived earlier in Eq. 4.10.5 that

$$\mathcal{V}^T S \mathcal{V} = \sum_{I=1}^{N} \lambda_I \mathbf{e}_I \otimes \mathbf{e}_I^T = \begin{bmatrix} \lambda_1 & 0 & .. & 0 \\ 0 & \lambda_2 & .. & 0 \\ \vdots & & & \\ 0 & 0 & .. & \lambda_N \end{bmatrix} = \mathcal{D}$$

Comparing the result obtained above with Eq. 4.13.3, we can conclude that the diagonalization of matrix S is a rotation of the original basis states by the transformation matrix $\mathcal{V}$. As given in Eqs. 4.13.9 and 4.13.10 the transformed basis states are given by

$$\mathbf{e}_I = \mathcal{V}^T \mathbf{v}_I \; ; \; \mathbf{e}_I^T = \mathbf{v}_I^T \mathcal{V} \tag{4.13.11}$$

4.13.2 Diagonalization and Rotation

Consider the symmetric matrix

$$A = \begin{bmatrix} 3 & 1 \\ 1 & 3 \end{bmatrix} \; : \; \det(A) = 8$$

Since the determinant is positive, the matrix can be diagonalized by a rotation. Diagonalizing A yields

$$A = R \begin{bmatrix} 2 & 0 \\ 0 & 4 \end{bmatrix} R^T \; ; \; R = \frac{1}{\sqrt{2}} \begin{bmatrix} 1 & 1 \\ -1 & 1 \end{bmatrix}$$

Hence

$$R^T A R = \begin{bmatrix} 2 & 0 \\ 0 & 4 \end{bmatrix} = \sum_I \lambda_I \mathbf{e}_I \otimes \mathbf{e}_I^T$$

Writing A in the eigenvector basis yields

$$A = \sum_{ij} A_{ij} \mathbf{e}_i \otimes \mathbf{e}_j^T = \sum_I \lambda_I \mathbf{v}_I \otimes \mathbf{v}_I^T$$

Applying the rotation on matrix A yields the diagonalization

$$R^T A R = \sum_I \lambda_I R^T \mathbf{v}_I \otimes \mathbf{v}_I^T R = \sum_I \lambda_I \mathbf{e}_I \otimes \mathbf{e}_I^T$$

The rotation matrix in E_2 is given by

$$\mathcal{O}(\theta) = \begin{bmatrix} \cos(\theta) & \sin(\theta) \\ -\sin(\theta) & \cos(\theta) \end{bmatrix} \; ; \; \mathcal{O}\mathcal{O}^T = \mathbb{I}$$

Hence, R is a rotation matrix given by

$$\sin(\frac{\pi}{4}) = \frac{1}{\sqrt{2}} = \cos(\frac{\pi}{4}) \;\Rightarrow\; R = \mathcal{O}(\frac{\pi}{4})$$

4.13.3 Rotation and Inversion

Consider matrix A is given by

$$A = \begin{bmatrix} 0 & 1 \\ 1 & 0 \end{bmatrix} \ : \ \det(A) = -1$$

The determinant of A is -1, which implies (in E_2) that A reflects one of the coordinate system, mapping say $\mathbf{e}_y$ to $-\mathbf{e}_y$. A rotation cannot change the sign of a coordinate, and hence the diagonalization of A needs to be a product of a rotation and an reflection.[4]

As discussed in Sect. 3.8, the eigenvectors of A, given by $\mathbf{v}_\pm$, are linearly independent, and their span is the entire E_2 space. Hence, these eigenvectors can also be the basis states for E_2. Consider the following basis states

$$\lambda_+ = +1 \ : \ \mathbf{v}_+ = \mathbf{b}_1 = \frac{1}{\sqrt{2}} \begin{bmatrix} 1 \\ 1 \end{bmatrix} \ ; \ \lambda_- = -1 \ : \ \mathbf{v}_- = \mathbf{b}_2 = \frac{1}{\sqrt{2}} \begin{bmatrix} 1 \\ -1 \end{bmatrix}$$

The basis vectors are related by a linear transformation R given in Eq. 4.10.8. By inspection, we obtain

$$R = \frac{1}{\sqrt{2}} \begin{bmatrix} 1 & 1 \\ 1 & -1 \end{bmatrix} \ ; \ \mathbf{b}_1 = R \begin{bmatrix} 1 \\ 0 \end{bmatrix} = R\mathbf{e}_1 \ ; \ \mathbf{b}_2 = R \begin{bmatrix} 0 \\ 1 \end{bmatrix} = R\mathbf{e}_2$$

As expected, the R matrix diagonalizes matrix A and yields the following

$$R^T A R = \begin{bmatrix} \lambda_+ & 0 \\ 0 & \lambda_- \end{bmatrix} = \begin{bmatrix} 1 & 0 \\ 0 & -1 \end{bmatrix}$$

an angle θ, and is given by

The matrix R is equal to $\mathcal{O}$, with $\theta = \pi/4$, times a reflection—since $\det R = -1$; hence

$$R = \frac{1}{\sqrt{2}} \begin{bmatrix} 1 & 1 \\ 1 & -1 \end{bmatrix} = \begin{bmatrix} 1 & 0 \\ 0 & -1 \end{bmatrix} \frac{1}{\sqrt{2}} \begin{bmatrix} 1 & 1 \\ -1 & 1 \end{bmatrix} = P\mathcal{O}(\frac{\pi}{4}) \ : \ P = \begin{bmatrix} 1 & 0 \\ 0 & -1 \end{bmatrix}$$

Note

$$A = R \begin{bmatrix} \lambda_+ & 0 \\ 0 & \lambda_- \end{bmatrix} R^T = PO(\frac{\pi}{4}) \begin{bmatrix} \lambda_+ & 0 \\ 0 & \lambda_- \end{bmatrix} O(\frac{\pi}{4})^T P$$

[4] A reflection relates the left hand to the right. As one can see, one cannot rotate the left hand into the right hand, but instead one can reflect the left hand into the right by standing in front of a mirror.

Hence, when A acts on E_2, the matrix P reflects $\mathbf{e}_y$ to $-\mathbf{e}_y$ and then A rotates E_2 by $O(\frac{\pi}{4})^T$.

4.13.4 Hermitian Matrix: Change of Basis

From Eq. 4.11.3, for the Hermitian case

$$\mathcal{U}^\dagger S \mathcal{U} = \sum_{I=1}^{N} \lambda_I \mathbf{e}_I \otimes \mathbf{e}_I^\dagger = \begin{bmatrix} \lambda_1 & 0 & .. & 0 \\ 0 & \lambda_2 & .. & 0 \\ \vdots & & & \\ 0 & 0 & .. & \lambda_N \end{bmatrix} = \mathcal{D}$$

From Eq. 4.13.11, for the Hermitian case

$$\mathbf{e}_I = \mathcal{U}^\dagger \mathbf{u}_I \ ; \ \mathbf{e}_I^\dagger = \mathbf{u}_I^\dagger \mathcal{U} \ \Rightarrow \ \mathcal{U}\mathbf{e}_I = \mathbf{u}_I \ ; \ \mathbf{e}_I^\dagger \mathcal{U}^\dagger = \mathbf{u}_I^\dagger \tag{4.13.12}$$

Hence, the equations above yield

$$S = \sum_{I=1}^{N} \lambda_I \mathcal{U}\mathbf{e}_I \otimes \mathbf{e}_I^\dagger \mathcal{U}^\dagger = \sum_{I=1}^{N} \lambda_I \mathbf{u}_I \otimes \mathbf{u}_I^\dagger \tag{4.13.13}$$

Similar to the symmetric matrices, for any matrix function $F(S)$, we have the matrix equation

$$F(S) = \sum_{I=1}^{N} F(\lambda_I) \mathbf{u}_I \otimes \mathbf{u}_I^\dagger \tag{4.13.14}$$

The spectral decomposition and diagonalization is derived for the Hermitian matrix using the components of the matrix $\mathcal{U}$; the symmetric case is obtained by making all elements real. The definition of $\mathcal{U}$ is given by

$$\mathcal{U} = [\mathbf{u}_1, \mathbf{u}_2, \cdots, \mathbf{u}_N] \ ; \ \mathbf{u}_I^\dagger \cdot \mathbf{u}_J = \sum_k u_I^*(k) u_J(k) = \delta_{I-J}$$

In components

$$\mathcal{U} = \sum_{ij} u_j(i) \mathbf{e}_i \otimes \mathbf{e}_j^\dagger \ \Rightarrow \ \mathcal{U}_{ij} = u_j(i) \equiv u_{ji} \ ; \ \mathcal{U}_{ij}^\dagger = (\mathcal{U}_{ji})^* = u_i^*(j) \equiv u_{ji}^*$$

The matrix $\mathcal{U}$ is unitary since

$$\sum_k \mathcal{U}^\dagger_{ik}\mathcal{U}_{kj} = \sum_k u^*_{ki}u_{kj} = \mathbf{u}^\dagger_i \cdot \mathbf{u}_j = \delta_{i-j}$$

The transformation of basis states is given by

$$\mathcal{U}^\dagger \mathbf{u}_K = \sum_{ij} u^*_{ij}\mathbf{e}_j \otimes (\mathbf{e}^\dagger_i \cdot \mathbf{u}_K)$$
$$= \sum_{ij} u^*_{ij}u_{ik}\mathbf{e}_j = \mathbf{e}_K \Rightarrow \mathbf{u}^\dagger_K\mathcal{U} = \mathbf{e}^\dagger_K$$

4.14 Problems

1. From the definition of the inverse of a matrix in terms of its cofactors, find the inverse of
$$M = \begin{vmatrix} 2 & 3 & 5 \\ 3 & 1 & 2 \\ 1 & 4 & 3 \end{vmatrix}$$
2. Prove that the eigenvalues of a Hermitian matrix are all real.
3. Let U be a unitary matrix and let $\mathbf{v}_1, \mathbf{v}_2$ be its eigenvectors with eigenvalues λ_1, λ_2, respectively. Show the following
   - $|\lambda_1| = 1 = |\lambda_2|$
   - If $\lambda_1 \neq \lambda_2$, then $\mathbf{v}^\dagger_1 \cdot \mathbf{v}_2$=0
4. Can the following matrix non-symmetric matrix be diagonalized?
$$M = \begin{bmatrix} 1 & 0 & 2 \\ 0 & 2 & 2 \\ 0 & 1 & 1 \end{bmatrix}$$
5. Find the eigenvalues and left and right eigenvectors of the following matrix non-symmetric
$$M = \begin{bmatrix} 1 & 0 & 2 \\ 0 & 2 & 2 \\ 0 & 1 & 1 \end{bmatrix}$$
Verify that the left and right eigenvectors are orthonormal.
6. Consider the symmetric matrix
$$M = \begin{bmatrix} a & b \\ b & c \end{bmatrix}$$
Find the all the matrix elements of the exponential of M, given by $\exp\{M\}$.

7. Consider the Markov matrix

$$M = \begin{bmatrix} a & 1-b \\ 1-a & b \end{bmatrix}$$

Find the eigenvalues and the left and right eigenvectors. Verify that the left and right eigenvectors are orthonormal.

8. Using the alternating-sign method, find the determinant of the following matrices

$$M = \begin{bmatrix} 7 & -1 & 4 \\ -3 & 0 & 5 \\ 2 & 1 & 1 \end{bmatrix} \; ; \; N = \begin{bmatrix} 2 & 1 & 3 \\ 1 & 0 & 2 \\ 2 & 0 & -2 \end{bmatrix}$$

9. Find the inverse of

$$A = \begin{bmatrix} 3 & 2 & 7 \\ 8 & 4 & 1 \\ 2 & 6 & 5 \end{bmatrix} \; ; \; N = \begin{bmatrix} 4 & 2 & 1 \\ 5 & 8 & 2 \\ 0 & 0 & 3 \end{bmatrix}$$

Part III
Calculus

Chapter 5
Integration

Abstract Integration naturally follows from the idea of summation, and this route is chosen to introduce the idea of the integral. Integration follows from the concept of an infinite convergent sum, and for this reason it is discussed before differentiation. The continuum limit is taken to show one obtains a well defined limit. Definite and indefinite integrals, as well multiple integrals are defined. The integration of special functions is obtained from first principles. Gaussian integration for single and N-variables are studied. Examples from economics and finance are analyzed.

5.1 Introduction

Consider the rate of investment I in a company. Suppose the capital of the company is C and that investments are made regularly over time; with the increase in C being due to the investments being made. Consider two instants, say t and $t + \Delta t$; the change in the capital stock formation, denoted by ΔC over the time interval Δt is given by

$$\Delta C = C(t + \Delta t) - C(t) = I(t)\Delta t$$

Suppose the rate of investments is $I(t) = \alpha t^{1/2}$. Then one can ask: what is the value of the company $C(T)$ at time T? Clearly, one has to *add* (sum) all the investments made over each small time interval Δt to find the value of $C(T)$. Suppose the investments start at $t = 0$ and are made at fixed instants of time given by $t_n = n\Delta t$; in total N investments are made, with last payment at time T; hence $T = N\Delta t$. The value of the company at time T is then given by

$$C(T) = C(0) + \sum_{n=1}^{N} I(t_n)\Delta t \tag{5.1.1}$$

where $C(0)$ was the value of the company at $t = 0$.

So far there is no need for calculus. Now suppose one wants to know what is $C(T)$ in the limit $\Delta t \to 0$, in other words when the investments are made continuously?

B. E. Baaquie, *Mathematical Methods and Quantum Mathematics for Economics and Finance*, https://doi.org/10.1007/978-981-15-6611-0_5

In this case $N = T/\Delta t \to \infty$. To find $C(T)$ for the continuous case, an *integration* is required, and this is where one needs calculus.

One can see from Eq. 5.1.1 that to evaluate $C(T)$ for continuous time, an infinite summation has to be taken. Recall that in Sect. 1.5, a number of infinite series have been discussed, including a brief discussion on the convergence of these series. The discussion of integration is best started by studying the properties of some simple examples of infinite summations.

Noteworthy 5.1: Summing integers

The finite sum of integers is given by

$$S_N = 1 + 2 + \cdots + N$$

Writing this sum in reverse order yields

$$S_N = N + (N-1) + (N-2) + \cdots + 1$$

Adding the two yields

$$2S_N = N(N+1) \Rightarrow S_N = \frac{1}{2}N(N+1)$$

A more formal way of deriving above results is to consider the following

$$(n+1)^2 - n^2 = 2n + 1$$

Summing above expression from $n = 1$ to $n = N$ leads, on the left hand side, to a telescopic sum with all terms except the first and last canceling out. This yields

$$\sum_{n=1}^{N}\{(n+1)^2 - n^2\} = (N+1)^2 - 1 = 2\sum_{n=1}^{N} n + N \Rightarrow \sum_{n=1}^{N} n = \frac{1}{2}N(N+1)$$

Similarly, consider

$$(n+1)^3 - n^3 = 3n^2 + 3n + 1$$

Summing from $n = 1$ to $n = N$ and with the telescopic sum on the left hand side yields

$$\sum_{n=1}^{N}\{(n+1)^3 - n^3\} = (N+1)^3 - 1$$

and hence

$$(N+1)^3 - 1 = 3\sum_{n=1}^{N} n^2 + \frac{3}{2}N(N+1) + N \;\Rightarrow\; 3\sum_{n=1}^{N} n^2 = N^3 + \frac{3}{2}N^2 + \frac{1}{2}N$$

Hence

$$\sum_{n=1}^{N} n^2 = \frac{1}{6}N(2N+1)(N+6) \simeq= \frac{1}{3}N^3$$

Similarly

$$\sum_{n=1}^{N} n^3 = \frac{1}{4}N^2(N+1)^2$$

It can be shown, using the Binomial theorem, that

$$\sum_{n=1}^{N} n^k = \frac{1}{k+1}N^{k+1} + O(N^k)$$

Consider the area under a curve; let us start with a triangle; of course we know the answer, but the technique that will be employed can be generalized to any curve. The curve is given by $y = x$. Let the length of the base of the curve be U, which is divided it into N segments, with each segment having a length of U/N. The points on the line can be numbered by an integer n such that the distance of a point x_n from the origin is given by

$$y \to x_n = \frac{U}{N}n \;:\; \Delta x \equiv x_{n+1} - x_n = \frac{U}{N} \;;\; n = 1, 2, \cdots, N \tag{5.1.2}$$

As shown in Fig. 5.1, the area under the curve has been broken up into small segments, which are rectangles with base of size Δx and height x_n; each segment's area is $y\Delta x \simeq x_n \Delta x$; if all the small rectangles can be summed up, an approximate estimate will be found for the the full area. The sum of all the segments is given by the following

$$I_N = \Delta x \sum_{n=1}^{N} x_n = (\Delta x)^2 \sum_{n=1}^{N} n = \frac{U^2}{N^2} \cdot \frac{1}{2}N(N+1)$$

The remarkable property of I_N is that the series

$$I_1, I_2, \cdots, I_N$$

obeys Cauchy's condition for convergence. In the limit of

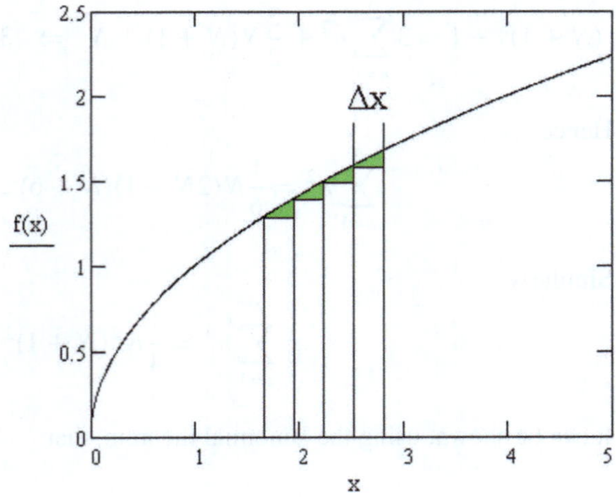

Fig. 5.1 The area under the curve, made out of a sum of rectangles with base Δx and height $f(x)$. Published with permission of © Belal E. Baaquie 2020. All Rights Reserved

$$N \to \infty \;\Rightarrow\; \Delta x = \frac{U}{N} \to 0$$

and the sum converges to the following

$$\lim_{\Delta x \to 0} I_N = \frac{1}{2}U^2 + \frac{1}{2}\Delta x U \to I = \frac{1}{2}U^2$$

What the result states is that one can compose the area of a triangle by an infinite sum over rectangles that are infinitesimally small.

The reason that Δx has to be made vanishingly small is because the error in the computation, shown by small shaded triangles beneath the curve shown in Fig. 5.1, goes to zero as $\Delta x \to 0$.

Consider a parabola, given by the equation $y = x^2$; let us try and evaluate the area under a parabola over the interval $[0, U]$. The curve is given by

$$y^2 \to x_n^2 \;;\; y\Delta x \to x_n^2 \Delta x$$

Summing all the small rectangles yields[1]

$$I'_N = \Delta x \sum_{n=1}^{N} x_n^2 = (\Delta x)^3 \sum_{n=1}^{N} n^2 = \frac{U^3}{N^3} \cdot \frac{1}{6} N(N+1)(2N+1)$$

[1]The sum of the square of integers is given by

$$1^2 + 2^2 + 3^2 + \cdots + N^2 = \frac{1}{6}N(N+1)(2N+1)$$

.

Hence

$$\lim_{\Delta x \to 0} I'_N = \frac{1}{3}U^3 + \frac{1}{2}\Delta x U^2 + \frac{1}{6}(\Delta x)^2 U \to I' = \frac{1}{3}U^3$$

Finally, consider the area under the curve $y = x^3$. The sum of all the small rectangles yields[2]

$$I''_N = \Delta x \sum_{n=1}^{N} x_n^3 = \frac{U^4}{N^4} \cdot \frac{1}{4} N^2 (N+1)^2$$

Hence

$$\lim_{\Delta x \to 0} I''_N = \frac{1}{4}U^4 + \frac{1}{2}\Delta x U^3 + \frac{1}{4}(\Delta x)^2 U^2 \to I'' = \frac{1}{4}U^4$$

There is a pattern. In general it can be shown that

$$I = \lim_{N \to \infty} \Delta x \sum_{n=1}^{N} x_n^k = \frac{1}{k+1} U^{k+1} \tag{5.1.3}$$

This is the remarkable result of calculus, that there are infinite summations that yield convergent results—results that are independent of the underlying summation.[3]

5.2 Sums Leading to Integrals

The results obtained can be interpreted directly, without the need to look at the results of the convergent infinite series, and without any reference to the underlying limiting process. To do so, the simplest case is examined

$$I_N = \Delta x \sum_{n=1}^{N} x_n$$

The line segment is the base of a rectangle that has a height of x_n; I is the area that is the sum of contiguous rectangles, as shown in Fig. 5.2. What does $N \to \infty$ imply for x_n? Recall

---

[2]The sum of the first N cubes is given by

$$1^3 + 2^3 + 3^3 + \cdots + N^3 = \frac{1}{4} N^2 (N+1)^2$$

.

[3]Of course not all infinite summations are convergent, and so we need to decide what is the criterion of convergence, which will be addressed later.

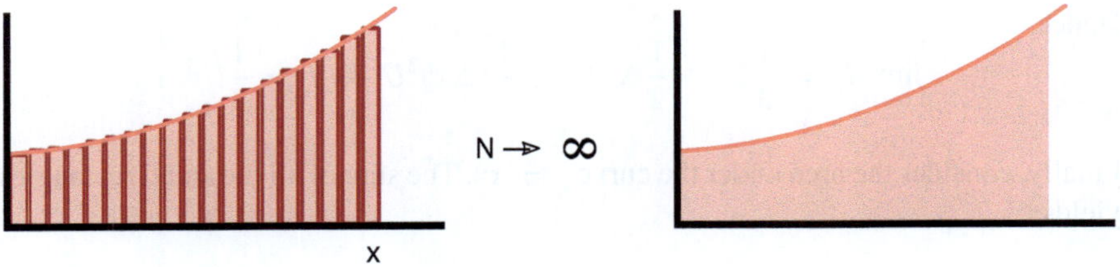

Fig. 5.2 The area under the curve x^k. Published with permission of © Belal E. Baaquie 2020. All Rights Reserved

$$x_n = \Delta x_n = \frac{U}{N} n \; ; \; n = 1, 2, \cdots, N \;\Rightarrow\; \lim_{N\to\infty} x_n \to x : \text{ Real number } \in [0, U]$$

What is U/N? From the discussion in the Appendix on real numbers, we have

$$\lim_{N\to\infty} \Delta x = \lim_{N\to\infty} \frac{U}{N} \to \epsilon$$

Consider the sum; at the lower and upper point, x_n takes the following values

$$\sum_{n=1}^{N} \; ; \; x_1 = \frac{U}{N} \to 0 \; ; \; x_N = \frac{U}{N} N = U$$

The infinitesimal is given by

$$\lim_{N\to\infty} \Delta x = \epsilon = dx : \text{ dimensionof } x$$

The limit yields

$$\lim_{N\to\infty} \Delta x \sum_{n=1}^{N} = \epsilon \sum_{n=1}^{N} \to \int_0^U dx : \text{integral}$$

Hence, the final result is given by

$$\lim_{N\to\infty} \Delta x \sum_{n=1}^{N} x_n \to \int_0^U dx \; x = \frac{1}{2} U^2$$

The integral is the area under the curve $y = x$ from 0 to U, as shown for the general case in Fig. 5.2.

In general, the result obtained in Eq. 5.1.3 is the area under the curve $y = x^k$ and is given by

$$\int_0^U dx \; x^k = \frac{1}{k+1} U^{k+1}$$

Note the result above is dimensionally consistent.

The main takeaway from the infinite summations is that, if one takes the continuum limit given by $\Delta x \to 0$, or equivalently $N \to \infty$, one finds that hidden in the infinite summations is a *finite continuum limit* of the infinite summation. Calculus has its own rules. The cardinal rule of calculus is that all of calculus is based on first computing with $\epsilon = \Delta x \neq 0$, and then taking the continuum limit of $\epsilon = \Delta x \to 0$.

5.3 Definite and Indefinite Integrals

The result of an integration can also be written as an *indefinite integral*. An indefinite integral is defined upto a constant and without the limits being defined; for arbitrary function $f(x)$, in most table of integrals, it is written as

$$g(x) = \int dx f(x)$$

What the above equation actually means is the following

$$g(x) = \int_L^x dy f(y)$$

The variable y is the integration variable and the function $f(y)$ is the integrand; the indefinite integral is the result of integrating the integrand; the lower limit not being specified means that the function $g(x)$ is defined only up to a constant.

The integration of the integrand x^α, where α is any real number, is given by

$$\int dx x^\alpha = \frac{1}{\alpha+1} x^{\alpha+1} + C$$

The special case of $\int dx/x$ is obtained by taking the following limit

$$\begin{aligned}
\int \frac{dx}{x} &= \lim_{\alpha \to -1} \int dx x^\alpha \\
&= \lim_{\alpha \to -1} \frac{1}{\alpha+1} (1 + (\alpha+1)\ln(x) + O(\alpha+1)^2) + C \\
\Rightarrow \quad \int \frac{dx}{x} &= \ln(x) + C
\end{aligned} \tag{5.3.1}$$

An integral is a *linear function* of the integrand; for functions $f(x)$, $g(x)$ integration, linearity yields

$$\int dx[af(x) + bg(x)] = a \int dx f(x) + b \int dx g(x)$$

Consider an arbitrary function

$$f(x) = \sum_{n=-\infty}^{+\infty} c_n x^n \ ; \ n \neq -1$$

where c_n are constants. The linearity of the integral yields

$$\int dx f(x) = \sum_{n=-\infty}^{+\infty} c_n \int dx x^n = \sum_{n=-\infty}^{+\infty} c_n \frac{1}{n+1} x^{n+1} \ ; \ n \neq -1$$

Exponential

$$\begin{aligned}
\int dx e^{\alpha x} &= \sum_{n=0}^{\infty} \frac{\alpha^n}{n!} \int dx x^n \\
&= \sum_{n=0}^{\infty} \frac{\alpha^n}{n!} \frac{1}{n+1} x^{n+1} = \frac{1}{a} \sum_{n=1}^{\infty} \frac{(\alpha x)^n}{n!} = \frac{1}{a}(e^{\alpha x} - 1) \\
\Rightarrow \int dx e^{\alpha x} &= \frac{1}{a} e^{\alpha x} + C
\end{aligned} \tag{5.3.2}$$

Sine, cosine

$$\begin{aligned}
\int dx \sin(\alpha x) &= \sum_{n=1}^{\infty} (-1)^{n-1} \frac{\alpha^{2n-1}}{(2n-1)!} \int dx x^{2n-1} \\
&= \frac{1}{\alpha} \sum_{n=1}^{\infty} (-1)^{n-1} \frac{(\alpha x)^{2n}}{(2n-1)!(2n)} = \frac{-1}{\alpha} \sum_{n=0}^{\infty} (-1)^n \frac{(\alpha x)^{2n}}{(2n)!} + \frac{1}{\alpha} \\
\Rightarrow \int dx \sin(\alpha x) &= -\frac{1}{\alpha} \cos(\alpha x) + C
\end{aligned} \tag{5.3.3}$$

Similarly

$$\int dx \cos(\alpha x) = \frac{1}{\alpha} \sin(\alpha x) + C$$

Logarithm
Note that

$$\lim_{\alpha \to 0} \frac{1}{\alpha}(x^\alpha - 1) = \ln x$$

Using the linearity of integration yields

$$\int dx \ln x = \lim_{\alpha\to 0}\frac{1}{\alpha}\Big(\frac{1}{1+\alpha}x^{\alpha+1} - x\Big)$$
$$\simeq \lim_{\alpha\to 0}\frac{x}{\alpha}((1-\alpha+\cdots)(1+\alpha\ln x+\cdots)-1)$$
$$= \lim_{\alpha\to 0}\frac{x}{\alpha}(\alpha\ln x - \alpha) = x\ln x - x$$

Examples

$$I = \int dx(10e^x - x^3 + \frac{4}{x}) = 10e^x - \frac{1}{4}x^4 + 4\ln x$$

$$I = \int \frac{dx}{(a+x)(b+x)} = \frac{1}{ab}\int dx\left(\frac{1}{b+x} - \frac{1}{a+x}\right)$$
$$= \frac{1}{ab}\left(\ln(b+x) - \ln(a+x)\right) = \frac{1}{ab}\ln\left(\frac{b+x}{a+x}\right)$$

Definite Integral

In general, a definite integral has a lower and upper limit and, unlike an indefinite integral, does not include an arbitrary constant. It is denoted by

$$\int_a^b dx f(x)$$

The constant in the indefinite integral always cancels out when forming the definite integral.

For example, the definite integral of x^n is given by

$$\int_a^b dx x^n = \frac{1}{n+1}x^{n+1}\Big|_a^b = \frac{1}{n+1}(b^{n+1} - a^{n+1})$$

A definite integral obeys the rule

$$\int_a^b dx f(x) = \int_a^c dx f(x) + \int_c^b dx f(x)$$

A special case of the rule yields

$$\int_a^b dx f(x) + \int_b^a dx f(x) = \int_a^a dx f(x) = 0 \Rightarrow \int_b^a dx f(x) = -\int_a^b dx f(x)$$

Example
Consider the indefinite integral

$$\int dx \frac{3x-5}{x^{1/3}} = \int dx 3x^{2/3} - \int dx 5x^{-1/3} = \frac{9}{5}x^{5/3} - \frac{15}{2}x^{2/3} = I(x)$$

The definite integral is given by

$$\int_a^b dx \frac{3x-5}{x^{1/3}} = I(b) - I(a)$$

Definite integrals can be infinite and hence undefined if the integrand becomes infinite in the range of integration. Consider the following definite integral

$$n \geq 1 \ : \quad \int_0^b \frac{dx}{x^n} = \frac{x^{-n+1}}{-n+1}\Big|_0^b = \frac{1}{-n+1}\left(\frac{1}{b^{n-1}} - \frac{1}{0^{n-1}}\right) = -\infty$$

Hence, if $f(x) = x^n, n \geq 1$ in the interval (a, b), then using the result above it follows that

$$\int_a^b \frac{dx}{f(x)} \to \infty$$

In general, to analytically find the integral of an arbitrary function is almost impossible. For example, it is almost impossible to exactly evaluate

$$I(w, \gamma, \beta; a, b) = \int_a^b dx e^{wx} \frac{1}{1 + \tan(\gamma x) + x^{5/3}} \frac{1}{\sqrt{\sin(\beta x) + x^3}}$$

However, using any of the common software, an integral such as $I(w, \gamma, \beta; a, b)$, with one integration variable x, can be obtained numerically within seconds.

5.4 Applications in Economics

Net investment I is defined as the rate of change in capital stock formation C over time t. If the process of capital formation is continuous over time, $dC(t) = I(t)dt$. Hence, at time t, the total capital of any enterprise is given by

$$\int_0^T dt I(t) = \int_0^T dC = C(T) - C(0)$$

The marginal cost M is the change in the total cost of production T for an additional quantity of input Q. Suppose the input quantity Q is increased by dQ; then the change in the total cost of production is given by the marginal cost of production by the following

$$T(Q + dQ) - T(Q) = dT = M(Q)dQ$$

Hence, the increase in the total cost of production is given by

$$T = \int M(Q)dQ + c$$

Example 1 Suppose the in-flow of investment capital is given by $I(t) = 3t^{1/3}$. What is the stock of capital today, at time T?

$$C(T) = \int_0^T dt I(t) + C(0) = \frac{3}{4}T^{4/3} + C(0)$$

Example 2 ABC company produces boats. Its marginal cost function is $M = x^2 + 10x + 200$, where x is the number of boats produced. The company's fixed costs are \$500. Find the total cost T of producing x boats.

$$\begin{aligned} T &= \int M dx + c \\ &= \int (x^2 + 10x + 200) dx + c \\ &= \frac{1}{3}x^3 + 5x^2 + 200x + c \end{aligned}$$

Substituting fixed cost= 500 for c yields the total cost of production

$$T = \frac{1}{3}x^3 + 5x^2 + 200x + 500$$

5.5 Multiple Integrals

Consider two independent integration variables x, y and a function of two variables $f(x, y)$. One can define a multiple integral by the product of successive single variable integrations. For now consider indefinite integrals so that we need not specify the domain over which the variables x, y take values. One possible definition is the following

$$I = \int dy \left[\int dx f(x, y) \right] = \int dy g(y) \ : \ g(y) = \int dx f(x, y)$$

The integral can also be evaluated in the following manner

$$J = \int dx \left[\int dy f(x, y) \right] = \int dx h(x) \ : \ h(x) = \int dy f(x, y)$$

Multiple integration is independent of the order in which the integration is performed (except for some pathological functions that are of very little interest). Hence, the multiple integral is defined without specifying the order in which the integration is carried out and yields

$$I = \int dx dy f(x, y) \equiv \int \int dy dx f(x, y)$$

Example To illustrate the integration can be performed in any order, the following are evaluated.

$$(i) \int_1^4 \int_{-1}^2 dx dy (2x + 6x^2 y); \ \ (ii) \int_{-1}^2 \int_4^1 dy dx (2x + 6x^2 y)$$

(i) Consider the integral

$$\int_1^4 dx \left[\int_{-1}^2 dy (2x + 6x^2 y) \right] = \int_1^4 dx \left[2xy + 3x^2 y^2 \right]_{-1}^2$$
$$= \int_1^4 dx (6x + 9x^2) = \left[3x^2 + 3x^3 \right]_1^4 = 234$$

(ii) Performing the integration in the other order yields the integral

$$\int_{-1}^2 dy \left[\int_1^4 dx (2x + 6x^2 y) \right] = \int_{-1}^2 dy \left[x^2 + 2x^3 y \right]_1^4$$
$$= \int_{-1}^2 dy (126y + 15) = \left[63y^2 + 15y \right]_{-1}^2 = 234$$

A definite multiple integral is defined by the integration variables x, y taking values over a fixed range. The range can be set independently for each variable such that

$$I(a, b, c, d) = \int_a^b dx \int_c^d dy f(x, y)$$

The definite integral in general is given by

$$I(r) = \int_{\mathcal{D}(x,y)} dx dy f(x, y)$$

The domain $\mathcal{D}$ can also be any subspace of $\Re_2$ and the integration variables lie in the domain; hence

$$\{x, y|x, y \in \mathcal{D}(x, y)\}$$

For example, the domain can be defined by the disk

$$\mathcal{D}(x, y) : r^2 - (x^2 + y^2) \geq 0$$

For N-independent integration variables $\mathbf{x}$

$$I_N = \int_{\mathcal{D}(\mathbf{x})} dx_1 dx_2 \cdots dx_N f(\mathbf{x})$$

Taking the limit of $N \to \infty$ yields I_N to be a functional integral, and is discussed in Sect. 13.10.

5.5.1 Change of Variables

Consider the linear transformation

$$\begin{bmatrix} y_1 \\ y_2 \end{bmatrix} = \begin{bmatrix} a & c \\ b & d \end{bmatrix} \begin{bmatrix} x_1 \\ x_2 \end{bmatrix} \Rightarrow \begin{bmatrix} dy_1 \\ dy_2 \end{bmatrix} = \begin{bmatrix} a & c \\ b & d \end{bmatrix} \begin{bmatrix} dx_1 \\ dx_2 \end{bmatrix} \; ; \; d\mathbf{y} = Ld\mathbf{x}$$

The linear transformation yields a change of area given by the determinant of the linear transformation, as discussed in Sect. 3.5. The infinitesimal area $dy_1 dy_2$ changes by an amount that is given by the determinant of L. Hence

$$dy_1 dy_2 = |\det L| dx_1 dx_2$$

For the transformation of N-independent integration variables $\mathbf{y}$ to $\mathbf{x}$, the change of variables for the linear case yields, for the N-dimensional volume element $dy_1 dy_2 \cdots dy_N$, the following

$$\mathbf{y} = L\mathbf{x} \Rightarrow d\mathbf{y} = Ld\mathbf{x} \Rightarrow dy_1 dy_2 \cdots dy_N = |\det L| dx_1 dx_2 \cdots dx_N$$

The domain also transforms under a change of variables and yields

$$\int_{\mathcal{D}} dy_1 dy_2 \cdots dy_N f(\mathbf{y}) = \int_{L(\mathcal{D})} dx_1 dx_2 \cdots dx_N f(L\mathbf{x})$$

5.6 Gaussian Integration

The mathematical framework of white noise and stochastic processes is based on the normal random variable, which is turn is described using Gaussian integration. Gaussian integration also plays a key role in studying and employing functional integrals.

The main results of Gaussian integration that will be employed later are derived. The basic Gaussian integral is given by

$$Z = \int_{-\infty}^{+\infty} dx e^{-\frac{1}{2}\lambda x^2 + jx} = \sqrt{\frac{2\pi}{\lambda}} e^{\frac{1}{2\lambda} j^2} \tag{5.6.1}$$

A proof is given on Eq. 5.6.1. Completing the square yields

$$-\frac{1}{2}\lambda x^2 + jx = -\frac{1}{2}\lambda\left(x - \frac{j}{\lambda}\right)^2 + \frac{1}{2}\frac{j^2}{\lambda}$$

Introduce new variable y by

$$y = x - \frac{j}{\lambda} \quad \Rightarrow \quad -\infty \leq y \leq +\infty$$

Equation 5.6.1 is given by

$$Z = \int_{-\infty}^{+\infty} dx e^{-\frac{1}{2}\lambda x^2 + jx} = e^{\frac{1}{2\lambda} j^2} \int_{-\infty}^{+\infty} dy e^{-\frac{1}{2}\lambda y^2} = e^{\frac{1}{2\lambda} j^2} I \tag{5.6.2}$$

It is shown in Eq. 6.8.4 that

$$I = \int_{-\infty}^{+\infty} dy e^{-\frac{1}{2}\lambda y^2} = \sqrt{\frac{2\pi}{\lambda}} \tag{5.6.3}$$

5.6.1 Gaussian Integration for Options

A major application of Gaussian integration is in evaluating the price of an option. This integral is discussed again in obtaining Eq. 11.10.3, and the main steps required for carrying out the integration are shown below.

Consider the integral

$$
\begin{aligned}
I &= \frac{1}{\sqrt{2\pi\sigma^2}} \int_{-\infty}^{+\infty} dx e^{-\frac{1}{2\sigma^2}(x-a)^2} [e^x - K]_+ \\
&= \frac{e^a}{\sqrt{2\pi\sigma^2}} \int_{-\infty}^{+\infty} dx e^{-\frac{1}{2\sigma^2}x^2} [e^x - e^{-a} K]_+ \\
&= \frac{e^a}{\sqrt{2\pi}} \int_{-\infty}^{+\infty} dx e^{-\frac{1}{2}x^2} [e^{\sigma x} - e^{-a} K]_+ \\
&= \frac{e^a}{\sqrt{2\pi}} \int_{-d}^{+\infty} dx e^{-\frac{1}{2}x^2} [e^{\sigma x} - e^{-a} K] \\
&= \frac{e^a}{\sqrt{2\pi}} \int_{-d}^{+\infty} dx e^{-\frac{1}{2}x^2} e^{\sigma x} - K I_1
\end{aligned}
\tag{5.6.4}
$$

where

$$
d = -\frac{\ln(e^{-a}K)}{\sigma} = \frac{\ln(e^a/K)}{\sigma}
$$

The integral I_1 is given by

$$
I_1 = \frac{1}{\sqrt{2\pi}} \int_{-d}^{+\infty} dx e^{-\frac{1}{2}x^2} = \frac{1}{\sqrt{2\pi}} \int_{-\infty}^{d} dx e^{-\frac{1}{2}x^2} = N(d)
$$

and the cumulative distribution function for the Gaussian random variable is given by

$$
N(z) = \frac{1}{\sqrt{2\pi}} \int_{-\infty}^{z} dx e^{-\frac{1}{2}x^2}
$$

The remaining integral is given by

$$
\begin{aligned}
I_2 &= \frac{1}{\sqrt{2\pi}} \int_{-d}^{+\infty} dx e^{-\frac{1}{2}x^2} e^{\sigma x} = \frac{e^{\sigma^2/2}}{\sqrt{2\pi}} \int_{-\infty}^{d} dx e^{-\frac{1}{2}(x+\sigma)^2} \\
&= \frac{e^{\sigma^2/2}}{\sqrt{2\pi}} \int_{-\infty}^{d+\sigma} dx e^{-\frac{1}{2}x^2}
\end{aligned}
$$

Hence

$$
I = e^a I_2 - K I_1 = e^{a+\sigma^2/2} N(d+\sigma) - K N(d)
$$

5.7 N-Dimensional Gaussian Integration

The moment generating function for the N-dimensional Gaussian random variable is given by

$$Z[j] = \int_{-\infty}^{+\infty} dx_1 \cdots dx_N e^S \tag{5.7.1}$$

where the 'action' is given by

$$S = -\frac{1}{2}\sum_{i,j=1}^{N} x_i A_{ij} x_j + \sum_i J_i x_i \tag{5.7.2}$$

Let A_{ij} be a positive definite symmetric matrix; it can be diagonalized and has only positive eigenvalues. Furthermore, A_{ij} can be diagonalized by an orthogonal matrix M

$$A = M^T \begin{pmatrix} \lambda_1 & & 0 \\ 0 & \ddots & 0 \\ 0 & & \lambda_N \end{pmatrix} M \ ; \ M^T M = I \ : \text{Orthogonal matrix}$$

Define new variables

$$z_i = M_{ij} x_j \quad ; \quad x_i = M^T_{ij} z_j$$
$$\prod_{i=1}^{N} dz_i = \det M \prod_{i=1}^{N} dx_i = \prod_{i=1}^{N} dx_i \tag{5.7.3}$$

Hence, from Eq. 5.6.1

$$Z[j] = \prod_i^N \int dz_i e^{-\frac{1}{2}\lambda_i z_i^2 + (J^T M^T)_i z_i} = \prod_{i=1}^{N} \left[\sqrt{\frac{2\pi}{\lambda_i}} e^{\frac{1}{2\lambda_i}(J^T M^T)_i (J^T M^T)_i} \right]$$

In matrix notation

$$\sum_i \frac{1}{\lambda_i}(J^T M^T)_i (J^T M^T)_i = \sum_{ijk} J_j M^T_{ji} \frac{1}{\lambda_i} J_k M^T_{ki} = \sum_{jk} J_j \Big(\sum_i M^T_{ji} \frac{1}{\lambda_i} M_{ik} \Big) J_k$$
$$= \sum_{jk} J_j (\frac{1}{A})_{jk} J_k = J^T A^{-1} J \tag{5.7.4}$$

The pre-factor is given by

$$\prod_{i=1}^{N} \sqrt{\frac{2\pi}{\lambda_i}} = (2\pi)^{N/2} \frac{1}{\sqrt{\det A}}$$

Hence

$$Z[J] = \frac{(2\pi)^{N/2}}{\sqrt{\det A}} \exp\left\{\frac{1}{2} J A^{-1} J\right\} \tag{5.7.5}$$

$Z[J]$ is called the generating function since all the correlators can be evaluated from it. In particular, the correlator of two of the variables is given by

$$E[x_i x_j] \equiv \frac{1}{Z} \int Dx\, x_i x_j e^S \;\; ; \;\; Z = Z[0]$$

$$E[x_i x_j] = \overbrace{x_i x_j} = \frac{\partial^2}{\partial J_i \partial J_j} \ln Z[J]\Big|_{J=0} \tag{5.7.6}$$

$$= A^{-1}_{ij} \quad : \text{contraction} \tag{5.7.7}$$

The correlation of N variables is given by

$$\langle x_{i_1} x_{i_2} x_{i_3} \cdots x_{i_{N-1}} x_{i_N} \rangle = \overbrace{x_{i_1} x_{i_2}} \cdots \overbrace{x_{i_{N-1}} x_{i_N}} + \text{all possible permutations.}$$

5.8 Problems

1. Evaluate the integral

$$I = \int dx \frac{1}{a + bx^2} \;\; ; \;\; ab > 0$$

2. Prove that integrations commute by showing that

$$\int_1^4 dx \int_0^5 dy (7x^2 y + 5x - 2y) = \int_0^5 dy \int_1^4 dx (7x^2 y + 5x - 2y)$$

3. A region is bounded by the three curves

$$y = \frac{1}{\sqrt{(x^2+1)}} \;\text{ and }\; x = 1 \;\text{ and }\; y = 0$$

(a) Compute volume of the region around x-axis. (b) Compute volume around y-axis

4. Find the volume of a pyramid of height h whose base is an equilateral triangle of length L.
5. Consider a triangle bounded by $y = -5x + 3$, and the x and y-axes. A cone is formed by rotating the triangle around the y-axis. Find its volume.

6. What's the present value of a house with a loan for which the annual interest payment is $\$30 \times 10^5$, to be paid for 30 years, with an interest rate of 5%? And, compare that to the present value of a house for another loan with interest rate of 6.5%.
7. An equipment factory depreciates rapidly in its first and second year and slower in its third and fourth year. The depreciation's rate over the years of $V(t)$ is $250(t-3)$. The equipment's price is \$8 millions. Find the following
   - The value of the equipment V (t).
   - The total amount of depreciation in the first two years
   - The total amount of depreciation in the following two years.
8. A company's marginal cost of production is MC=$3x^2 - 4x + 400$. What is the total cost of producing 5 extra units if 10 units are presently being produced?
9. (optional) An object's acceleration is given by $a(t) = \cos(\pi t)$, and it's velocity at time $t = 0$ is $1/(2\pi)$. Find the total distance traveled in the first 20 seconds.
10. (optional) An observation deck on 88th floor is 983 feet above the ground. If a steel ball is dropped, its velocity at time t is approximately $v(t) = -32t$ feet per second. Find the average speed between the time it is dropped and the time it hits the ground, and find its speed when it hits the ground.
11. Evaluate the following integral using the technique of partial fractions

$$\int dx \frac{1+x}{x^2+x-2}$$

12. Use a change of variables to evaluate the following integral

$$\int dx \frac{1+x}{x^2+x+5}$$

13. Use integration by parts to evaluate

$$\int dx x e^x$$

14. Express in terms of the normal cumulative distribution $N(z)$, the following integral

$$\frac{1}{\sqrt{2\pi}} \int_a^b dx e^{-\frac{1}{2\sigma^2}x^2}$$

15. Evaluate

$$\int_a^b \frac{dx}{x^2+2x-24}$$

Is the integral finite for i) a=1,b=3 and ii) a=1, b=5? If not, why?

Chapter 6
Differentiation

Abstract Differentiation is the inverse of integration, and its salient features are developed. Differentiation is defined by comparing two integrals that differ infinitesimally—in contrast to starting with a continuous curve and defining differentiation as the tangent to the curve. Taylor expansion is defined and used for analyzing the maximum and minimum of functions. The Hessian matrix is defined to study various types of optimization problems that occur in economics; in particular, the Cobb-Douglas production function is used for analyzing the maximization of a firm's profit. Constrained optimization using the Lagrange multiplier is studied.

6.1 Introduction

Differentiation is widely used in economics and finance. Two examples of marginal cost and rate of capital formation were utilized in the earlier discussion on integration, where the concept of differentiation was indirectly used. The marginal cost M is the change in the total cost of production T for an additional quantity of input Q. For a change of dQ in the input quantity Q, the change in the total cost of production dT is given by the marginal cost of production $M(Q)$, which is given by

$$dT = M(Q)dQ \;\Rightarrow\; M(Q) = \frac{dT}{dQ}$$

Net investment I is defined as the rate of change in capital stock formation C over time t. If the process of capital formation is continuous over time, the rate of capital formation $I(t)$ is given by

$$dC(t) = I(t)dt \;\Rightarrow\; I(t) = \frac{dC}{dt}$$

In both the examples, the generic operation of differentiation denoted by df/dx, was defined intuitively. This is now studied more carefully, starting with the relation of differentiation to integration.

B. E. Baaquie, *Mathematical Methods and Quantum Mathematics for Economics and Finance*, https://doi.org/10.1007/978-981-15-6611-0_6

6.2 Differentiation: Inverse of Integration

Differentiation is the inverse of integration, and has rules of its own. Integration depends on the value of a function $f(x)$ over the range it is being integrated whereas differentiation depends only on the value of $f(x)$ at the point x, and the ϵ neighborhood of x.

Consider the integral $I(x)$ given in Fig. 6.1a

$$I(x) = \int_a^x dyf(y)$$

For a point close to x given by $x + \epsilon$, the integral, as shown in Fig. 6.1b, yields

$$I(x+\epsilon) = \int_a^{x+\epsilon} dyf(y) = \int_a^x dyf(y) + \int_x^{x+\epsilon} dyf(y)$$

The second integral on the right hands is an infinitesimal rectangle, with base ϵ and height $f(x)$. Hence

$$I(x+\epsilon) - I(x) = \int_x^{x+\epsilon} dyf(y) = \epsilon f(x)$$

Note that the *ratio* of two vanishing small terms, $I(x+\epsilon) - I(x)$ and ϵ, has a *finite value* given by $f(x)$. Define the derivative of $I(x)$ at the point x by the following

$$\frac{dI(x)}{dx} = \lim_{\epsilon \to 0} \frac{I(x+\epsilon) - I(x)}{\epsilon} = f(x) \tag{6.2.1}$$

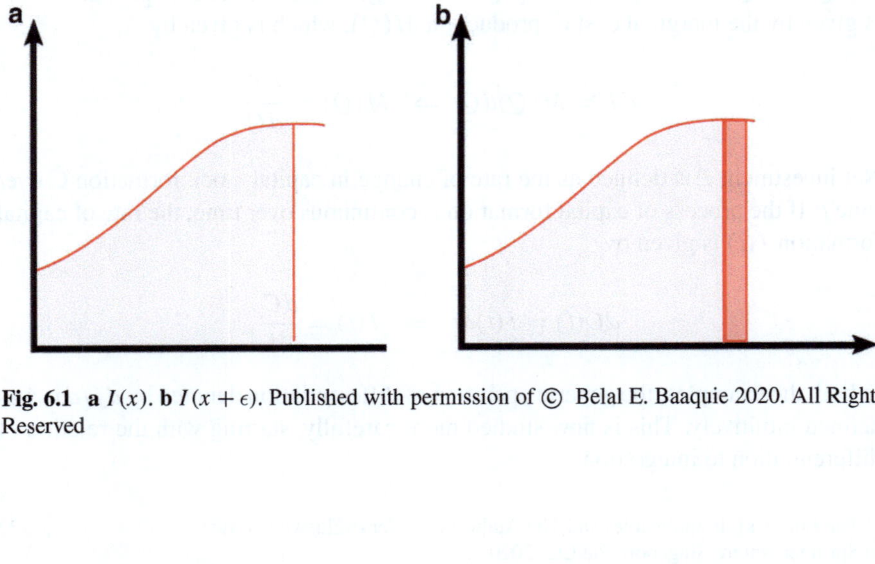

Fig. 6.1 **a** $I(x)$. **b** $I(x+\epsilon)$. Published with permission of © Belal E. Baaquie 2020. All Rights Reserved

To summarize, the following relation, called the *first fundamental theorem of calculus*, is

$$\frac{dI(x)}{dx} = \frac{d}{dx}\int_a^x dyf(y) = \frac{d}{dx}\int^x dyf(y) = f(x)$$

Since differentiation compares only two nearby values of $I(x)$, it does not matter whether one differentiates a definite or an indefinite integral. The result is the same.

Differentiating $I(x)$ is the inverse of the operation of integration and yields the integrand $f(x)$. The second fundamental theorem of calculus states the following

$$I = \int_a^b dx\frac{dh(x)}{dx} = \int_{h(a)}^{h(b)} dh = h(b) - h(a) \tag{6.2.2}$$

The proof of the second theorem follows from the definition of an integral. Discretize the x-axis into a lattice of points such that

$$x_n = a + n\epsilon \ : n = 1, 2, \ldots, N \ ; \ \epsilon = \frac{(b-a)}{N}$$

The integral I is given by the limit of the following telescopic summation

$$I = \lim_{N\to\infty} I_N = \sum_{n=1}^{N}(x_{n+1} - x_n)\frac{h(x_{n+1}) - h(x_n)}{(x_{n+1} - x_n)} = h(x_{N+1}) - h(x_1)$$

Taking the limit of $N \to \infty$ yields

$$I = \int_a^b dx\frac{dh(x)}{dx} = \int_{h(a)}^{h(b)} dh = h(b) - h(a)$$

The special case of $h = x$ yields the expected result

$$\int_a^b dx = b - a$$

6.3 Rules of Differentiation

The rule for differentiation, as given in Eq. 6.2.1, yields

$$\frac{df(x)}{dx} = \lim_{\epsilon\to 0}\frac{f(x+\epsilon) - f(x)}{\epsilon} \ : \text{right derivative}$$

This is called the right derivative since the the value at x is being compared to the value at $x + \epsilon$. The left derivative is defined by comparing the function at x with the point $x - \epsilon$ to the left

$$\frac{df(x)}{dx} = \lim_{\epsilon \to 0} \frac{f(x) - f(x - \epsilon)}{\epsilon} \quad : \text{ left derivative}$$

A symmetric derivative is given by

$$\frac{df(x)}{dx} = \lim_{\epsilon \to 0} \frac{f(x + \epsilon/2) - f(x - \epsilon/2)}{\epsilon} \quad : \text{ symmetric derivative}$$

Function $f(x)$ is a curve when plotted against x, as shown in Fig. 6.2. The derivative df/dx is the tangent to the curve at the point x; this can be seen from the geometrical interpretation of

$$\frac{df(x)}{dx} = \frac{\Delta f(x)}{\Delta x} = \tan \theta$$

where $\tan \theta$ is shown in Fig. 6.2.

For $f(x)$ being a 'smooth' function of x, the symmetric, left and right derivatives are all equal. However, if the function has a 'kink' or is discontinuous at point x, the left and right derivatives are not equal. Figure 6.3 shows typical cases when left and right derivatives are not equal.

For most cases of interest, the left and right derivatives are equal. However, there is an important case in stochastic calculus in which the future and past time derivative, the analogy of left and right derivatives, are not equal. To avoid arbitrage, the correct time derivative for option pricing is the *past time derivative* given by

$$\frac{df(t)}{dt} = \lim_{\epsilon \to 0} \frac{f(t) - f(t - \epsilon)}{\epsilon} \quad : \text{ past derivative}$$

Let f, g be smooth functions of x and a, b constants. The following are the rules for differentiation.

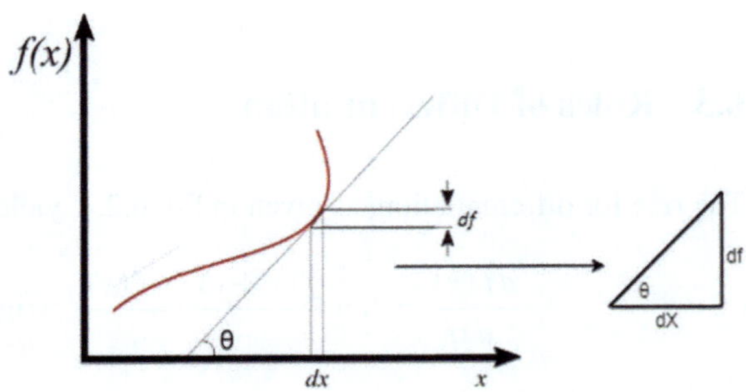

Fig. 6.2 The derivative df/dx. Published with permission of © Belal E. Baaquie 2020. All Rights Reserved

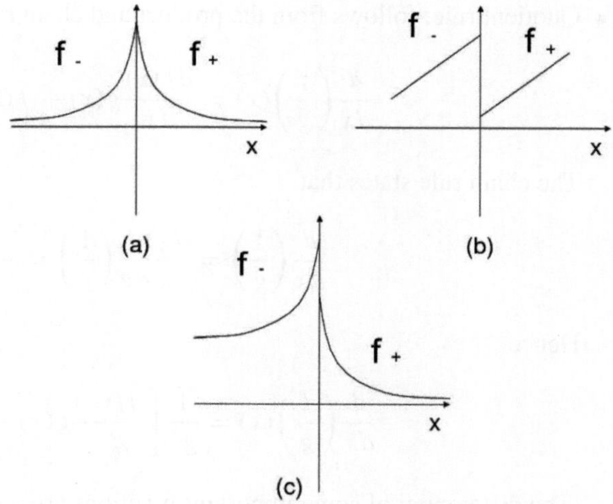

Fig. 6.3 Discontinuous derivatives. Published with permission of © Belal E. Baaquie 2020. All Rights Reserved

- Linearity

$$\frac{d}{dx}(af(x) + bg(x)) = a\frac{df(x)}{dx} + b\frac{dg(x)}{dx}$$

- Product rule

$$\frac{d}{dx}(fg)(x) = \frac{df(x)}{dx}g(x) + f(x)\frac{dg(x)}{dx}$$

A derivation is the following

$$\delta f(x) = f(x+\epsilon) - f(x) = \epsilon df(x)/dx \;\; ; \;\; \delta g(x) = \epsilon dg(x)/dx$$

It follows that

$$\begin{aligned} \frac{d}{dx}(fg)(x) &= \frac{1}{\epsilon}\big[(fg)(x+\epsilon) - (fg)(x)\big] \\ &= \frac{1}{\epsilon}\big[\delta f(x)g(x) + f(x)\delta f(x) + \delta f(x)\delta g(x)\big] \\ \Rightarrow \frac{d}{dx}(fg)(x) &= \frac{df(x)}{dx}g(x) + f(x)\frac{dg(x)}{dx} + O(\epsilon) \end{aligned}$$

- Chain rule

$$\frac{df(g(x))}{dx} = \frac{dg(x)}{dx}\frac{df(g)}{dg}$$

- Quotient rule; follows from the product and chain rule

$$\frac{d}{dx}\left(\frac{f}{g}\right)(x) = \frac{df(x)}{dx}g(x) + f(x)\frac{d}{dx}\left(\frac{1}{g}\right)$$

The chain rule states that

$$\frac{d}{dx}\left(\frac{1}{g}\right) = \frac{dg}{dx}\frac{d}{dg}\left(\frac{1}{g}\right) = -\frac{1}{g^2}\frac{dg}{dx}$$

Hence

$$\frac{d}{dx}\left(\frac{f}{g}\right)(x) = \frac{1}{g^2}\left[\frac{df(x)}{dx}g(x) - f(x)\frac{dg}{dx}\right]$$

The differential of some important functions are the following.

- $f(x) = x^n$

$$\Rightarrow \quad \frac{df(x)}{dx} = \frac{1}{\epsilon}[(x+\epsilon)^n - x^n] = \frac{1}{\epsilon}[(x^n + n\epsilon x^{n-1}) - x^n] = nx^{n-1}$$

- $f(x) = e^{ax}$

$$\Rightarrow \quad \frac{df(x)}{dx} = \frac{1}{\epsilon}[e^{a(x+\epsilon)} - e^{ax}] = e^{ax}[1 + a\epsilon - 1] = ae^{ax}$$

- $f(x) = \ln(ax) = \ln(a) + \ln(x)$

$$\Rightarrow \quad \frac{df(x)}{dx} = \frac{1}{\epsilon}[\ln(x+\epsilon) - \ln(x)] = \frac{1}{\epsilon}[\ln(1 + \frac{\epsilon}{x})] = \frac{1}{x}$$

- $f(x) = e^{i\theta}$

$$\frac{df(\theta)}{d\theta} = ie^{i\theta} = i(\cos(\theta) + i\sin(\theta)) = \frac{d\cos(\theta)}{d\theta} + i\frac{d\sin(\theta)}{d\theta}$$

Equating real and imaginary part of both sides yields

$$\frac{d\cos(\theta)}{d\theta} = -\sin(\theta) \ ; \quad \frac{d\sin(\theta)}{d\theta} = \cos(\theta)$$

Example: $f(x) = (x^3 + 3)(6x^2 + 5x - 3)$

$$\Rightarrow \quad \frac{df(x)}{dx} = 3x^2(6x^2 + 5x - 3) + (x^3 + 3)(12x + 5)$$

Example: $f(x) = g + 7g^3$; $g = -3x^2 + 4x$

$$\Rightarrow \quad \frac{df(g(x))}{dx} = \frac{dg(x)}{dx}\frac{df(g)}{dg} = (-6x+4)(1+21g^2)$$

6.4 Integration by Parts

The product rule of differentiation yields a useful transformation of integrals, called integration by parts. Consider applying the product rule to the indefinite integral

$$\int dx f\frac{dg}{dx} = \int dx\left[\frac{d(fg)}{dx} - \frac{df}{dx}g\right]$$
$$= \int d(fg) - \int dx\frac{df}{dx}g = (fg)(x) - \int dx\frac{df}{dx}g$$

The definite integral yields

$$\int_a^b dx f\frac{dg}{dx} = \int_a^b dx\left[\frac{d(fg)}{dx} - \frac{df}{dx}g\right]$$
$$= \int_a^b d(fg) - \int_a^b dx\frac{df}{dx}g = (fg)(b) - (fg)(a) - \int_a^b \frac{df}{dx}g dx$$

Consider the example

$$\int dx\frac{\ln(x)}{x^2} = -\int dx \ln(x)\frac{d}{dx}\frac{1}{x}$$
$$= -\int d(\frac{\ln(x)}{x}) + \int dx\frac{1}{x^2} = -\frac{\ln(x)}{x} - \frac{1}{x}$$

Differentiating the left and right hand side yields

$$LHS = \frac{\ln(x)}{x^2} \;;\; RHS = -\frac{1}{x^2} + \frac{\ln(x)}{x^2} + \frac{1}{x^2}$$

thus verifying the result.

6.5 Taylor Expansion

Suppose x and a are nearby points. One would expect that the value of $f(x)$ should be computable if $df(a)/dx$ is known. From the definition of a derivative given in Eq. 6.2.1, one approximately has

$$\frac{df(a)}{dx} \simeq \frac{f(a+\epsilon) - f(a)}{\epsilon}$$

Let $\epsilon = x - a$; then

$$\frac{df(a)}{dx} \simeq \frac{f(x) - f(a)}{x - a} \Rightarrow f(x) = f(a) + (x-a)\frac{df(a)}{dx} + \cdots$$

Taylor expansion allows one to calculate a function at nearby points more and more accurately, the higher the derivatives one uses.[1]

To evaluate the full Taylor expansion, note that the second fundamental theorem of calculus given in Eq. 6.2.2 yields the recursion equation

$$\int_a^x dt \frac{d^{n+1} f(t)}{dt^{n+1}} = \int_a^x dt \frac{d}{dt}\left(\frac{d^n f(t)}{dt^n}\right) = \frac{d^n f(x)}{dt^n} - \frac{d^n f(a)}{dt^n}$$

Hence

$$\frac{d^n f(x)}{dt^n} = \frac{d^n f(a)}{dt^n} + \int_a^x dt \frac{d^{n+1} f(t)}{dt^{n+1}} \tag{6.5.1}$$

For $n = 0$, Eq. 6.5.1 yields

$$f(x) = f(a) + \int_a^x dt_1 \frac{df(t_1)}{dt_1}$$

Using the recursion equation for $n = 1$ yields

$$f(x) = f(a) + \int_a^x dt_1 \left[\frac{df(a)}{dt_1} + \int_a^{t_1} dt_2 \frac{d^2 f(t_2)}{dt_2^2}\right]$$

Define the notation

$$\frac{d^n f(a)}{dx^n} = f^{(n)}(a) \; ; \quad f^{(0)}(a) = f(a)$$

Continuing with the recursion above yields

[1] What is the maximum distance that the point a can be from point x and still have a valid Taylor expansion? To answer this one needs to look at the behavior of the function by first complexifying the argument x and then determining the function's analytic structure.

$$
\begin{aligned}
f(x) &= f(a) + \int_a^x dt_1 \frac{df(a)}{dt_1} + \int_a^x dt_1 \int_a^{t_1} dt_2 \frac{d^2 f(a)}{dt_2^2} + \cdots \\
&\quad \cdots + \int_a^x dt_1 \int_a^{t_1} dt_2 \cdots \int_a^{t_{n-1}} dt_n \frac{d^n f(t_n)}{dt_n^n} + \cdots \\
&= f(a) + f^{(1)}(a) \int_a^x dt_1 + f^{(2)}(a) \int_a^x dt_1 \int_a^{t_1} dt_2 + \cdots \\
&\quad \cdots + f^{(n)}(a) \int_a^x dt_1 \int_a^{t_1} dt_2 \cdots \int_a^{t_{n-1}} dt_n
\end{aligned}
\tag{6.5.2}
$$

One has the following coefficients

$$
\int_a^x dt_1 \int_a^{t_1} dt_2 \cdots \int_a^{t_{n-1}} dt_n = \frac{1}{n!} \left(\int_a^x dt_1 \right)^n = \frac{1}{n!}(x-a)^n
$$

Hence, the Taylor expansion is given by

$$
\begin{aligned}
f(x) &= f(a) + (x-a) f^{(1)}(a) + \frac{1}{2!} f^{(2)}(a) + \cdots + \frac{1}{n!}(x-a)^n f^{(n)}(a) + \cdots \\
&= f(a) + (x-a) \frac{df(a)}{dx} + \frac{(x-a)^2}{2!} \frac{d^2 f(a)}{dx^2} + .. + \frac{(x-a)^n}{n!} \frac{d^n f(a)}{dx^n} + .. \\
&= \sum_{n=0}^{\infty} \frac{(x-a)^n}{n!} \frac{d^n f(a)}{dx^n}
\end{aligned}
\tag{6.5.3}
$$

To check Taylor expansion, consider the case when $a = 0$; define

$$
d_n = \frac{1}{n!} \frac{d^n f(0)}{dx^n}
$$

Hence, from Eq. 6.5.3

$$
f(x) = \sum_{n=0}^{\infty} d_n x^n
$$

The polynomial expansion for $f(x)$ is given by

$$
f(x) = \sum_{n=0}^{\infty} c_n x^n \tag{6.5.4}
$$

This yields

$$
\frac{d^n f(0)}{dx^n} = n! c_n \;\Rightarrow\; c_n = d_n
$$

and hence Taylor expansion agrees with the expansion given in Eq. 13.8.8, as expected.

6.6 Minimum and Maximum

A minimum and maximum is a feature of functions having a value at some point—usually denoted by x_c—that is less or greater than its value at all the points neighboring x_c. The proof usually uses the geometrical approach of studying the plot of $f(x)$ against x, which will be discussed later; the problem is first examined analytically.

Suppose the first derivative of $f(x)$ is zero at some point x_c; in other words $df(x_c)/dx = 0$; the point x_c is called a critical point. What is the significance of the following:

$$\frac{df(x_c)}{dx} = 0$$

Using the left and right definition of the derivative yields

$$f(x_c + \epsilon) - f(x_c) = 0 = f(x_c) - f(x_c - \epsilon)$$
$$\Rightarrow \; f(x_c) = f(x_c + \epsilon) = f(x_c - \epsilon) : \text{ stationary}$$

In other words, there is no change in $f(x)$ in the near neighborhood of x_c. As one moves away from x_c, the function can have the following behavior.

- The value of $f(x)$ can *decrease* in both directions, giving rise to x_c being a maxima—as shown in Fig. 6.4a.
- The value of $f(x)$ can *increase* in both directions, giving rise to a minima—as shown in Fig. 6.4b.
- The value of $f(x)$ can increase in one directions and decrease in the other direction, giving rise to an inflexion point, and shown in Fig. 6.5.

Whether a critical point is a maxima, minima or an inflexion point depends on the *sign* of $d^2 f(x_c)/dx^2$. To see this, consider the Taylor expansion of $f(x)$ in the neighborhood of the point x_c, which yields, from Eq. 13.8.10, the following

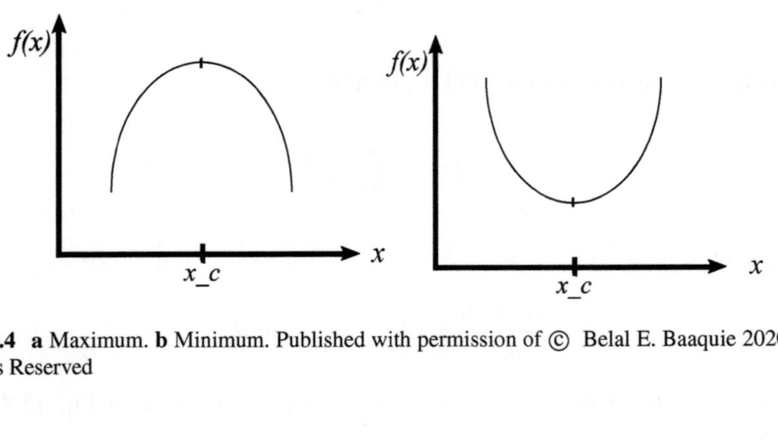

Fig. 6.4 **a** Maximum. **b** Minimum. Published with permission of © Belal E. Baaquie 2020. All Rights Reserved

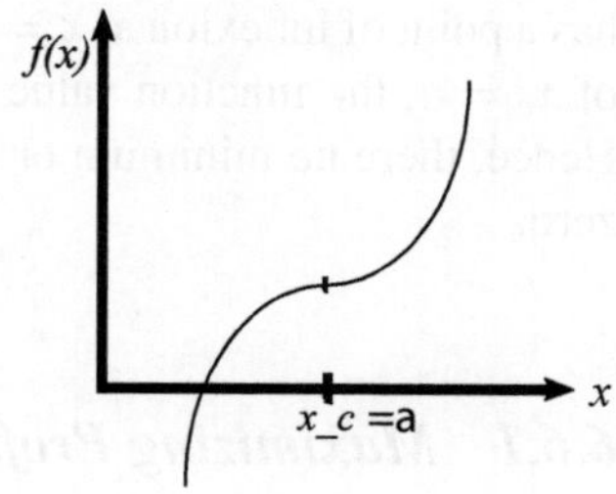

Fig. 6.5 The point $x = a$ is a point of inflexion. Published with permission of © Belal E. Baaquie 2020. All Rights Reserved

$$f(x) = f(x_c) + \frac{(x - x_c)^2}{2!}\frac{d^2 f(x_c)}{dx^2} + O(x - x_c)^3$$

Since $(x - x_c)^2 > 0$, the behavior of the function in the neighborhood of x is entirely determined by the sign of $d^2 f(x_c)/dx^2$.

Consider the two cases, for which the variable x takes a *range of values* in the neighborhood of x_c, denoted by $\mathcal{D}$.

$$f(x) = f(x_c) + \frac{(x - x_c)^2}{2!}\left|\frac{d^2 f(x_c)}{dx^2}\right| + O(x - x_c)^3 \;\Rightarrow\; f(x_c)\;:\;\text{minimum}$$
$$f(x) = f(x_c) - \frac{(x - x_c)^2}{2!}\left|\frac{d^2 f(x_c)}{dx^2}\right| + O(x - x_c)^3 \;\Rightarrow\; f(x_c)\;:\;\text{maximum} \quad (6.6.1)$$

Function with a maximum and minimum are shown in Fig. 6.4. In summary, from Eq. 6.6.1

- $$\frac{d^2 f(x_c)}{dx^2} = -\left|\frac{d^2 f(x_c)}{dx^2}\right| < 0 \Rightarrow f(x) < f(x_c);\ \forall x \in \mathcal{D}\ :\ f(x_c)\ :\ \text{maximum}$$
- $$\frac{d^2 f(x_c)}{dx^2} = +\left|\frac{d^2 f(x_c)}{dx^2}\right| > 0 \Rightarrow f(x) > f(x_c);\ \forall x \in \mathcal{D}\ :\ f(x_c)\ :\ \text{minimum}$$

For example, the function

$$f(x) = \frac{1}{2}(x - a)^2$$

has a minimum at $x = x_c = a$.

For the special case of $d^2 f(x_c)/dx^2 = 0$, one needs to find out the first non-zero higher derivative to gauge the nature of the critical point. An inflexion point needs certain conditions to be obeyed by the third derivative. The function

$$f(x) = b(x - a)^3$$

has a point of inflexion at $x = x_c = a$, as shown in Fig. 6.5. As one moves to the left of $x_c = a$, the function value decreases and as one moves to the right it increases. Hence, there no minimum or maximum at $x = a$ even though the first derivative is zero.

6.6.1 Maximizing Profit

Differentiation has many applications in economics and finance, and the concept of profit maximization is studied with a simple model of revenue and costs. Let Q be the quantity of a commodity produced by an enterprise. Suppose the revenue $R(Q)$ and cost $C(Q)$ accruing from producing quantity Q is given by

$$R(Q) = 1200 - 2Q^2 \ ; \ C(Q) = Q^3 - 61.25Q^2 + 1528.5Q + 2000$$

The profit function $\pi(Q)$ is given by

$$\pi(Q) = R(Q) - C(Q) = -Q^3 + 59.25Q^2 - 328.5Q - 2000$$

Maximizing profit requires

$$\frac{d\pi}{dQ} = -3Q^2 + 118.5Q - 328.5 = 0 \ \Rightarrow \ Q_c = 3 \text{ or } 36.5$$

To determine whether the critical value of Q_c maximizes profit, the second derivative of π is required for the critical values. Hence

$$\frac{d^2\pi}{dQ^2} = -6Q + 118.5 = \begin{cases} > 0 \ Q_c = 3 \\ < 0 \ Q_c = 36.5 \end{cases}$$

Hence, the quantity that maximizes profit is $Q_c = 36.5$, with the other value minimizing profit. The maximum profit is

$$\pi^* = \pi_m(36.5) = 16,318.44$$

Note the maximizing condition can also be written as follows

$$\frac{d\pi}{dQ} = \frac{d\pi}{dQ} - \frac{dC}{dQ} = 0 \ \Rightarrow \ \frac{dR}{dQ} = \frac{dC}{dQ}$$

In other words, for determining the quantity Q_c that yields the maximum profit, the marginal revenue dR/dQ must be equal to the marginal cost dC/dQ.

6.7 Integration; Change of Variable

Consider the following integration of two functions f, g.

$$I = \int_a^b dx f(x) \frac{dg(x)}{dx}$$

A change of variables yields the following

$$y = g(x) \ \Rightarrow \ dy = \frac{dg(x)}{dx} dx \ ; \ x = g^{-1}(y)$$

The limits yield the following

$$x = a \Rightarrow y = g(a) \ ; \ x = b \Rightarrow y = g(b)$$

The change of variables yields the integral

$$I = \int_a^b dx f \frac{dg}{dx} = \int_{g(a)}^{g(b)} dy f(x) = \int_{g(a)}^{g(b)} dy f(g^{-1}(y))$$

The second fundamental theorem of calculus follows from a change of variable. Consider the change of variable

$$f(x) = \frac{dh(x)}{dx} \ ; \ y = h(x) \ \Rightarrow \ dy = \frac{dh(x)}{dx} dx$$

that yields the result given earlier in Eq. 6.2.2

$$I = \int_a^b dx f(x) = \int_a^b dx \frac{dh(x)}{dx} = \int_{h(a)}^{h(b)} dy = h(b) - h(a)$$

Example:

$$I = \int_{-\infty}^{+\infty} dx \frac{e^x}{1 + e^{2x}}$$

Perform the change of variable

$$y = e^x \ : \ dy = e^x dx \ ; \ x = \ln(y)$$

The limits are given by

$$x = -\infty \Rightarrow y = 0 \ ; \ x = \infty \Rightarrow y = \infty$$

Hence

$$I = \int_0^{+\infty} \frac{dy}{1+y^2}$$

Consider another change of variables

$$y = \tan(\theta) \ : \ dy = \frac{d\theta}{\cos^2(\theta)} \ ; \ y \in [0, \infty] \ \Rightarrow \ \theta \in [0, \frac{\pi}{2}]$$

Hence,

$$I = \int_0^{\pi/2} \frac{d\theta}{\cos^2(\theta)} \frac{1}{1+\tan^2(\theta)} = \int_0^{\pi/2} d\theta = \frac{\pi}{2}$$

6.8 Partial Derivatives

Consider a function of N-variables x_i given by

$$f(\mathbf{x}) = f(x_1, x_2, \ldots, x_N) \ ; \ \mathbf{x} = \sum_i x_i \mathbf{e}_i$$

A partial derivative with respect to one of the variable x_i is defined by

$$\begin{aligned}\frac{\partial f}{\partial x_i} &= \lim_{\epsilon \to 0} \frac{1}{\epsilon}[f(\mathbf{x} + \epsilon \mathbf{e}_i) - f(\mathbf{x})] \\ &= \lim_{\epsilon \to 0} \frac{1}{\epsilon}[f(x_1, \ldots, x_i + \epsilon, \ldots, x_N) - f(x_1, x_2, \ldots, x_N)]\end{aligned}$$

Partial derivatives commute and yield

$$\frac{\partial^2 f}{\partial x_i \partial x_j} = \frac{\partial^2 f}{\partial x_j \partial x_i}$$

Suppose all the variables $\mathbf{x}$ depend on some parameter t, and the function f also has an explicit dependence on t. The total change in f on varying t is given by

$$\begin{aligned}\frac{df(\mathbf{x}(t); t)}{dt} &= \frac{1}{\epsilon}[f(\mathbf{x}(t+\epsilon); t+\epsilon) - f(\mathbf{x}(t); t)] \\ &= \frac{1}{\epsilon}\left[f(\mathbf{x}(t) + \epsilon\frac{d\mathbf{x}(t)}{dt}; t+\epsilon) - f(\mathbf{x}(t); t)\right] \\ &= \frac{\partial f(\mathbf{x}(t); t)}{\partial t} + \sum_i \frac{dx_i(t)}{dt} \frac{\partial f(\mathbf{x}(t); t)}{\partial x_i} \\ &\equiv \frac{\partial f(x(t); t)}{\partial t} + \frac{d\mathbf{x}(t)}{dt} \cdot \nabla f(x(t); t)\end{aligned}$$

where the gradient operator is defined by

$$\nabla = \sum_i \mathbf{e}_i \frac{\partial}{\partial x_i}$$

Example: Let $f = x^2 - 9xy + 11$; $x = t^2 + 5$; $y = 3t - 7$; then

$$\frac{df(x(t); y(t))}{dt} = \frac{dx(t)}{dt}\frac{\partial f(x(t); y(t))}{\partial x} + \frac{dy(t)}{dt}\frac{\partial f(x(t); y(t))}{\partial y}$$
$$= 2t(2x - 9y) + (3)(-9x) = 4t^3 - 73t^2 + 146t - 135$$

Example: Let $f(x, y) = 3x^3y^2 - 9x^2 + 7y$

$$\frac{\partial^2 f}{\partial y \partial x} = 18x^2 y = \frac{\partial^2 f}{\partial x \partial y}$$

6.8.1 Cobb-Douglas Production Function

To produce a quantity Q of commodities, the Cobb-Douglas production function postulates that the capital K and workers L required is given by

$$Q = AL^{\alpha}K^{\beta}$$

and where A is determined by the technology of the production process. To evaluate the partial derivatives, note that

$$dQ = \frac{\partial Q}{\partial L}dL + \frac{\partial Q}{\partial K}dK \tag{6.8.1}$$

and hence

$$\frac{\partial Q}{\partial L} = \alpha A L^{\alpha-1}K^{\beta} \; ; \quad \frac{\partial Q}{\partial K} = \beta A L^{\alpha}K^{\beta-1}$$

Suppose $Q = Q_0$ is *fixed*; then K, L are no longer independent; for example, the change in K if one changes L is given by

$$\frac{\partial K}{\partial L}$$

From Eq. 6.8.1

$$0 = dQ_0 = \frac{\partial Q}{\partial L}dL + \frac{\partial Q}{\partial K}dK$$

Hence

$$\frac{\partial K}{\partial L} = -\frac{\partial Q}{\partial L} / \frac{\partial Q}{\partial K} = -\frac{\alpha K}{\beta L} < 0$$

If Q is not a constant, then L, K are independent variables with $\partial K/\partial L = 0$.

6.8.2 Chain Rule; Jacobian

The chain rule has a generalization for the case of many variables. Consider a change of variables from $\mathbf{x}$ to $\mathbf{y}$ given by $\mathbf{x} = \mathbf{x}(\mathbf{y})$. For a function $f(\mathbf{x})$, the chain rule yields

$$\frac{\partial f(\mathbf{x})}{\partial y_i} = \sum_j \frac{\partial x_j}{\partial y_i} \frac{\partial f(\mathbf{x})}{\partial x_j}$$

Consider a change in the differential of a coordinate dx_i; the chain rule yields

$$dx_i = \sum_{j=1}^{N} \frac{\partial x_j}{\partial y_i} dy_j \equiv \sum_{j=1}^{N} J_{ij} dy_j$$

where the Jacobian matrix J_{ij} of the transformation is given by

$$J_{ij} = \frac{\partial x_j}{\partial y_i} = J_{ij}(\mathbf{y})$$

The Jacobian measures how much the original volume, in terms of the original coordinates x, changes from point to point in going to the new coordinate system y.

A multiple integral, under the change of variables, transforms in the following manner

$$I = \int_{\mathcal{D}} d\mathbf{x} f(\mathbf{x}) = \int_{\mathcal{D}'} d\mathbf{y} J(\mathbf{y}) f(\mathbf{x}(\mathbf{y}))$$

where $J(\mathbf{y})$ is the mod of the Jacobian matrix and is given by

$$J(\mathbf{y}) = \left| \det[J_{ij}(\mathbf{y})] \right|$$

A linear change of variables, discussed in in Sect. 5.5.1, is an examples of how the Jacobian appears in a change of variables.

6.8.3 Polar Coordinates; Gaussian integration

Consider the change of variables

$$x = r\cos\theta \ ; \ \ y = r\sin\theta \ : \ 0 \le r \le \infty \, ; \, 0 \le \theta \le 2\pi$$

The partial derivatives are given by

$$\frac{\partial x}{\partial r} = \cos\theta \, ; \ \frac{\partial y}{\partial r} = \sin\theta \, ; \ \frac{\partial x}{\partial \theta} = -r\sin\theta \, ; \ \frac{\partial y}{\partial \theta} = r\cos\theta$$

The Jacobian is given by

$$J = \begin{bmatrix} \frac{\partial x}{\partial r} & \frac{\partial x}{\partial \theta} \\ \frac{\partial y}{\partial r} & \frac{\partial y}{\partial \theta} \end{bmatrix} = \begin{bmatrix} \cos\theta & -r\sin\theta \\ \sin\theta & r\cos\theta \end{bmatrix} \tag{6.8.2}$$

$$\Rightarrow \ \det J = r(\cos^2\theta + \sin^2\theta) = r$$

Hence

$$dxdy = \det J dr d\theta = r dr d\theta \tag{6.8.3}$$

The change of variables for the integral over x, y is given by

$$\int_{-\infty}^{+\infty} dxdy f(x, y) = \int_{0}^{+\infty} dr \int_{0}^{2\pi} d\theta J(r, \theta) f(r, \theta) = \int_0^{+\infty} dr r \int_0^{2\pi} d\theta f(r, \theta)$$

From Eq. 5.6.3

$$I = \int_{-\infty}^{+\infty} dy e^{-\frac{1}{2}\lambda y^2} = \sqrt{\frac{2\pi}{\lambda}}$$

Hence

$$I^2 = \int_{-\infty}^{+\infty} dy e^{-\frac{1}{2}\lambda y^2} \int_{-\infty}^{+\infty} dz e^{-\frac{1}{2}\lambda z^2} = \int_{-\infty}^{+\infty} dy dz e^{-\frac{1}{2}\lambda(y^2+z^2)}$$

Define new variables by the polar coordinates r, θ such that

$$y = r\cos\theta \, , \ z = r\sin\theta \ ; \ \ 0 \le r \le +\infty \ ; \ \ 0 \le \theta \le 2\pi$$

The change of variable, from Eq. 6.8.3, yields the following for the square of the integral I

$$I^2 = \int_0^{2\pi} d\theta \int_0^{+\infty} rdre^{-\frac{1}{2}\lambda r^2} = 2\pi \int_0^{+\infty} rdre^{-\frac{1}{2}\lambda r^2}$$

Define new variable

$$w = r^2 \ , \ dw = 2rdr \ ; \ 0 \le w \le +\infty$$

Hence

$$I^2 = 2\pi \int_0^{+\infty} \frac{dw}{2} e^{-\frac{1}{2}\lambda w} = \pi \cdot \frac{2}{\lambda}$$

and yields the final answer

$$I = \sqrt{\frac{2\pi}{\lambda}} \tag{6.8.4}$$

The result above and Eq. 5.6.2 yields the expression given in Eq. 5.6.3.

6.9 Hessian Matrix: Critical Points

The Taylor expansion for N-variables is obtained by using vectors $\mathbf{a}$ and $\mathbf{x}$; Eq. 6.5.3 yields

$$f(\mathbf{x}) = f(\mathbf{a}) + \sum_i (\mathbf{x} - \mathbf{a})_i \frac{\partial f(\mathbf{a})}{\partial x_i} + \frac{1}{2!} \sum_{ij} (\mathbf{x} - \mathbf{a})_i (\mathbf{x} - \mathbf{a})_j \frac{\partial^2 f(\mathbf{x})}{\partial x_i \partial x_j} + \cdots \tag{6.9.1}$$

The real symmetric $N \times N$ Hessian matrix H is given by

$$H_{ij} = \frac{\partial^2 f(\mathbf{a})}{\partial x_i \partial x_j} \ : \ \text{Hessian matrix}$$

and has real eigenvalues.

The critical point $\mathbf{a} = \mathbf{x}_c$ in N-dimensions is determined by

$$\frac{\partial f(\mathbf{x}_c)}{\partial x_i} = 0 \ : \ i = 1, 2, \ldots, N$$

Hence, near the critical point $\mathbf{x}_c = (x_c, y_c)$

$$f(\mathbf{x}) = f(\mathbf{x}_c) + \frac{1}{2!} \sum_{ij} (\mathbf{x} - \mathbf{x}_c)_i (\mathbf{x} - \mathbf{x}_c)_j \frac{\partial^2 f(\mathbf{x}_c)}{\partial x_i \partial x_j} + \cdots \tag{6.9.2}$$

Similar to the case of the single variable given in Eq. 6.6.1, the nature of the critical point is determined by the eigenvalues the Hessian matrix. At each critical point, the Hessian matrix is evaluated, and the value of eigenvalues at the critical points determine the nature of the critical points.

Let the Hessian be diagonalized by an orthogonal transformation $\mathcal{O}$; for $N = 2$, in matrix notation, the Hessian matrix H is given by

$$H = \mathcal{O}^T \begin{bmatrix} \lambda_1 & 0 \\ 0 & \lambda_2 \end{bmatrix} \mathcal{O} \ ; \ \mathbf{z}_I = \mathcal{O}_{Ij}(\mathbf{x} - \mathbf{x}_c)_j$$

Then, from Eq. 6.9.1

$$f(\mathbf{x}) = f(\mathbf{x}_c) + \lambda_1 \mathbf{z}_1^2 + \lambda_2 \mathbf{z}_2^2 + \cdots$$

Hence, after rotating the coordinate system by $\mathcal{O}$, it is seen that the maximum or minimum of the function $f(\mathbf{x})$ is determined by the eigenvalues of the Hessian matrix H. The generalization for the case of N variables is given by

$$f(\mathbf{x}) = f(\mathbf{x}_c) + \lambda_1 \mathbf{z}_1^2 + \lambda_2 \mathbf{z}_2^2 + \cdots + \lambda_N \mathbf{z}_N^2$$

For simplicity consider two independent variables x, y and let the critical point $\mathbf{x}_c = (x_c, y_c)$ be defined by

$$\frac{\partial f(x_c, y_c)}{\partial x} = 0 = \frac{\partial f(x_c, y_c)}{\partial y}$$

For two independent variables x, y the Hessian matrix is given by

$$H = \begin{bmatrix} \frac{\partial^2 f(x_c,y_c)}{\partial x^2} & \frac{\partial^2 f(x_c,y_c)}{\partial x \partial y} \\ \frac{\partial^2 f(x_c,y_c)}{\partial y \partial x} & \frac{\partial^2 f(x_c,y_c)}{\partial y^2} \end{bmatrix} = H^T \qquad (6.9.3)$$

Let the two eigenvalues of the Hessian matrix be given by λ_1, λ_2. The classification of the critical point $\mathbf{x}_c = (x_c, y_c)$ is given by the following.

- $\lambda_1, \lambda_2 > 0$: $f(x_c, y_c)$ is a minimum.
- $\lambda_1, \lambda_2 < 0$: $f(x_c, y_c)$ is a maximum.
- $\lambda_1 > 0, \lambda_2 < 0$ or $\lambda_1 < 0, \lambda_2 > 0$: $f(x_c, y_c)$ is a saddle point.
- $\lambda_1 = 0 = \lambda_2$: no information; need higher derivatives to characterize the critical point.

Noteworthy 6.1: Two dimensional Hessian

From Noteworthy 4.7, consider the following 2×2 real symmetric matrix

$$S = \begin{bmatrix} \alpha & \beta \\ \beta & \gamma \end{bmatrix} = S^T \ ; \ \alpha, \beta, \gamma : \text{real}$$

From Eq. 4.7.3, the eigenvalues are given by

$$\lambda_\pm = \frac{1}{2}(\alpha + \gamma) \pm \sqrt{\frac{1}{4}(\alpha - \gamma)^2 + \beta^2} : \text{real}$$

- Consider the case when $\alpha, \gamma < 0$; then $\lambda_- < 0$; for a *maximum* one needs

$$\lambda_+ = \frac{1}{2}(\alpha + \gamma) + \sqrt{\frac{1}{4}(\alpha - \gamma)^2 + \beta^2} < 0 \ \Rightarrow \ \det(S) = \alpha\gamma - \beta^2 > 0$$

- Consider the case when $\alpha, \gamma > 0$; then $\lambda_+ > 0$; for a *minimum* one needs $\lambda_- > 0$ or equivalently

$$\det(S) = \alpha\gamma - \beta^2 > 0$$

From Noteworthy 4.7, for the two-dimensional case, the determinant of the Hessian, given by

$$\det H = \lambda_1 \lambda_2$$

contains the following information. One needs to specify the critical point, and at the critical point one then evaluates the determinant of the Hessian matrix.

- $\det H > 0$: either a maximum or a minimum
- $\det H < 0$: saddle point
- $\det H = 0$: no information

From Noteworthy 6.1, one also has the following

- $H_{11}, H_{22} < 0$; $\det H > 0$: maximum
- $H_{11}, H_{22} > 0$; $\det H > 0$: minimum

Example 1 Consider an enterprise that produces two commodities that can be substituted for each other. Hence their demands are coupled. Let the prices and quantities of the commodities, denoted by P_1, P_2 and Q_1, Q_2, obey the following equations

$$P_1 = 55 - Q_1 - Q_2$$
$$P_2 = 70 - Q_1 - 2Q_2$$

The total revenue of the enterprise is given by

$$R = P_1 Q_1 + P_2 Q_2 = 55Q_1 + 70Q_2 - 2Q_1Q_2 - Q_1^2 - 2Q_2^2$$

Let the cost function be given by

$$C = Q_1^2 + Q_1Q_2 + Q_2^2$$

Hence the profit function is given by

$$\pi = R - C = 55Q_1 + 70Q_2 - 3Q_1Q_2 - 2Q_1^2 - 3Q_2^2$$

Prices are fixed by the market, so the only choice the enterprise has is to decide on the quantities Q_1, Q_2 that should be produced to have maximum profit; for this, the following minimum are required

$$\frac{\partial \pi}{\partial Q_1} = 0 = 55 - 3Q_2 - 4Q_1 \ ; \ \frac{\partial \pi}{\partial Q_2} = 0 = 70 - 3Q_1 - 6Q_2$$

Solving the linear equation for the quantities yields

$$Q_1^* = 8 \ ; \ Q_2^* = 7.666..$$

Note that

$$\frac{\partial^2 \pi}{\partial Q_1^2} = -4 \ ; \ \frac{\partial^2 \pi}{\partial Q_2^2} = -6 \ ; \ \frac{\partial^2 \pi}{\partial Q_1 \partial Q_2} = -3$$

Since the profit function is quadratic, there is only one critical point given by Q_1^*, Q_2^*; the Hessian matrix at the critical point—and every other point in the Q_1, Q_2 space—is a constant given by

$$H = \begin{bmatrix} -4 & -3 \\ -3 & -6 \end{bmatrix} \tag{6.9.4}$$

The two eigenvalues of H are

$$\lambda_1 = -5 + \sqrt{12} \ ; \ \lambda_2 = -5 - \sqrt{12}$$

Hence, $\lambda_1, \lambda_2 < 0$, and the value of $\pi^* = \pi(Q_1^*, Q_2^*)$ is a maximum. Substituting the critical values of the two quantities yields

$$P_1^* = 39.333.. \ ; \ P_2^* = 46.666.. \ ; \ \pi^* = 488.333..$$

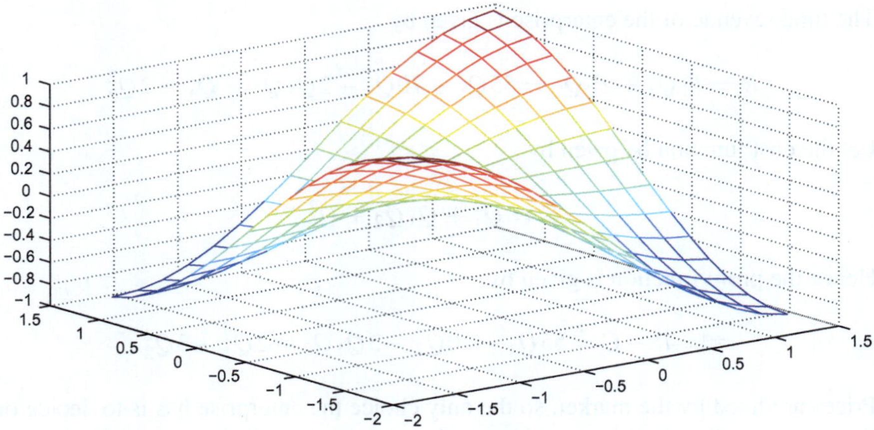

Fig. 6.6 Function $\sin(x)\sin(y)$. Published with permission of © Belal E. Baaquie 2020. All Rights Reserved

Example 2 Consider the function, shown in Fig. 6.6

$$f(x, y) = \sin(x)\sin(y)$$

The critical points x_c, y_c of the function are given by

$$\cos x_c \sin y_c = \frac{\partial f}{\partial x} = 0 = \frac{\partial f}{\partial y} = \sin x_c \cos y_c$$

The following are the three distinct critical points

$$I: \ x_c = 0 = y_c \ ; \ II: \ \ x_c = \pi/2 = y_c \ ; \ III: \ \ x_c = \pi/2 = -y_c$$

From Eq. 6.8.2, the Hessian matrix is given by

$$H = \begin{bmatrix} -\sin x \sin y & \cos x \cos y \\ \cos x \cos y & -\sin x \sin y \end{bmatrix} \tag{6.9.5}$$

Table 6.1 Critical points and Hessian eigenvalues

Critical point	λ_+	λ_-	Classification
$x_c = 0 = y_c$	$+1$	-1	Saddle point
$x_c = \pi/2 = y_c$	-1	-1	Maximum
$x_c = \pi/2 = -y_c$	1	1	Minimum

The eigenvalue equation

$$\det \begin{vmatrix} -\sin x \sin y - \lambda & \cos x \cos y \\ \cos x \cos y & -\sin x \sin y - \lambda \end{vmatrix} = 0$$

yields

$$\lambda_{\pm}(x, y) = -\sin x \sin y \pm \cos x \cos y$$

Given in Table 6.1 are the eigenvalues at the critical points. The result of the classification is seen to be consistent with the shape of the function $f(x, y) = \sin x \sin y$ given in Fig. 6.6.

6.10 Firm's Profit Maximization

The profit of a firm is given by

$$\pi = \mathcal{R} - wL - K$$

where w is wages, L is the number of workers employed, K is the capital input required and $\mathcal{R}$ is revenue. For a quantity of commodity Q produced by the firm and sold at market price P, the revenue is

$$\mathcal{R} = PQ$$

The firm has to decide how much of commodities Q should it produce to maximize its profits. For this, one needs to know how much capital K and workers L are required to produce quantity Q of commodities. The Cobb-Douglas production function postulates that

$$Q = AL^{\alpha}K^{\beta}$$

and hence

$$\pi = APL^{\alpha}K^{\beta} - wL - K$$

In general, $\alpha \neq \beta$. For profit that scales with input, one has $\beta = 1 - \alpha$.

The marginal productivity of labor is the increase in output for a unit increase of worker and is given by

$$\frac{\partial n}{\partial L} = \alpha APL^{\alpha-1}K^{\beta} - w$$

The marginal productivity of capital is given by

$$\frac{\partial n}{\partial K} = \beta APL^{\alpha}K^{\beta-1} - 1$$

The amount of capital and labor to be employed by the firm is given by the following first-order condition for profit maximization

$$\frac{\partial n}{\partial L} = 0 = \frac{\partial n}{\partial K}$$

The solution is given by

$$L = L_* \ : \ K = K_*$$

and yields

$$\alpha APL_*^{\alpha-1}K_*^{\beta} - w = 0 = \beta APL_*^{\alpha}K_*^{\beta-1} - 1 \tag{6.10.1}$$

To ascertain conditions for the critical point L_*, K_* to be a maxima one needs to obtain the Hessian matrix for the profitability of the firm. The Hessian matrix is given by

$$\begin{aligned} H &= \begin{bmatrix} \frac{\partial^2 \pi}{\partial L^2} & \frac{\partial^2 \pi}{\partial L \partial K} \\ \frac{\partial^2 \pi}{\partial K \partial L} & \frac{\partial^2 \pi}{\partial K^2} \end{bmatrix} \\ &= AP \begin{bmatrix} \alpha(\alpha-1)L^{\alpha-2}K^{\beta} & \alpha\beta L^{\alpha-1}K^{\beta-1} \\ \alpha\beta L^{\alpha-1}K^{\beta-1} & \beta(\beta-1)L^{\alpha}K^{\beta-2} \end{bmatrix} \end{aligned} \tag{6.10.2}$$

The second order derivatives are all evaluated at the critical point L_*, K_*. From Eq. 6.9, maximum requires

$$H_{11}(L_*, K_*), H_{22}(L_*, K_*) < 0 \ ; \ \det H(L_*, K_*) > 0 : \text{maximum}$$

The condition H_{11}, $H_{22} < 0$, from Eq. 6.10.2, requires

$$\alpha < 1 \ ; \ \beta < 1$$

The determinant is given by

$$\begin{aligned} \det H &= (AP)^2 L^{2\alpha-2} K^{2\beta-2} \alpha\beta \Big\{(\alpha-1)(\beta-1) - \alpha\beta\Big\} \\ &= (AP)^2 L^{2\alpha-2} K^{2\beta-2} \alpha\beta \Big\{1 - (\alpha+\beta)\Big\} \end{aligned} \tag{6.10.3}$$

Hence, from Eq. 6.10.3

$$\det H(L_*, K_*) > 0 \ \Rightarrow \ \alpha + \beta < 1$$

Note that if $\alpha + \beta = 1$, the critical point is a saddle point and there is no value of L, K for which the firm can maximize its profit.

Noteworthy 6.2: Scaling of Profit and Inputs

Let λ scale the inputs; this yields

$$\pi(\lambda) = AP(\lambda L)^{\alpha}(\lambda K)^{\beta} - w\lambda L - \lambda K = \lambda\Big[AP\lambda^{\alpha+\beta-1}(L)^{\alpha}(K)^{\beta} - wL - K\Big]$$

Since $\beta + \alpha < 1$ for obtaining a maxmimum for the profit function, increasing the inputs by scaling the input with $\lambda > 1$ may reduce the maximum profit of the firm depending on the value of the terms in the bracket.

For the special case of $\beta = 1 - \alpha$, the profit scales if the number of workers and capital are increased by the same scale.

$$\pi(\lambda) = AP(\lambda L)^{\alpha}(\lambda K)^{1-\alpha} - w\lambda L - \lambda K = \lambda\pi$$

The case when the profit scales results in no unique values for L, K that yield the maximum profit: the reason being that one can obtain a higher profit as long as one keeps on increasing λ. This is the reason that the Hessian is zero for $\beta = 1 - \alpha$.

For α, β satisfying the conditions above, the firm's profit is maximized by Eq. 6.10.1, which yields the following

$$L_* = \frac{w\beta}{\alpha}K_* \Rightarrow \alpha AP(\frac{w\beta}{\alpha}K_*)^{\alpha-1}(K_*)^{\beta} = w \tag{6.10.4}$$

Hence, from Eq. 6.10.4

$$\begin{aligned} K_* &= \left(\frac{\alpha AP}{w}\right)^{1/(1-\alpha-\beta)}\left(\frac{\alpha}{w\beta}\right)^{(1-\alpha)/(1-\alpha-\beta)} \\ \Rightarrow L_* &= \frac{w\beta}{\alpha}\left(\frac{\alpha AP}{w}\right)^{1/(1-\alpha-\beta)}\left(\frac{\alpha}{w\beta}\right)^{(1-\alpha)/(1-\alpha-\beta)} \end{aligned} \tag{6.10.5}$$

The quantity of produced commodities that maximize the firm's profit is given by

$$\begin{aligned} Q_* &= AL_*^{\alpha}K_*^{\beta} = A\left(\frac{w\beta}{\alpha}K_*\right)^{\alpha}K^{\beta} \\ &= A\left(\frac{w\beta}{\alpha}\right)^{\alpha}\left(\frac{\alpha AP}{w}\right)^{(\alpha+\beta)/(1-\alpha-\beta)}\left(\frac{\alpha}{w\beta}\right)^{(\alpha+\beta)(1-\alpha)/(1-\alpha-\beta)} \end{aligned}$$

and maximum profit is

$$\pi_* = PQ_* - wL_* - K_*$$

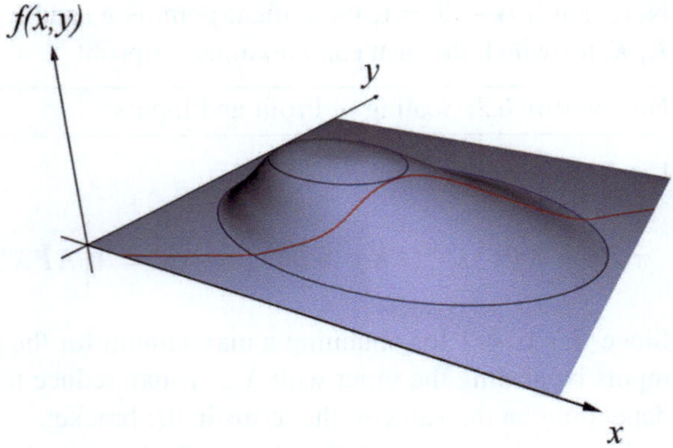

Fig. 6.7 Constrained optimization. Image credit: By Nexcis (Own work) [Public domain], via Wikimedia Commons

6.11 Constrained Optimization: Lagrange Multiplier

The extremal value of a function can be found by finding its critical point $\mathbf{x}_c$ given by

$$\frac{\partial f(\mathbf{x}_c)}{\partial x_i} = 0 \ : \ i = 1, 2, \ldots, N$$

The Hessian matrix determines whether the critical point is a maxima or a minima. Often, the critical point has to be found subject an additional constraint on the underlying variables.

Consider a two dimensional system; one needs to find the maximum of the function $f(x, y)$ subject to the constraint that $g(x, y) = c$; the problem can be stated as finding the values of (x_c, y_c) such the following three equations are satisfied

$$\frac{\partial f(x_c, y_c)}{\partial x} = 0 = \frac{\partial f(x_c, y_c)}{\partial y} \ ; \ g(x_c, y_c) = c$$

An example of a typical function and constraint are shown in Figs. 6.7 and 6.8.

The constraint relation $g(x, y) = c$ is a line in two dimensions. The contours of $f(x, y)$=constant are also lines in the xy-plane. The contour lines and constraint are shown in Fig. 6.8.

Suppose one is at the critical point of $f(x, y)$ that lies on the constraint equation and denoted by (x_c, y_c); then if one moves infinitesimally along the constraint equation, the value of $f(x, y)$ should not change since otherwise one is not at a maximum. One way the value of $f(x, y)$ does not change is because one is moving along the contour line of $f(x, y)$ on which the value of $f(x, y)$ is fixed.

Since one is moving along the line for which the constraint equation is valid, the value of $g(x, y)$ *also* does not change. We conclude that for *both* of these conditions to hold, the required critical point (x_c, y_c) is the point where the contour lines of $f(x, y)$ and $g(x, y)$ are *parallel*; being parallel is equivalent to the line perpendicular to the two contour lines be proportional. Hence, the gradient of the two contour lines are

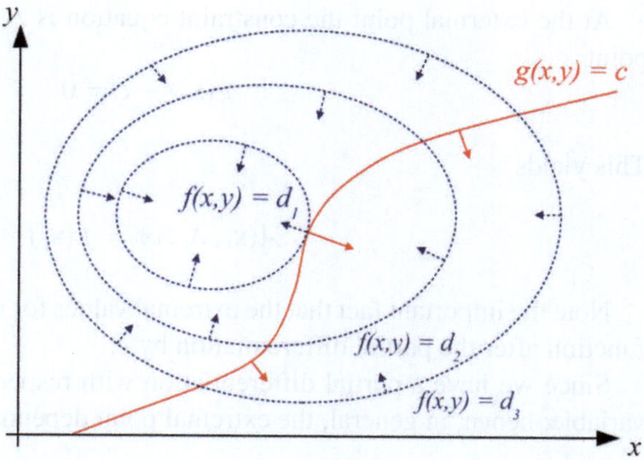

Fig. 6.8 The two dimensional contours of $f(x, y) = d_i$ and constraint equation $g(x, y) = c$. The gradients of $f(x, y)$ and $g(x, y)$ at the extremal point are proportional. Image credit: By Nexcis (Own work) [Public domain], via Wikimedia Commons

proportional, as shown in Fig. 6.8. Define the gradient operator ∇ by the following

$$\nabla = \mathbf{e}_x \frac{\partial}{\partial x} + \mathbf{e}_y \frac{\partial}{\partial y}$$

The requirement that the gradients be proportional yields

$$\nabla f(x_c, y_c) = \lambda \nabla[g(x_c, y_c) - c]$$

$$\Rightarrow \frac{\partial f(x_c, y_c)}{\partial x} = \lambda \frac{\partial[g(x_c, y_c) - c]}{\partial x} \; ; \quad \frac{\partial f(x_c, y_c)}{\partial y} = \lambda \frac{\partial[g(x_c, y_c) - c]}{\partial y}$$

where λ is a constant of proportionality. The reasoning holds in N-dimensions; define an auxiliary function, also called the Lagrange function, by the following

$$\mathcal{A}(\mathbf{x}, \lambda, c) = f(\mathbf{x}) - \lambda[g(\mathbf{x}) - c] \tag{6.11.1}$$

The constrained minimization determines the constrained critical point $\mathbf{x}_c$ and λ_c by simultaneously solving the following equations

$$\frac{\partial \mathcal{A}(\mathbf{x}_c, \lambda_c, c)}{\partial \lambda} = 0 \; ; \quad \frac{\partial \mathcal{A}(\mathbf{x}_c, \lambda_c, c)}{\partial x_i} = 0 \; : \; i = 1, 2, \ldots, N \tag{6.11.2}$$

6.11.1 Interpretation of λ_c

From Eq. 6.11.1

$$\frac{\partial \mathcal{A}(\mathbf{x}, \lambda, c)}{\partial c} = \lambda \tag{6.11.3}$$

At the extermal point the constraint equation is zero and hence, at the extremal point

$$g(\mathbf{x}_c) - c = 0$$

This yields

$$\mathcal{A}(\mathbf{x}_c, \lambda_c, c) = f(\mathbf{x}_c) \tag{6.11.4}$$

Note the important fact that the extremal values for $\mathbf{x}_c$, λ_c are put into the auxiliary function after the partial differentiation by c.

Since we have a partial differentiation with respect to c, let the value of c be a variable; hence, in general, the extremal point depends on c, and one has

$$\mathbf{x}_c = \mathbf{x}_c(c) \ ; \ \ \lambda_c = \lambda_c(c)$$

The extremal point is a function of c and the maximum value is given by

$$M(c) = f(\mathbf{x}_c(c)) = \mathcal{A}(\mathbf{x}_c(c), \lambda_c(c), c)$$

Consider the total variation of $M(c)$ as a function of c; this yields, using the optimization equations, the following

$$\begin{aligned}\frac{dM(c)}{dc} = \frac{d\mathcal{A}(\mathbf{x}_c, \lambda_c, c)}{dc} &= \frac{\partial \mathcal{A}(\mathbf{x}_c, \lambda_c, c)}{\partial x}\frac{\partial x_c(c)}{\partial c} + \frac{\partial \mathcal{A}(\mathbf{x}_c, \lambda_c, c)}{\partial y}\frac{\partial y_c(c)}{\partial c} \\ &+ \frac{\partial \mathcal{A}(\mathbf{x}_c, \lambda_c, c)}{\partial \lambda}\frac{\partial \lambda_c(c)}{\partial c} + \frac{\partial \mathcal{A}(\mathbf{x}_c, \lambda_c, c)}{\partial c}\frac{\partial c}{\partial c} = \frac{\partial \mathcal{A}(\mathbf{x}_c, \lambda_c, c)}{\partial c}\end{aligned} \tag{6.11.5}$$

Hence, from Eqs. 6.11.3 and 6.11.5 above yields

$$\Rightarrow \quad \frac{dM(c)}{dc} = \frac{df(\mathbf{x}_c(c))}{dc} = \frac{\partial \mathcal{A}(\mathbf{x}_c, \lambda_c, c)}{\partial c} = \lambda_c \tag{6.11.6}$$

The result given in Eq. 6.11.6 means that the left hand side must have a result that obeys the result obtained earlier in Eq. 6.11.3. Hence, Eqs. 6.11.4 and 6.11.6 yield the final result

$$\lambda_c = \frac{df(\mathbf{x}_c(c))}{dc} = \frac{df(c)}{dc} \tag{6.11.7}$$

In general, the (complicated) dependence of $\mathbf{x}_c$ on c is not known. The derivation, instead of attacking the problem directly, uses the property of the auxiliary function. Hence, there is no need to explicitly compute the dependence of the critical point $\mathbf{x}_c(c)$ on c. Instead, we have the important result is that λ_c is the change in the function that is being optimized due to a change in the constraint condition. The examples considered below are the constrained maximization involving two commodities. To determine whether the critical point is a maximum or minimum requires the analysis of a 3×3 Hessian matrix, and is beyond the scope of this book.

Example 1 Suppose the profit from making a steel product is given by

$$f(h, s) = 200h^{\frac{2}{3}}s^{\frac{1}{3}}$$

where s is the amount of steel (in tons) used and h is the hours of labor expended.

The budget constraint limits the amount of steel and labor that can be paid for; suppose the price of steel per ton is 170 and the cost of labor per hour is 20; the total budget c is given by 20000 and is the budget constraint. Hence, the constraint equation is given by

$$g(h, s) = c \;\Rightarrow\; g(h, s) - c = 20h + 170s - 20000$$

Define the auxiliary function by

$$\mathcal{A}(h, s, \lambda) = 200h^{2/3}s^{1/3} - \lambda(20h + 170s - 20000)$$

The maximization of profit, given the budget constraint, is the following

$$\frac{\partial \mathcal{A}(h_c, s_c, \lambda_c)}{\partial h} = 0 \;\Rightarrow\; 200 \cdot \frac{2}{3} \cdot \left(\frac{s_c}{h_c}\right)^{1/3} = 20\lambda_c$$
$$\frac{\partial \mathcal{A}(h_c, s_c, \lambda_c)}{\partial s} = 0 \;\Rightarrow\; 200 \cdot \frac{1}{3} \cdot \left(\frac{h_c}{s_c}\right)^{2/3} = 170\lambda_c$$
$$\frac{\partial \mathcal{A}(h_c, s_c, \lambda_c)}{\partial \lambda} = 0 \;\Rightarrow\; 20h + 170s - 20000 = 0$$

Equations above yield

$$\frac{h_c}{s_c} = 17 \;\Rightarrow\; \lambda_c = \frac{20}{51}(17)^{2/3} = 2.592.. \tag{6.11.8}$$

Using the constraint equation yields

$$20h_c + 170s_c = 20000 \;\Rightarrow\; s_c = 392.157.. \;;\; h_c = 6666.666..$$

The profit is

$$f(h_c, s_c) = 518140$$

The quantities s_c, h_c are the amount of steel and labor used—given the constraint on the budget—and $f(h_c, s_c)$ is the profit.

To apply (and verify) Eq. 6.11.7 requires the constraint equation to be solved with a varying budget. Let the budget (constraint) be given by From Eq. 6.11.8

$$340s + 170s = c \;\Rightarrow\; s_c(c) = \frac{c}{510} \;;\; h_c(c) = \frac{c}{30}$$

Hence,

$$f(h_c(c)s_c(c)) = 200\left(\frac{c}{510}\right)^{2/3}\left(\frac{c}{30}\right)^{1/3} = \frac{20}{51}(17)^{2/3}c$$

From Eq. 6.11.7, the Lagrange multiplier λ_c is given by

$$\lambda_c = \frac{df((h_c(c)s_c(c))}{dc} = \frac{20}{51}(17)^{2/3} \tag{6.11.9}$$

and the result obtained earlier in Eq. 6.11.8 has been recovered.

Example 2 Suppose the profit from making a steel product is given by

$$f(h, s) = a^2 h^{2/3} s^{1/3}$$

where s is the amount of steel (in tons) used and h is the hours of labor expended.

The budget constraint limits the amount of steel and labor that can be paid for, and is given by

$$g(h, s) = \frac{1}{2}\alpha h^2 + \frac{1}{2}\beta s^2 = c$$

and yields the following auxiliary function

$$\mathcal{A}(h, s, \lambda) = a^2 h^{2/3} s^{1/3} + \lambda\Big(\frac{1}{2}\alpha h^2 + \frac{1}{2}\beta s^2 - c\Big)$$

The optimization equation yields

$$\frac{\partial \mathcal{A}(h_c, s_c, \lambda_c)}{\partial h} = 0 \;\Rightarrow\; a^2 \cdot \frac{2}{3} \cdot \left(\frac{s_c}{h_c}\right)^{1/3} = \alpha h_c \lambda_c$$
$$\frac{\partial \mathcal{A}(h_c, s_c, \lambda_c)}{\partial s} = 0 \;\Rightarrow\; a^2 \cdot \frac{1}{3} \cdot \left(\frac{h_c}{s_c}\right)^{2/3} = \beta s_c \lambda_c$$
$$\frac{\partial \mathcal{A}(h_c, s_c, \lambda_c)}{\partial \lambda} = 0 \;\Rightarrow\; \frac{1}{2}\alpha h_c^2 + \frac{1}{2}\beta s_c^2 = c$$

From above equations

$$h_c = \sqrt{\frac{2\beta}{\alpha}} s_c \;;\; \lambda_c = \frac{a^2}{3\beta s_c}\left(\frac{h_c}{s_c}\right)^{2/3}$$

From constraint equation

$$s_c = \sqrt{\frac{2c}{3\beta}} \Rightarrow \lambda_c = \frac{a^2}{\sqrt{3\alpha c}}\left(\frac{1}{2\beta}\right)^{1/6}$$

The total profit as a function of c is given by

$$f = a^2 h_c^{2/3} s_c^{1/3} = a^2(\frac{2\beta}{\alpha})^{1/3} s_c = \frac{a^2}{\sqrt{3\alpha}}\left(\frac{1}{2\beta}\right)^{1/6} 2\sqrt{c}$$

To verify the result given in Eq. 6.11.7 note, as expected from Eq. 6.11.7

$$\frac{df}{dc} = \frac{a^2}{\sqrt{3\alpha c}}\left(\frac{1}{2\beta}\right)^{1/6} = \lambda_c$$

In both these examples, the amount λ_c is the change in the profit due to a change in the total budget c. If one increases the budget by one unit, then the increase in profit is λ_c, and vice versa. In microeconomics, λ_c is often called the shadow price of the commodity being produced. The shadow price is the profit gained (lost) by increasing (decreasing) the total budget.

6.12 Line Integral; Exact and Inexact Differentials

Consider a $f(x, y)$, a function of two variables. One can integrate the function along a line (contour) $\mathcal{C}$ in the xy-plane, as shown in Fig. 6.9. Let the *arc length* of the line segment at point x, y be denoted by ds. The value of the coordinates on the contour are given by $x(s), y(s)$. The line integral is given by

$$I = \int_{\mathcal{C}} ds f(\mathbf{x})$$

The line element ds can change from point to point as one integrates along $\mathcal{C}$; change to a new variable t that does not depend on the position along $\mathcal{C}$. As shown in Fig. 6.9.

$$ds = |\mathbf{x}(t + dt) - \mathbf{x}(t)| = |\frac{d\mathbf{x}}{dt}|dt$$

Let the parameter $t \in [a, b]$, where the curve beings at $\mathbf{x}(a)$ and ends at $\mathbf{x}(b)$. Then

$$I = \int_{\mathcal{C}} ds f(\mathbf{x}) = \int_a^b dt|\frac{d\mathbf{x}}{dt}|f(\mathbf{x}(t))$$

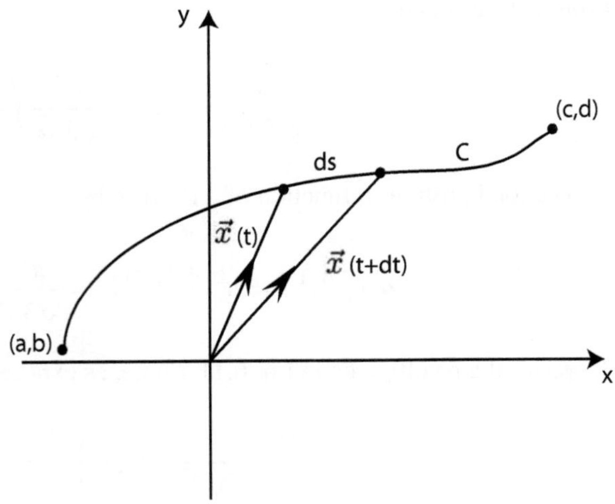

Fig. 6.9 The two dimensional contours C_1, C_2, C_3, C_4. Published with permission of © Belal E. Baaquie 2020. All Rights Reserved

Example: Let the contour C be a circle S^1 with radius r. The dependence of the coordinates on the parameter θ is given by

$$x = r\cos\theta \ ; \quad y = r\sin\theta \ ; \quad \left|\frac{d\mathbf{x}}{d\theta}\right| = 1 \ ; \quad \theta \in [0, 2\pi]$$

The line integral is given by

$$I = \int_{S^1} ds f(\mathbf{x}) = \int_0^{2\pi} d\theta f(\mathbf{x}(\theta))$$

Let

$$f(\mathbf{x}(\theta)) = x^2 = x^2(\theta) = r^2\cos^2\theta$$

Then

$$I = \int_0^{2\pi} d\theta r^2 \cos^2\theta = \frac{1}{2}r^2$$

Consider a $f(x, y)$, a function of two variables. Then, from the chain rule

$$df(x, y) = \frac{\partial f}{\partial x}dx + \frac{\partial f}{\partial y} = Mdx + Ndy$$

One then has

$$\frac{\partial^2 f}{\partial y \partial x} = \frac{\partial^2 f}{\partial x \partial y} \quad \Rightarrow \quad \frac{\partial M}{\partial y} = \frac{\partial N}{\partial x} \tag{6.12.1}$$

Consider a differential dg

$$dg(x, y) = a(x, y)dx + b(x, y)dy$$

From Eq. 6.12.1, dg is an exact differential if

$$\frac{\partial a}{\partial y} = \frac{\partial b}{\partial x}$$

and one can conclude that dg is the result of taking the differential of some function $g(x, y)$. However consider a differential

$$đH(x, y) = \alpha(x, y)dx + \beta(x, y)dy \;\Rightarrow\; \frac{\partial \alpha}{\partial y} \neq \frac{\partial \beta}{\partial x}$$

$đH(x, y)$ is called an *inexact differential*, with the differential denoted by$đ$. There does **not** exist a function $F(x, y)$ such that $dF(x, y) = đH(x, y)$.

Consider a path in the xy-plane that goes from point a, b to the endpoint c, d; let the path be denoted by $\mathcal{C}(a, b; c, d)$. The difference between an exact and inexact differential lies in the nature of the line integration of these two cases. Consider a line integral taken along the path $\mathcal{C}(a, b; c, d)$.

For the exact differential, the value of the integral depends only on the endpoints and yields

$$I = \int_{\mathcal{C}(a,b;c,d)} dh = h(c, d) - h(a, b)$$

The value of the line integral for an inexact differential depends on path chosen—and not only on the end points.

$$J = \int_{\mathcal{C}(a,b;c,d)} đH(x, y) = J[\mathcal{C}(a, b; c, d)]$$

Example: Consider the exact differential

$$f = 3x^2 - xy \;\Rightarrow\; df = (6x - y)dx - xdy$$

Consider integrating from point (0,0) to point (L, L) along two different paths, given in Fig. 6.10 as $\mathcal{C}_1 + \mathcal{C}_2$ and $\mathcal{C}_3 + \mathcal{C}_4$. First one goes along the x-axis and then y-axis as in $\mathcal{C}_1 + \mathcal{C}_2$. We have the following.

$$I_1 = \int_0^L dx(6x - 0) + \int_0^L dy(-L) = 3L^2 - L^2 = 2L^2$$

The other line first goes along the y-axis and then along the x-axis as in $\mathcal{C}_3 + \mathcal{C}_4$.

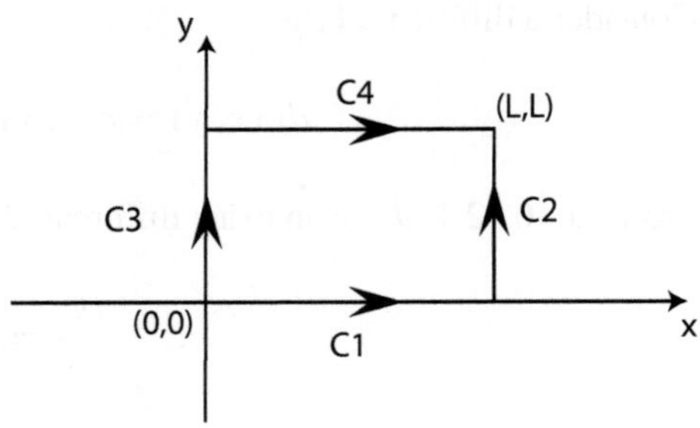

Fig. 6.10 The two dimensional contours C_1, C_2, C_3, C_4. Published with permission of © Belal E. Baaquie 2020. All Rights Reserved

$$I_2 = \int_0^L dy(-0) + \int_0^L dx(6x - L) = 3L^2 - L^2 = 2L^2$$

Hence, the value of an exact differential integrated along a line depends only on the endpoints.

Consider the inexact differential

$$đH = xydx + xy^2dy$$

The inexact case is integrated along the same lines as for the exact case. The first line goes along the x-axis and then y-axis as in $C_1 + C_2$.

$$J_1 = \int_0^L dx(0) + \int_0^L dy(Ly^2) = \frac{1}{3}L^4$$

The other line first goes along the y-axis and then along the x-axis as in $C_3 + C_4$.

$$J_2 = \int_0^L dy(0) + \int_0^L dx(xL) = \frac{1}{2}L^3 \neq J_1$$

Hence, the value of an inexact differential integrated along a line depends on the path taken.

6.13 Problems

1. Evaluate the integral

$$\int_0^{\pi/2} dx \frac{\cos x}{(2 - \cos^2 x)}$$

(Hint: Use the identity $\sin^2 x + \cos^2 x = 1$.)

2. Find dw/dt if

$$w = x^3 - y^3 \ ; \ x = \frac{1}{1+t} \ , \ y = \frac{t}{1+t}$$

3. Find the partial derivatives and the second partial derivatives of the following function

$$f(x, y) = x^2 + 4xy + y^2$$

4. The Cobb-Douglas production function is given by

$$Q = AL^{\alpha} L^{\beta}$$

Find the first and second order partial derivatives and show that mixed derivatives are equal.
Consider the case where $Q = Q_0 =$ constant; find

$$\frac{\partial^2 K}{\partial L^2} \ ; \ \frac{\partial^2 L}{\partial K^2}$$

5. Consider the implicit definition of $y = y(x)$ by the equation

$$e^{xy} + 2(x + y) = 5$$

Show that

$$\frac{dy}{dx} = -\frac{ye^{xy} + 2}{xe^{xy} + 2}$$

6. Consider the change of variables from x_1, x_2 to y_1, y_2 given by

$$y_1 = 2x_1 + 3x_2 \ ; \ y_2 = 4x_1^2 + 12x_1x_2 + 9x_2^2$$

Show that the Jacobian of the transformation is zero.
7. Find the extrema and saddle points of

$$f(x, y) = x^2 - 3xy - y^2 + 2y - 6x$$

8. Consider the function

$$f(x, y) = x^2 - 3xy - y^2 + 2y - 6x$$

Find the critical point(s). Evaluate the Hessian and determine whether the critical point(s) are maximum, minimum or saddle points.
9. Find the exterma of

$$f(x, y) = 4x^2 - 4xy + y^2$$

subject to the constraint

$$x^2 + y^2 = 1$$

10. Consider the utility function given by

$$\mathcal{U}(x_1, x_2) = 2\ln(x_1) + \ln(x_2)$$

with budget constraint

$$p_1x_1 + p_2x_2 = M$$

Evaluate quantity x_1^*, x_2^* that maximizes the utility function, subject to the budget constraint; hence, determine the demand function

$$\mathcal{D}(p_1, xp_2) = \mathcal{U}(x_1^*, x_2^*)$$

11. The profit from making a steel product is given by

$$f(h, s) = 200h^{2/3}s^{1/3}$$

where s is the amount of steel (in tons) used and h is the hours of labor expended. Suppose the price of steel per ton is 170 and the cost of labor per hour is 20; the total budget constraint is given by

$$20h + 170s = 20000$$

Find the maximum profit subject to the budget constraint.

Chapter 7
Ordinary Differential Equations

Abstract Ordinary differential equations—both first and second order differential equations—are studied with emphasis on linear differential equations. A number of distinct cases of the first order differential equations are studied. Second order linear differential equations, both the homogeneous and inhomogeneous cases, are studied is some detail. Applications to economics are considered; in particular, a general solution is provided of the Swan-Solo equation.

7.1 Introduction

Consider the price of a commodity P. Let the demand for the commodity be denoted by Q_d and quantity in supply for the commodity be denoted by Q_s. One expects the demand for the quantity to fall with its price and the supply to increase with an increase in its price. Hence, a simple model for supply and demand is given

$$Q_d = a - bP \ ; \ Q_s = -c + dP \tag{7.1.1}$$

One expects the equilibrium price P_* to be given when the quantity in demand for the commodity is equal to its supply, which yields

$$a - bP_* = -c + dP_* \ \Rightarrow \ P_* = \frac{a+c}{b+d} \tag{7.1.2}$$

What happens if, at time t, the price of the commodity $P(t)$ is not equal to P_*? One expects that the effect of supply and demand should change the price $P(t)$ and move it towards P_*. Hence, one needs to know how $P(t)$ changes with time, and any description should encode the fact that the price should evolve to P_* as $t \to \infty$.

Let us assume that change in price is proportional to the excess demand in the market; Eq. 7.1.1 yields

$$\frac{dP}{dt} = \alpha(Q_d - Q_s) = k - \mu P \ : \ k = \alpha(a+c) \ ; \ \mu = \alpha(b+d) \tag{7.1.3}$$

B. E. Baaquie, *Mathematical Methods and Quantum Mathematics for Economics and Finance*, https://doi.org/10.1007/978-981-15-6611-0_7

Equation above is a linear first order differential equation

$$\frac{dP}{dt} + \mu P = k \tag{7.1.4}$$

Only after one obtains a solution of Eq. 7.1.4 can one conclude that $P(t)$ converges to P_* as $t \to \infty$. This will be discussed later.

7.2 Separable Differential Equations

The usual notation for studying differential equations is that one takes x to the independent variable, and $y = y(x)$ the dependent variable that is defined by the differential equation. Consider the general separable first order differential equation given by

$$\frac{dy}{dx} = f(y)g(x) \;\Rightarrow\; \int \frac{dy}{f(y)} = \int dx g(x) \tag{7.2.1}$$

Both the integrations above are indefinite, and there is one constant that is not determined; for this reason, one boundary condition is required to fully specify the solution. This entails specifying the value of y at some special point of x, say x_0; hence

$$\text{Boundary condition}: y(x_0) = y_0 \;:\; \text{given}$$

A special case of Eq. 7.2.1 is when $f(y) = 1$; in that case

$$\frac{dy}{dx} = g(x) \;\Rightarrow\; y(x) = \int_a^x dx' g(x') + c \tag{7.2.2}$$

with

$$\text{Boundary condition}: y(a) = c \;:\; \text{given}$$

Example

$$f(y) = e^y \;;\; g(x) = x \;;\; y(0) = -1$$

From Eq. 7.2.1

$$\int dy e^{-y} = \int dx x \;\Rightarrow\; -e^{-y} = \frac{1}{2}x^2 + c \;\Rightarrow\; y = -\ln\left(-c - \frac{1}{2}x^2\right)$$

The boundary condition yields

$$-1 = -\ln(-c) \;\Rightarrow\; c = -e$$

and the solution is given by

$$y = -\ln\left(e - \frac{1}{2}x^2\right)$$

If one requires that y is real, it gives the additional restriction that

$$-\sqrt{2e} \leq x \leq \sqrt{2e}$$

7.3 Linear Differential Equations

Equation 7.1.4 is a special case of linear differential equations that are defined by

$$\frac{dy}{dx} + f(x)y(x) = h(x) \tag{7.3.1}$$

Define the integrating factor by

$$e^{P(x)} = \exp\{\int_a^x dx' f(x')\} \;;\; P(a) = 0 \tag{7.3.2}$$

Using the chain rule, Eq. 7.3.1 yields

$$\begin{aligned} &e^{-P(x)}\frac{d}{dx}\left(e^{P(x)}y(x)\right) = h(x) \\ \Rightarrow &\int_a^x dx' \frac{d}{dx'}\left(e^{P(x')}y(x')\right) = \int_a^x dx' e^{P(x')}h(x') \end{aligned} \tag{7.3.3}$$

Hence, from Eqs. 7.2.2 and 7.3.3, one obtains the solution

$$e^{P(x)}y(x) - y(a) = \int_a^x dx' e^{P(x')}h(x')$$

$$y(x) = ce^{-P(x)} + e^{-P(x)}\int_a^x dx' e^{P(x')}h(x') \;:\; y(a) = c \tag{7.3.4}$$

Consider the special case of $f(x) = \mu$ is a constant

$$\frac{dy}{dx} + \mu y(x) = h(x) \tag{7.3.5}$$

Hence, from Eq. 7.3.2

$$P(x) = e^{\mu(x-a)}$$

and

$$y(x) = ce^{-\mu(x-a)} + e^{-\mu(x-a)} \int_a^x dx' e^{\mu(x'-a)} h(x') \tag{7.3.6}$$

7.3.1 Dynamics of Price: Supply and Demand

From Eq. 7.1.4

$$\frac{dP}{dt} + \mu P = k \ ; \quad P(0) \ : \ \text{initial value}$$

Similar to Eqs. 7.3.3, 7.1.4 yields

$$e^{-\mu t}\frac{d}{dt}(e^{\mu t}P) = k$$

$$\Rightarrow \int_0^t d(e^{\mu t}P) = k\int_0^t dt' e^{\mu t'} = \frac{k}{\mu}(e^{\mu t} - 1)$$

and we obtain the special case of Eq. 7.3.6

$$P(t) = e^{-\mu t}P(0) + \frac{k}{\mu}(1 - e^{-\mu t})$$

As $t \to \infty$ above equation yields, from Eqs. 7.1.2 and 7.1.3

$$P(t) \to \frac{k}{\mu} = P_*$$

The solution for the time-dependent price $P(t)$ can be re-written as

$$P(t) = e^{-\mu t}(P(0) - P_*) + P_*$$

The result shows that the assumptions on how $P(t)$ evolves in time is correct, since the market price $P(t)$ is seen to converge to the equilibrium price P_*; the time taken to converge is given by $t_* \simeq 1/\mu$.

Noteworthy 7.1: Formal solution

Solution for $y(x)$ given in Eq. 7.3.4 can be written in a more formal manner that leads to an important generalization for second order differential equations. Equation 7.3.1 can be written as

$$(D + f)y = h \ ; \ \ D = \frac{d}{dx}$$

Then

$$y = y_H + \left[\frac{1}{(D + f)}\right] h$$

Comparing with Eq. 7.3.4 yields

$$(D + f)y_H = 0 \ ; \ \ \left[\frac{1}{(D + f)}\right] h = e^{-P(x)} \int_a^x dx' e^{P(x')} h(x') \tag{7.3.7}$$

7.4 Bernoulli Differential Equation

The Bernoulli equation is a special case of a nonlinear differential equation that can be reduced to a linear differential equation. The Bernoulli equation is defined by

$$\frac{dy}{dx} + f(x)y(x) = h(x)y^{\alpha}(x) \tag{7.4.1}$$

where α is any real number and not necessarily an integer. Consider the change of variables

$$u = y^{1-\alpha} \ \ \Rightarrow \ \ y = u^{\frac{1}{1-\alpha}} \ \ \Rightarrow \ \ \frac{dy}{dx} = \frac{du}{dx}\frac{dy}{du} = \frac{1}{1-\alpha} u^{\frac{\alpha}{1-\alpha}} \frac{du}{dx}$$

Hence, from Eq. 7.4.1

$$\frac{1}{1-\alpha} u^{\frac{\alpha}{1-\alpha}} \frac{du}{dx} + f(x) u^{\frac{1}{1-\alpha}} = h(x) u^{\frac{\alpha}{1-\alpha}}$$

$$\Rightarrow \ \ \frac{du}{dx} + (1-\alpha) f(x) u(x) = (1-\alpha) h(x) \tag{7.4.2}$$

Equation 7.4.2 is a linear differential equation for $u(x)$, with its solution given by Eq. 7.3.4. On obtaining $u(x)$, the solution for $y(x)$ is given by

$$y(x) = u^{\frac{1}{1-\alpha}}(x)$$

Example
Consider the Bernoulli equation

$$\frac{dy}{dx} = xy + y^2$$

Defining

$$u = y^{-1}$$

yields

$$\frac{du}{dx} = -xu - 1 \Rightarrow e^{-x^2/2}\frac{d}{dx}\left(e^{x^2/2}u\right) = -1$$

Hence

$$u(x) = e^{-x^2/2}\left(u(0) - \int_0^x dte^{t^2/2}\right)$$

and the final result is

$$y(x) = \frac{1}{u(x)} = y(0)\frac{e^{x^2/2}}{1 - y(0)\int_0^x dte^{t^2/2}}$$

7.5 Swan-Solow Model

In economics the Swan-Solow growth model is an example of the Bernoulli equation. Let K be the required capital for producing output Q that has a market price of P and yields revenue Y. The rate of change of capital, given by dK/dt, is determined by the reinvestment in capital goods by the firm as well as changes in K due to depreciation of capital goods, improvement in technology and growth of the working force. Hence

$$\frac{dK}{dt} = fY - (a + b + t)K$$

where f is the reinvestment, a is the rate of population growth, b is the rate of capital depreciation, and t is technological innovations. Suppose Q is determined the Cobb-Douglas production function and yields

$$Y = PAL^{\alpha}K^{\beta}$$

Hence, for $s = fPAL^{\alpha}, q = a + b + t$, we obtain the Bernoulli equation

$$\frac{dK}{dt} + qK = sK^{\beta} \;\; ; \;\; \beta < 1 \tag{7.5.1}$$

In Eq. 7.4.1 $f(x) = q, h(x) = s$, and from Eq. 7.4.2 the Swan-Solow equation has the following representation as a first order ordinary linear differential

$$\frac{du}{dx} + \gamma q u(x) = \gamma s \;\; ; \;\; K = u^{1/\gamma} \;\; ; \;\; \gamma = 1 - \beta > 0 \tag{7.5.2}$$

From Eq. 7.3.6 , the solution is given by

$$u(t) = u(0)e^{-\gamma qt} + \frac{\gamma s}{\gamma q}\left(1 - e^{-\gamma qt}\right) = \frac{s}{q} + \left(u(0) - \frac{s}{q}\right)e^{-\gamma qt} \tag{7.5.3}$$

$$\Rightarrow \;\; K(t) = u(t)^{1/\gamma} = \left\{\frac{s}{q} + \left(u(0) - \frac{s}{q}\right)e^{-(1-\beta)qt}\right\}^{1/(1-\beta)} \tag{7.5.4}$$

Hence, as $t \to \infty$

$$u(t) \to u_* = \frac{s}{q} \;\; \Rightarrow \;\; K_* = u_*^{1/\gamma} = \left(\frac{s}{q}\right)^{1/(1-\beta)}$$

The value of critical capital goods K_* also directly follows from Eq. 7.5.1 where

$$\left.\frac{dK}{dt}\right|_{K_*=0} = 0 \;\; \Rightarrow \;\; qK_* = sK_*^{\beta} \;\; \Rightarrow \;\; K_* = \left(\frac{s}{q}\right)^{1/(1-\beta)}$$

The critical capital goods K_* is shown in Fig. 7.1a.

Note that for

$$K < K_* \;\; \Rightarrow \;\; \frac{dK}{dt} > 0 \;\; ; \;\; K > K_* \;\; \Rightarrow \;\; \frac{dK}{dt} < 0$$

In other words, for K less than the critical capital goods K_*, dK/dt has to grow and hence $dK/dt > 0$, whereas for K greater than the critical capital goods K_*, dK/dt slows down and hence $dK/dt < 0$. The rate of change of capital K reaches equilibrium when $K = K_*$, and yields $dK/dt = 0$. See Fig. 7.1b.

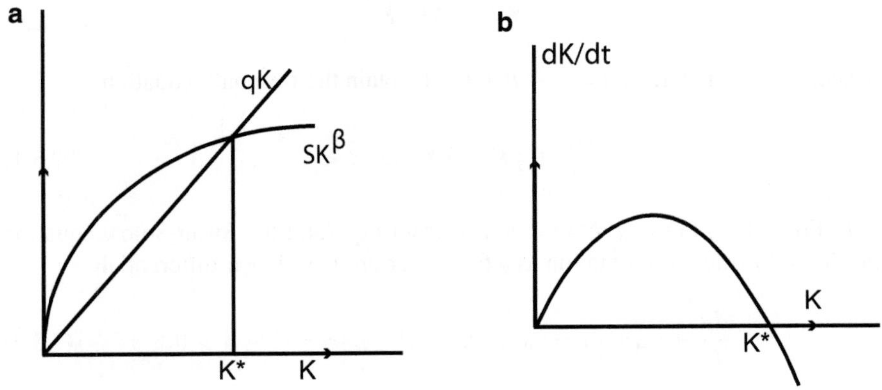

Fig. 7.1 **a** qK versus sK^{β}; the critical value K_{*} is indicated. **b** The different domains of dK/dt. Published with permission of © Belal E. Baaquie 2020. All Rights Reserved

7.6 Homogeneous Differential Equation

Consider a first order equation

$$M(x, y) + N(x, y)\frac{dy}{dx} = 0 \tag{7.6.1}$$

A homegeneous differential equation is defined by the coefficients satisfying

$$M(tx, ty) = t^{\alpha}M(x, y) \ ; \ \ N(tx, ty) = t^{\alpha}N(x, y)$$

Hence, taking $t = 1/x$ yields

$$\frac{M(x, y)}{N(x, y)} = \frac{M(tx, ty)}{N(tx, ty)} = \frac{M(1, y/x)}{N(1, y/x)} = f(y/x) = f(u)$$

where $u = y/x$. The chain rule yields

$$\frac{dy}{dx} = \frac{d(ux)}{dx} = x\frac{du}{dx} + u$$

Hence, from Eq. 7.6.1

$$\frac{dy}{dx} + f(u) = 0 \ \Rightarrow \ x\frac{du}{dx} = -f(u) - u$$

This is a separable first order equation that can be solved by integration and yields

$$\int \frac{du}{f(u) + u} = -\ln(x)$$

Noteworthy 7.2: Market model for price trends

Traders may base the price they offer for a commodity depending on their expectation of future trends. In that case, both supply and demand become time dependent and an inter-temporal equation for equilibrium is needed for determining market prices. Consider the following demand and supply functions

$$Q_d = 15 - 4P + 2\frac{dP}{dt} + 6\frac{d^2P}{dt^2}$$
$$Q_s = 3 + 7P$$

The equilibrium market price is given by $Q_d = Q_s$ and yields

$$6\frac{d^2P}{dt^2} + 2\frac{dP}{dt} - 11P = -12$$

Equation above is an example of a second order differential equation. The general techniques for solving these equations are discussed in the following Sections.

7.7 Second Order Linear Differential Equations

A general linear second order differential equation is given by

$$y'' + f(x)y' + g(x)y = 0 \ ; \ y' = \frac{dy}{dx} \ ; \ y'' = \frac{d^2y}{dx^2} \tag{7.7.1}$$

Since Eq. 7.7.1 is a linear equation, the solutions obey the superposition principle; let y_1, y_2 be two distinct solutions; then the linear combination

$$y = Ay_1 + By_2 \tag{7.7.2}$$

is also a solution of Eq. 7.7.1.

The general case does not have any closed form solution. Instead, consider the case of constant coefficients given by

$$y'' + 2by' + cy = 0 \tag{7.7.3}$$

All linear differential equations, similar to Eq. 7.7.3, are solved by the following ansatz

$$y(x) = e^{\lambda x}$$

Substituting the ansatz into Eq. 7.7.3 yields

$$\lambda^2 + 2b\lambda + c = 0 \ : \ \lambda_1 = -b + \sqrt{b^2 - c} \ ; \ \lambda_2 = -b - \sqrt{b^2 - c}$$

The superposition principle given in Eq. 7.7.2 yields the general solution

$$y = Ae^{\lambda_1 x} + Be^{\lambda_2 x} = e^{-bx}\left[Ae^{Dx} + Be^{-Dx}\right] \ ; \ D = \sqrt{b^2 - c}$$

For complex roots, $D = iC$ and

$$y = Ae^{\lambda_1 x} + Be^{\lambda_2 x} = e^{-bx}\left[Ae^{iCx} + Be^{-iCx}\right] \ ; \ C = \sqrt{c - b^2}$$

If y is real, the boundary conditions yield the following form for the solution

$$y = e^{-bx}\left[\widetilde{A}\cos(Cx) + \widetilde{B}\sin(Cx)\right]$$

The two constants A, B (or $\widetilde{A}$, $\widetilde{B}$) require two boundary conditions on $y(x)$.

7.7.1 *Special Case*

For the special case of $b^2 = c$, the two eigenvalues both are equal to $-b$ and the differential equation is given by

$$y'' + 2by' + b^2 y = 0 \tag{7.7.4}$$

The general solution is

$$y = \Big(A + Bx\Big)e^{-bx}$$

To verify the solution, note that

$$y' = -b(A + Bx)e^{-bx} + Be^{-bx} \ ; \ y'' = b^2(A + Bx)e^{-bx} - 2bBe^{-bx}$$

Adding the result above yields

$$y'' + 2by' = -b^2 y$$

which is the equation given in Eq. 7.7.4.

7.8 Riccati Differential Equation

Consider the first order nonlinear differential equation

$$\frac{df}{dt} = 2\beta f - 2gf^2 + \sigma^2 \quad : \quad \text{Riccati equation} \tag{7.8.1}$$

For $\sigma^2 = 0$, the Ricatti equation reduces to the Bernoulli equation, and for $g = 0$ it is a linear first order inhomogeneous equation. For β, g, σ^2 being constants, the Ricatti equation can be solved exactly by the following change of variables. Let

$$f = \frac{1}{2g}\frac{\dot{u}}{u} \Rightarrow \dot{f} = \frac{1}{2g}\left(\frac{\ddot{u}}{u} - \left(\frac{\dot{u}}{u}\right)^2\right) \; ; \; \dot{x} \equiv \frac{dx}{dt}$$

Substituting the expression for $f, \dot{f}$ on both sides of the Riccati equation given in Eq. 7.8.1 yields

$$\frac{1}{2g}\left(\frac{\ddot{u}}{u} - \frac{\dot{u}^2}{u^2}\right) = 2\beta \cdot \frac{1}{2g}\frac{\dot{u}}{u} - \frac{1}{2g}\left(\frac{\dot{u}}{u}\right)^2 + \sigma^2$$

The nonlinear term $(\dot{u}/u)^2$ cancels in the above equation, and yields a homogeneous second order linear equation for $u(t)$ given by

$$\ddot{u} - 2\beta\dot{u} - 2g\sigma^2 u = 0$$

Using the ansatz $u(t) \simeq e^{\alpha t}$ yields

$$\alpha^2 - 2\beta\alpha - 2g\sigma^2 = 0 \Rightarrow \alpha = \beta \pm \sqrt{\beta^2 + 2\sigma^2 g} = \beta \pm \lambda$$

Hence the solution is given by

$$u(t) = ce^{\beta t}(e^{\lambda t} + ke^{-\lambda t})$$

where c, k are two integration constants. $f(t)$ is the solution of a first order differential equation and hence needs only a single initial value. The superfluous constant c cancels out, as is required for $f(t)$, in the following manner

$$f(t) = \frac{1}{2g}\frac{\dot{u}}{u} = \frac{1}{2g}\frac{(\beta + \lambda)e^{\lambda t} + k(\beta - \lambda)e^{-\lambda t}}{e^{\lambda t} + ke^{-\lambda t}}$$

where

$$f(0) = \frac{1}{2g}\frac{\beta + \lambda + k(\beta - \lambda)}{1 + k}$$

Equivalently, the constant k is given by $f(0)$ as follows

$$k = \frac{2gf(0) - \beta - \lambda}{\beta - \lambda - 2gf(0)}$$

Note

$$\lambda \equiv \beta\left(1 + \frac{2\sigma^2 g}{\beta^2}\right)^{1/2} \simeq \beta\left(1 + \frac{\sigma^2 g}{\beta^2}\right) \simeq \beta$$

Consider $1 \ll \beta t \ll \infty$, that is $1/\beta \ll t \ll \infty$. Let $f(0) \ll 1$; then, for infinite time, $f(t)$ is given by

$$\Rightarrow \lim_{t\to\infty} f(t) \cong \frac{1}{2g}\left[\frac{2\beta e^{\lambda t}}{e^{\lambda t} + ke^{-\lambda t}}\right] = \frac{\beta}{g}\frac{\frac{1}{k}e^{2\lambda t}}{1 + \frac{1}{k}e^{2\lambda t}} \to \frac{\beta}{g}$$

The cross-over of $f(t)$ from its initial value $f(0)$ to its equilibrium value β/g takes place at t_0 such that $f(t) \simeq \beta/g$. The cross-over time t_0 is given by

$$\frac{1}{k}e^{2\lambda t} \simeq 1 \;\Rightarrow\; t_0 \simeq \frac{1}{2\lambda}\ln(k).$$

7.9 Inhomogeneous Second Order Differential Equations

Consider the following inhomogeneous equation

$$\frac{d^2x}{dt^2} + \beta\frac{dx}{dt} + \omega^2 x = R \;;\; t \in [0, +\infty] \tag{7.9.1}$$

Let $D = d/dt$; then

$$(D^2 + \beta D + \omega^2)x = R \tag{7.9.2}$$

or

$$(D + r_1)(D + r_2)x = R \;;\; r_{1,2} = \frac{\beta}{2} \pm \sqrt{\frac{\beta^2}{4} - \omega^2} \tag{7.9.3}$$

Let x_H be the homogeneous solution of Eq. 7.9.1 with

$$(D + r_1)(D + r_2)x_H = 0 \tag{7.9.4}$$

Note the boundary values that specify a unique solution for $x(t)$ are carried by x_H. The complete solution of Eq. 7.9.1 is given by

$$x = x_H + \frac{1}{(D+r_1)(D+r_2)}\sigma R$$
$$= x_H + \frac{1}{r_2 - r_1}\left[\frac{1}{D+r_1} - \frac{1}{D+r_2}\right]\sigma R$$
$$= x_H + \sigma \int_0^t G(t,t')R(t')dt' \tag{7.9.5}$$
$$= x_H + x_{IH}$$

where the inhomgeneous solution is given by

$$x_{IH} = \sigma \int_0^t G(t,t')R(t')dt'$$

Equation 7.3.7 yields

$$\frac{1}{D+r_1}R = \int_0^t e^{-r_1(t-\xi)}R(\xi)d\xi \ ; \ x_H = Ae^{-r_1 t} + Be^{-r_2 t}$$

and hence

$$G(t,t') = \frac{1}{\sqrt{\beta^2 - 4\omega^2}}\left\{e^{-r_2(t-t')} - e^{-r_1(t-t')}\right\} \tag{7.9.6}$$

The coefficients A, B are determined from the initial conditions using $x(0) = x_0$, $\dot{x}(0) = v_0$, or any other specifications of the boundary conditions.

7.9.1 Special Case

From Eq. 7.9.1, consider the following inhomogeneous equation with $\omega = 0$, and which yields

$$\frac{d^2x}{dt^2} + \mu\frac{dx}{dt} = h \ ; \ t \in [a,b] \tag{7.9.7}$$

with initial values

$$\frac{dx}{dt}(a) = A \ ; \ x(b) = B$$

Define the following new variable

$$s = \frac{dx}{dt} \ ; \ s(a) = A$$

Equation 7.9.7 yields the following first order ordinary differential equation

$$\frac{ds}{dt} + \mu s = h \ ; \ t \in [a, b] \tag{7.9.8}$$

Using the solution for first order differential equation given in Eq. 7.3.6 yields

$$s(t) = Ae^{-\mu(t-a)} + e^{-\mu(t-a)} \int_a^t dt' e^{\mu(t'-a)} h(t')$$

$$\Rightarrow \frac{dx}{dt} = s = Ae^{-\mu(t-a)} + \frac{h}{\mu}(1 - e^{-\mu(t-a)}) \tag{7.9.9}$$

Performing an integration of Eq. 7.9.9 yields

$$x(t) = B + \int_a^t dt' \left(\frac{h}{\mu} + \frac{A\mu - h}{\mu} e^{-\mu(t'-a)} \right) = B + \frac{h}{\mu}(t-a) + \frac{A\mu - h}{\mu^2}(1 - e^{-\mu(t-a)})$$

The result can be written in terms of the homogeneous and inhomogeneous as follows

$$x(t) = B + \frac{A}{\mu}(1 - e^{-\mu(t-a)}) + \frac{h}{\mu}(t-a) - \frac{h}{\mu^2}(1 - e^{-\mu(t-a)})$$

where

$$x_H = B + \frac{A}{\mu}(1 - e^{-\mu(t-a)}) \ ; \ x_{IH} = \frac{h}{\mu}(t-a) - \frac{h}{\mu^2}(1 - e^{-\mu(t-a)})$$

Example
Consider the differential equation

$$L[x(t)] \equiv \frac{d^2x}{dt^2} - \frac{dx}{dt} - 2x = 4t^2 \ ; \ t \in [0, +\infty] \tag{7.9.10}$$

with the solution

$$x(t) = x_H(t) + x_{IH}(t) \tag{7.9.11}$$

The homogeneous solution is given by

$$r^2 - r - 2 = (r+1)(r-2) = 0 \ \Rightarrow \ r_1 = -1, r_2 = 2$$

and yields

$$x_H(t) = Ae^{-2t} + Be^{t}$$

The inhomogeneous solution, from Eq. 7.9.6, is given by

$$x_{IH}(t) = \frac{4}{3}\int_0^t dt' \left(e^{(t-t')} - e^{-2(t-t')}\right)(t')^2 \tag{7.9.12}$$

The following integral needs to be evaluated

$$\begin{aligned}
I(\alpha) &= \frac{4}{3}\int_0^t dt' e^{\alpha(t-t')}(t')^2 = \frac{4e^{\alpha t}}{3}\frac{\partial^2}{\partial\alpha^2}\int_0^t dt' e^{-\alpha t'} \\
&= \frac{4e^{\alpha t}}{3}\frac{\partial^2}{\partial\alpha^2}\left[\frac{1}{\alpha}(1-e^{-\alpha t})\right] = \frac{4e^{\alpha t}}{3}\frac{\partial}{\partial\alpha}\left[-\frac{1}{\alpha^2} + (\frac{1}{\alpha^2} + \frac{t}{\alpha})e^{-\alpha t}\right] \\
&= \frac{4e^{\alpha t}}{3}\left[\frac{2}{\alpha^3} - \left(\frac{2}{\alpha^3} - \frac{2t}{\alpha^2} + \frac{t^2}{\alpha}\right)e^{-\alpha t}\right] \\
&= \frac{8}{3}\left[\frac{e^{\alpha t}}{\alpha^3} - \left(\frac{1}{\alpha^3} - \frac{t}{\alpha^2} + \frac{t^2}{2\alpha}\right)\right]
\end{aligned} \tag{7.9.13}$$

Hence, from Eqs. 7.9.12 and 7.9.13

$$x_{IH}(t) = I(1) - I(-2) = \frac{8}{3}\left[e^t + \frac{e^{-2t}}{2^3}\right] + Q \tag{7.9.14}$$

Q is given by

$$Q = \frac{8}{3}\left[-\left(1 - t + \frac{t^2}{2}\right) + \left(-\frac{1}{2^3} - \frac{t}{2^2} - \frac{t^2}{4}\right)\right] = -2t^2 + 2t - 3$$

The terms below

$$\frac{8}{3}\left[e^t + \frac{e^{-2t}}{2^3}\right]$$

are annihilated by the differential operator L and can be added to $x_{IH}(t)$ without altering the result. These terms belong to $x_H(t)$ and contribute to fixing the boundary conditions and reflect the ambiguity in the Green's function mentioned in Eq. 13.8.19.

Hence, from Eq. 7.9.11

$$\begin{aligned}
x(t) &= x_H(t) + x_{IH}(t) \\
&= (A + \frac{1}{3})e^{-2t} + (B + \frac{8}{3})e^t - 2t^2 + 2t - 3 \\
&= \widetilde{A}e^{-2t} + \widetilde{B}e^t - 2t^2 + 2t - 3
\end{aligned}$$

where $\widetilde{A}$, $\widetilde{B}$ are fixed by the boundary conditions. It can be directly verified that the solution obtained above satisfies the inhomogeneous equation given in Eq. 7.9.10.

7.10 System of Linear Differential Equations

A homogeneous linear second order differential equation is equivalent to a system of two first order equations. Consider Eq. 7.9.1

$$\frac{d^2x}{dt^2} + \beta\frac{dx}{dt} + \omega^2 x = 0$$

It can re-written as follows

$$\frac{dx}{dt} = y \ ; \ \frac{dy}{dt} = -\omega^2 x - \beta y$$

In general, one can has the following system of linear differential equations

$$\frac{dx}{dt} = ax + cy \ ; \ \frac{dy}{dt} = bx + dy$$

In matrix notation

$$\frac{d}{dt}\begin{bmatrix} x \\ y \end{bmatrix} = \begin{bmatrix} a & c \\ b & d \end{bmatrix}\begin{bmatrix} x \\ y \end{bmatrix} = \begin{bmatrix} ax + cy \\ bx + dy \end{bmatrix} \tag{7.10.1}$$

In vector and matrix notation, an $N$the order linear differential equation can be written as

$$\frac{d\mathbf{x}}{dt} = A\mathbf{x} \tag{7.10.2}$$

where $\mathbf{x}$ is a N dimensional column vector and A is a $N \times N$ matrix. A formal solution of Eq. 7.10.2 is given by

$$\mathbf{x}(t) = e^{tA}\mathbf{x}(0)$$

where $\mathbf{x}(0)$ are the initial conditions for the elements of the column vector. The value of $x_1(t)$ is obtained by projecting it out using the canonical basis; hence

$$x_1(t) = \mathbf{e}_1^T \cdot \exp\{tA\} \cdot \mathbf{x}(0)$$

The inhomogeneous case yields the following

$$\frac{d\mathbf{x}}{dt} = A\mathbf{x} + \mathbf{J} \tag{7.10.3}$$

The solution is similar to the single variable case with

$$\mathbf{x}(t) = e^{tA}\mathbf{x}(0) + e^{tA}\int_0^t dt' e^{-t'A}\mathbf{J}(t')$$

Example
Consider from Eq. 4.11.1 the 2×2 real symmetric matrix given by

$$S = \begin{bmatrix} \alpha & \beta \\ \beta & \gamma \end{bmatrix} = S^T \ ; \ \alpha, \beta, \gamma : \text{real}$$

Solving the eigenvalue equation yields

$$\lambda_{\pm} = \frac{1}{2}(\alpha + \gamma) \pm \sqrt{\frac{1}{4}(\alpha - \gamma)^2 + \beta^2} : \text{real}$$

The (normalized) eigenstates of the symmetric matrix S, from Eq. 4.7.5, are given by

$$\eta_+ = N_+ \begin{bmatrix} 1 \\ \frac{-\alpha+\lambda_+}{\beta} \end{bmatrix} \ ; \ \eta_- = N_- \begin{bmatrix} \frac{-\gamma+\lambda_-}{\beta} \\ 1 \end{bmatrix} \Rightarrow \eta_\pm^T \cdot \eta_\pm = 1 \ ; \ \eta_+^T \cdot \eta_- = 0$$

$$N_+ = \sqrt{\frac{\beta^2}{(-\alpha + \lambda_+)^2 + \beta^2}} \ ; \ N_- = \sqrt{\frac{\beta^2}{(-\gamma + \lambda_-)^2 + \beta^2}}$$

From Eq. 4.7.6 the spectral decomposition is given by

$$S = \lambda_+ \eta_+ \otimes \eta_+^T + \lambda_- \eta_- \otimes \eta_-^T = \begin{bmatrix} \alpha & \beta \\ \beta & \gamma \end{bmatrix}$$

Consider a linear differential equation defined by

$$\frac{d\mathbf{x}}{dt} = S\mathbf{x}$$

with the solution given by

$$\mathbf{x}(t) = e^{tS}\mathbf{x}(0) = \left[e^{t\lambda_+}\eta_+ \otimes \eta_+^T + e^{t\lambda_-}\eta_- \otimes \eta_-^T\right]\mathbf{x}(0)$$

Since $\lambda_+ > \lambda_-$, for $t >> 1$, the solution is projected to the η_+ eigenvector, and given by

$$\mathbf{x}(t) \to e^{t\lambda_+}\eta_+(\eta_+^T \cdot \mathbf{x}(0))$$

If the initial state is an eigenvector of S, the differential equation yields

$$\mathbf{x}(0) = \eta_\pm \ \Rightarrow \ \mathbf{x}_\pm(t) = e^{tS}\eta_\pm = e^{t\lambda_\pm}\eta_\pm$$

And hence

$$\frac{d}{dt}\mathbf{x}_\pm(t) = S\mathbf{x}_\pm(t) = \lambda_\pm \mathbf{x}_\pm(t) \ \text{ Eigenfunction equation}$$

Example
Consider the differential equation, which can be solved by inspection

$$\frac{d^2x}{dt^2} = \beta^2 x \ \Rightarrow \ x(t) = Ae^{\beta t} + Be^{-\beta t}$$

Define

$$\frac{dx}{dt} = \beta y \ ; \ \frac{dy}{dt} = \beta x$$

In matrix notation

$$\frac{d}{dt}\begin{bmatrix} x \\ y \end{bmatrix} = S\begin{bmatrix} x \\ y \end{bmatrix} \ \Rightarrow \ \begin{bmatrix} x(t) \\ y(t) \end{bmatrix} = e^{tS}\begin{bmatrix} x_0 \\ y_0 \end{bmatrix} \tag{7.10.4}$$

with the symmetric matrix given by

$$S = \begin{bmatrix} 0 & \beta \\ \beta & 0 \end{bmatrix}$$

The eigenvalues and eigenvectors of S are given by

$$\lambda_\pm = \pm\beta \ ; \ \eta_+ = \frac{1}{\sqrt{2}}\begin{bmatrix} 1 \\ 1 \end{bmatrix} \ ; \ \eta_- = \frac{1}{\sqrt{2}}\begin{bmatrix} -1 \\ 1 \end{bmatrix}$$

The exponential of S is given by

$$e^{tS} = e^{\beta t}\eta_+ \otimes \eta_+^T + e^{-\beta t}\eta_- \otimes \eta_-^T = \begin{bmatrix} \cosh(\beta t) & \sinh(\beta t) \\ \sinh(\beta t) & \cosh(\beta t) \end{bmatrix} \tag{7.10.5}$$

The exponential of S can also be evaluated by noting the following identities

$$S^2 = \beta^2\mathbb{I} \ ; \ S^3 = \beta^2 S, \ \ldots$$

It can be verified by differentiating both sides of the above equation that, in matrix notation

$$\frac{d}{dt}e^{tS} = e^{tS}S$$

Equations 7.10.4 and and 7.10.5 yield

$$\begin{bmatrix} x(t) \\ y(t) \end{bmatrix} = e^{tS}\begin{bmatrix} x_0 \\ y_0 \end{bmatrix} = \begin{bmatrix} \cosh(\beta t) & \sinh(\beta t) \\ \sinh(\beta t) & \cosh(\beta t) \end{bmatrix}\begin{bmatrix} x_0 \\ y_0 \end{bmatrix} \tag{7.10.6}$$

In terms of the components of the above equation

$$\begin{aligned} x(t) &= \cosh(\beta t)x_0 + \sinh(\beta t)y_0 \\ y(t) &= \sinh(\beta t)x_0 + \cosh(\beta t)y_0 = \frac{1}{\beta}\frac{dx(t)}{dt} \end{aligned}$$

Furthermore, as expected

$$x(0) = x_0 \ ; \ \frac{dx(0)}{dt} = \beta y(0) = \beta y_0$$

7.11 Problems

1. Suppose the price of a commodity $y(t)$ satisfies

$$\frac{dy}{dt} = \frac{y^2+1}{2y}e^t$$

with initial condition $y(0) = 3$. Find $y(t)$.

2. Solve the differential equation

$$2t\frac{dy}{dt} + \frac{1}{y} = y$$

with initial condition $y(1) = 4$.

3. Solve the differential equation

$$\frac{dy}{dt} + 2y = \ln(t)$$

with initial condition $y(0) = a$.

4. Solve the differential equation

$$\frac{dy}{dt} + 2y = 5y^2$$

with initial condition $y(0) = a$.

5. Solve the homogeneous equation

$$\frac{d^2y}{dt^2} - 4\frac{dy}{dt} + 3y = 0$$

with $y(0) = 1, dy(0)/dt = 2$.

6. Solve the homogeneous equation

$$\frac{d^2y}{dt^2} - 4\frac{dy}{dt} + 13y = 0$$

with $y(0) = 1,\ dy(0)/dt = 0$.

7. Consider

$$\frac{d^2y}{dt^2} + \alpha\frac{dy}{dt} + \beta y = 0$$

with $\alpha^2 = 4\beta$. Find the general solution for $y(t)$.

8. Find the general solution for the equation

$$\frac{d^2y}{dt^2} - 7\frac{dy}{dt} + 6y = 12$$

with initial conditions $y(0) = a,\ dy(0)/dt = b$.

9. Find the solution for the equation

$$\frac{d^2y}{dt^2} + 4\frac{dy}{dt} + 4y = 8t^2 + 8$$

with initial conditions $y(0) = 1,\ dy(0)/dt = 3$.

10. Find the solution for the equation

$$\frac{d^2y}{dt^2} + 4\frac{dy}{dt} + 4y = te^t$$

with initial conditions $y(0) = 1,\ dy(0)/dt = 1$.

11. The prices of two commodities are given by the following system of differential equations

$$\frac{d}{dt}\begin{bmatrix} x \\ y \end{bmatrix} = \begin{bmatrix} 3 & 2 \\ 4 & 1 \end{bmatrix}\begin{bmatrix} x \\ y \end{bmatrix}$$

Find the solution with initial values given by $x(0) = a$, $y(0) = b$.

Part IV
Probability Theory

Chapter 8
Random Variables

Abstract The future is uncertain, and this uncertainty is fundamental to economics and finance since all policies and plans for the economy, as well as all cash flows of financial instruments, are impacted by an uncertain future. The mathematical description of uncertainty is the subject matter of probability theory, which is built on the concept of the random variable. This Chapter introduces some of the key ideas of probability and important examples of random variables, both discrete and continuous, are discussed in some detail.

8.1 Introduction: Risk

For any entrepreneur or investor, two considerations are of utmost importance: (a) the profit that can be made, and (b) the risk that is inherent in obtaining this profit. The trade-off between return and risk is the essence of any business strategy. Clearly, all entrepreneurs and investors would like to maximize returns and minimize risk. What constitutes profit is quite simple, but the definition of risk is more complex since it involves quantifying the uncertainties that the future holds.

8.1.1 Example

Consider the following scenario. Suppose one buys, at time t, a stock at price $S(t)$, holds it for a duration of time T with stock price having a terminal value of $S(t+T)$ and during this period the stock earns dividends worth d. The (fractional) rate of return R for the period T is given by

$$R = \frac{d + S(t+T) - S(t)}{S(t)}$$

and R/T is the instantaneous rate of return.

B. E. Baaquie, *Mathematical Methods and Quantum Mathematics for Economics and Finance*, https://doi.org/10.1007/978-981-15-6611-0_8

What are risks involved in this investment? The future value of the stock price may either increase of decrease, and it is this uncertainty regarding the future that introduces an element of risk in the investment. There are many possible scenarios for the stock price. One scenario is that there is a boom in the market with stock price increasing; or there is a downturn and stock prices plummet; or that the market is in doldrums with only small changes in the stock price. One can assign probabilities for each scenario, and this in turn gives the investor a way of gauging the uncertainties of the market. A typical example of the various scenarios for some security $S(t)$ are shown in Table 8.1.

Label each scenario by a discrete variable s, it's probability by $p(s)$, and it's return by $R(s)$. The expected return for the investment is the average (mean) value of the return given by

$$\bar{R} = \text{Expected Return} = \sum_i p(i)R(i) \equiv E[R] \tag{8.1.1}$$

where the notation $E[X]$ denotes the expectation value of some random quantity X.

The **risk** inherent in obtaining the expected return is clearly the possibility that the return may deviate from this value. From probability theory it is known that the standard deviation indicates by what amount the mean value of any given sample can vary from its expected value, that is

$$\text{Actual Return} = \text{Expected Return} \pm \text{Standard Deviation}$$

The precise amount by which the actual return will deviate from the expected return—and the likelihood of this deviation—can be obtained only if one knows the probability density $p(S)$ of the stock price $S(t)$. Standard deviation, denoted by σ, is the square root of the variance defined by

$$\sigma^2 = \sum_i p(i)\big(R(i) - \bar{R}\big)^2 \equiv E[\big(R - \bar{R}\big)^2]$$

The risk inherent in any investment is given by σ—the larger risk the greater σ, and vice versa. In the example considered in Table 8.1, at the end of one year the investor

Table 8.1 Possible Scenarios for the Annual Change in the Price of a Security $S(t)$ with Current Price of \$100

i	Scenario	S	$p(s)$:Likelihood	$R(s)$:Annual Return	Average $\bar{R}$	Risk σ
1	Doldrum	\$100	$p(1) = 0.70$	$R(1) = 0.00$		
2	Downturn	\$85	$p(2) = 0.20$	$R(2) = -0.15$	−0.025	0.06
3	Boom	\$105	$p(3) = 0.10$	$R(3) = 0.05$		

with an initial investment of \$100 will have an expected amount of cash given by $\$100 \times [1+\bar{R}] \pm 6 = \$97.50 \pm 6$.

8.2 Probability Theory

The example uses a number of key ideas of probability theory, which are discussed below.

1 There is an entity with an **uncertain** outcome—in the example the future value of a stock—that needs to be described and modeled. A quantity that is uncertain is said to be a *random* quantity.

2 **Sample space**. The sample space Ω, also called the probability space, is a set consisting of all possibilities—scenarios—for the random entity one is describing. The space can be discrete or continuous. In the example, the space consists of only three possibilities for the stock price, labeled by $S(1), S(2), S(3)$. In general, for a discrete sample space, one has $x_i,\ i = 1, 2, \ldots, N$ and the sample space is given by the set of points $\Omega = \{x_1, x_2, \ldots, x_N\}$; for a continuous sample space, one has in general $\Omega = E_N$ or a subspace of E_N.

3 **Events**. In our example, the stock price going up is an event, labeled by $S(3)$. In general, events are the specific elements of the probability space $x_i,\ i = 1, 2, \ldots, N$.

4 **Probability of events**. The occurrence of a specific event s has a probability $p(s)$ assigned to it. An event s_0 that is certain is assigned $p(s_1) = 1$; and event s_0 that is impossible is assigned $p(s_1) = 0$; in general, for any event s, its probability of occurrence is given by $0 \le p(s) \le 1$.
Let $x_i,\ i = 1, 2, \ldots, N$ be the allowed values with probability $P(x_i)$. The probability of every event is always positive, with an event not occurring given by zero probability. Hence $P(x_i) > 0$. The probabilities must add to one since every time a specific event occurs, it must take one of the allowed values x_i. This is expressed by

$$\sum_{i=1}^{N} P(x_i) = 1 \ ; \ \ P(x_i) > 0 \tag{8.2.1}$$

5 **Random variable**. The quantity s is modeled to be *uncertain*, taking *values* with some probability $p(s)$. In general, random variables can be discrete or continuous and have many varieties. The following notation for a random variable is used: the random variable is denoted by a capital letter, say X and its specific values by lower case letter x_i. For sample space

$$\Omega = \{x_1, x_2, \ldots, x_N\}$$

the random variable is represented by

$$X = \begin{bmatrix} x_1 & x_2 & .. & x_N \\ P(x_1) & P(x_2) & .. & P(x_N) \end{bmatrix}$$

6 **Functions of random variable** One can form functions of s; in the example, the return $R(X)$ is a function of the random variable X.

7 **Expectation values**. What one observes are the average—or expected values—of uncertain quantities. In the example, both $\bar{R}$ and σ are expectation values. In general, the expectation values are obtained by summing the uncertain function in question over all possible values of x_i, weighed by the probability of occurrence $P(x_i)$ of the specific value of—as in Eq. 8.1.1. Hence

$$E[R(X)] = \sum_i R(x_i)P(x_i)$$

8 **Moment generating function** The moments of a random variable X are defined by $E[X^n]$. All the moments can be obtained from the moment generating function defined by

$$E[e^{tX}] = \sum_i e^{tx_i} P(x_i) \Rightarrow E[X^n] = \frac{d^n}{dt^n} E[e^{tX}]\Big|_{t=0}$$

The general principles of probability theory are developed by analyzing some important discrete and continuous random variables. A discussion of the general principles is given as a summary later in Sect. 10.2.

8.3 Discrete Random Variables

Discrete valued random variables are among the simplest cases and clearly bring out many features of probability theory. The following discrete random variables are discussed.

- Bernoulli random variable
- Binomial random variable
- Poisson random variable

8.3.1 Bernoulli Random Variable

The simplest and one of the most important random variable is the Bernoulli random variable, denoted by X. It has two values, which can be denoted by 0, 1. It can also be denoted by T, H.

The outcome of tossing a coin is an example of an entity that is represented by the Bernoulli random variable. The outcome is fully known from the laws of physics. But the actual outcome is so difficult to predict that the outcome of the toss is taken to be random, meaning that both the $H$and T can occur with a certain likelihood. In other words, the theory of probability reflects our ignorance regarding the phenomenon—and is not intrinsic to the phenomenon itself.

For an unbiased coin the probability $H$and T is equal but for a biased coin, this is not the case, with T having probability say p and H having probability say q. From Eq. 8.2.1

$$p + q = 1 \ ; \ \ p, q > 0$$

The Bernoulli random variable has the representation

$$X = \begin{bmatrix} 0 & 1 \\ p & q \end{bmatrix} \ ; \ q = 1 - p \tag{8.3.1}$$

The expectation value of the random variable is given by

$$\begin{aligned} E[X^n] &= \sum_i x_i^n P(x_i) = 0 \times P(0) + P(1) \\ &= q = 1 - p \end{aligned} \tag{8.3.2}$$

Hence

$$E[1] = 1 \ ; \ \ E[X] = 1 - p \ ; \ \ E[X^2] = 1 - p$$

and

$$\sigma^2 = E[X^2] - E^2[X] = pq > 0$$

The moment generating function is given by

$$E[e^{tX}] = \sum_i e^{tx_i} P(x_i) = p + e^t q$$

Hence

$$E[X^n] = \frac{d^n}{dt^n} E[e^{tX}]\Big|_{t=0} = q$$

and we have recovered Eq. 8.3.2.

8.3.2 Binomial Random Variable

Consider tossing a coin N times. A typical outcome is

$$HHTHTTHH \cdots THHT$$

The Binomial random variable K is a representation of the outcome of H occurring k times (and T occurring $N - k$), given that a H or a T occurs for each throw with the probability of p and q, respectively, where $p + q = 1$. Each throw is the sampling of a Bernoulli random variable, and it is shown later that a Binomial random variable is equivalent to the sum of N identical Bernoulli random variables.

Noteworthy 8.1: Example of Binomial Distribution

Consider throwing the coin $N = 3$ times and asking how many times one has $k = 1$ T and $N - k = 2$ H's? The collection of all the three throws, **not** keeping track of the order in which the coins occur when the coins are thrown, is given by

$$HHH, HHT, HTH, THH, HTT, THT, TTH, TTT \ : \ 2^3 = 8$$

Consider choosing one T: it can be either the first, second or third outcome and yields three ways of occurring

$$THH, HTH, HHT \ : \ 3$$

Hence, the *number of ways* that two Hs and one T occur, regardless of the order, is given by

$$3 = \frac{3!}{1!(3-1)!} = \frac{N!}{k!(N-k)!}$$

Consider now, choosing the 2 H's first and then the 1 T; this is done in two steps. (a) choose the first head denoted by H_1 without specifying the other choices, indicated by $*$. (b) Then choose the second head, denoted by H_2. The subscript of H keeps track of the order in which the heads are chosen. This yields the following.

- There are three ways the first H_1 can occur, in the first throw, second throw or third throw. Hence one has
$$H_1 * *, *H_1*, * * H_1$$
- Choose second H_2; each choice made above has 2 further choices, given in a bracket
$$(H_1H_2*, H_1 * H_2), (H_2H_1*, *H_1H_2), (H_2 * H_1, *H_2H_1)$$
- The 6 choices above are due to keeping track of the *order* in which the H's are chosen. Note we have three choices multiplied by a factor of 2 if we are not interested in the order in which the H's are chosen.
- Eliminating the ordering of the H's and keeping the bracket to indicate various possible choices without the ordering reduces the chosen elements to (note the third bracket is empty)

$$(HH*, H * H), (*HH), () \equiv HH*, H * H, *HH$$

- A more systematic way of counting the possible outcomes is to group all elements that are identical when two H's are *interchanged*, since the order of choosing is not needed. This leads to the following new grouping of the six elements

$$(H_1H_2*, H_2H_1*), (H_1*H_2, H_2*H_1), (*H_1H_2, *H_2H_1) \equiv HH*, H*H, *HH$$

- Choosing 1 T yields the final result

$$HHT, HTH, THH$$

The computation of how 2 H's and 1 T occur, regardless of the order of occurrence, is given by $N = 3, k = 2$; hence number of ways this can occur is

$$\frac{N!}{k!(N-k)!} = \frac{3!}{2!(3-2)!} = \frac{6}{2} = 3$$

which is what was obtained earlier.

Consider throwing a coin N times. For every throw, there are 2 possible outcomes. Hence, for N throws there are in total 2^N possible outcomes; this total number of outcomes is based on *not* keeping track of the order in which the coins occur.

The probability of obtaining k H's in N throws is now calculated. The first H can be chosen in N ways—the first outcome, second outcome and so on; there are $N-1$ ways of choosing the second H, and so on till the last H, which can be chosen in $N-(k-1) = N-k+1$ ways; hence, the total number of ways of generating k coins with H from N throws—and keeping track of the order in which this choice is made—is

$$N(N-1)\cdots(N-k+1) = \frac{N!}{(N-k)!}$$

In obtaining the number of ways choosing k coins in the above procedure, the order has been included. Suppose one is only interested in how many H's occurs and not in what order. In other words: what is the probability of the occurrence of k H, *regardless* of the order in which H is chosen?

If the order in which the coins occur is ignored, then if one *interchanges* the k coins chosen one does *not* get a new outcome. The number of ways of interchanging k H coins, called the possible permutations, is given by first choosing one of the k H's, then having a choice of $k-1$ H's for the next coin and so and yields the total number of permutation to be

$$k(k-1)\cdots 1 = k!$$

Hence, the number of ways of choosing k H coins, by keeping track of the order of outcome was over-counted by $k!$. For obtaining the number of outcomes of H

regardless of the place in the sequence that they occur, it is given by dividing out $N!/(N-k)!$ by $k!$: the counting *without order* and is given by the binomial coefficient

$$\binom{N}{k} = \frac{N!}{k!(N-k)!}$$

If H occurs k times, then T must occur $N-k$ times. The two ways of counting, picking either T first or H first yields the identity

$$\binom{N}{k} = \frac{N!}{k!(N-k)!} = \binom{N}{N-k} = \frac{N!}{(N-k)!k!}$$

Assume each throw of the coin is independent of the other throws. Suppose the probability of a H occurring is q. Then the probability of the sequence occurring is given by

$$\underbrace{HHTHTTHH\cdots THH}_{\text{N throws } kH} \text{ probability } q^k p^{N-k}$$

The probability $P(N,k)$ of obtaining k H and $N-k$ T from N throws, without specifying the ordering, is given by the probability of the obtaining a single outcome for obtaining k H and $N-k$ T multiplied by how many times the outcome occurs, which is given by the Binomial coefficient. Hence

$$B(N,k) \equiv \text{Probability } (k\ H\ \&\ (N-k)\ T)$$
$$\Rightarrow \quad B(N,k) = \frac{N!}{k!(N-k)!} q^k p^{N-k} = \binom{N}{k} p^k q^{N-k} \tag{8.3.3}$$

In a simulation or an actual experiment of sampling a coin, *each* sample of the binomial random variable requires that the coin be thrown 2^N times, and the number of times an outcome occurs is recorded. The procedure needs to repeated M times to get a sample of size M. Hence, the coin needs to be thrown in total $M \times 2^N$ times. The probability of obtaining say k heads in N throws, denoted by $p_H(k,N)$, is given by

$$p_H(k,N) = \lim_{M\to\infty} \frac{1}{M} \sum_{i=1}^{M} \Theta(\text{k-heads}) = \lim_{M\to\infty} \frac{M(k)}{M}$$

where Θ is the number of times that k heads occur in a sample of 2^N throws. $M(k)$ is the number of times for which k heads occur. If the probability for H in a single throw is q, then the binomial random variable predicts that

$$p_H(k,N) = \frac{N!}{k!(N-k)!} q^k p^{N-k} \quad ; \quad p = 1-q$$

Noteworthy 8.2: Binomial theorem

The proof of

$$(x+y)^N = \sum_{k=0}^{N} \frac{N!}{k!(N-k)!} x^k y^{N-k}$$

is identical to the reasoning for the Binomial random variable. In expanding the left hand side, given by $(x+y)^N$, one can consider x to be represented by H and y to be represented by T; the order in which the choice of x and y are made is irrelevant since $x^n x^m = x^m x^n$; hence the number of ways a term $x^k y^{N-k}$ occurs on the right hand side is given by $\binom{N}{k}$ and yields

$$(x+y)^N = \sum_{k=0}^{N} \frac{N!}{k!(N-k)!} x^k y^{N-k}$$

The total probability, given by the summing over all the probabilities of some event occurring, must be 1 since some outcome taking place is certain. For the Binomial random variable $p+q=1$, and using the Binomial theorem yields the expected result that

$$\sum_{k=0}^{N} B(N,k) = \sum_{k=0}^{N} \binom{N}{k} q^k p^{N-k} = (p+q)^N = 1 \ ; \quad B(N,k) > 0$$

Furthermore, the total number of possible outcomes is given by

$$\sum_{k=0}^{N} \binom{N}{k} = (1+1)^N = 2^N$$

The integer valued binomial random variable K, which can be thought of how many H's occur in the outcome of N throws, takes values from $0, \cdots, N$. It has the following representation

$$K = \begin{bmatrix} 0 & 1 & .. & N \\ B(N,0) & B(N,1) & .. & B(N,N) \end{bmatrix}$$

The moment generating function is given by

$$Z(t) = E[e^{tX}] = \sum_i e^{tx_i} B(N, x_i) = \sum_{k=0}^{N} e^{tk} \binom{N}{k} q^k p^{N-k} = (p + e^t q)^N \quad (8.3.4)$$

It follows that

$$E[X] = \frac{\partial \ln Z(0)}{\partial t} = Nq \; ; \; \sigma^2 = E[X^2] - E^2[X] = \frac{\partial^2 \ln Z(0)}{\partial t^2} = Npq.$$

8.3.3 *Poisson Random Variable*

Consider a problem from queing theory: what is the length of a queue, given the rate at which customers arrive. This has important consequences for say a ticket counter, since to fix the average length of a queue, and the waiting time that the length of a queue entails, would require a computation of how many ticket counters should service the customers.

Suppose the average number of customers arriving in a given time interval, say one hour, at a ticket counter is given by λ. $P(n, \lambda)$ is the probability that n customers will arrive in one hour and is given by

$$P(n, \lambda) = \frac{\lambda^n}{n!} e^{-\lambda} \quad : \text{ Poisson distribution}$$

Note that

$$\sum_{n=0}^{\infty} P(n, \lambda) = \sum_{n=0}^{\infty} \frac{\lambda^n}{n!} e^{-\lambda} = e^{\lambda} e^{-\lambda} = 1 \; ; \; P(n, \lambda) > 0$$

The Poisson random variable has the following representation

$$X = \begin{bmatrix} 0 & 1 & .. & \infty \\ P(0, \lambda) & P(1, \lambda) & .. & P(\infty, \lambda) \end{bmatrix}$$

The moment generating function is given by

$$E[e^{tX}] = \sum_i e^{tx_i} P(x_i) = \sum_{n=0}^{\infty} \frac{(e^t \lambda)^n}{n!} e^{-\lambda} = \exp\{e^t \lambda - \lambda\}$$

It follows that

$$E[X] = \lambda \; ; \; \sigma^2 = E[X^2] - E^2[X] = \lambda$$

8.4 Continuous Random Variables

Continuous random variables have a continuous sample space, and can take values over a finite or infinite range. Three widely employed continuous random variables are the following

- Uniform random variable

- Exponential random variable
- Normal (Gaussian) random variable.

8.4.1 Uniform Random Variable

A uniform random variable is equally likely to take any value in the continuous interval $[a, b]$. Hence $P(u)$=constant and

$$1 = \int_a^b du P(u) = P(u)(b-a) \;\Rightarrow\; P(u) = \frac{1}{(b-a)}$$

The uniform random variable is denoted by $U[a, b]$ and is represented as follows

$$U[a, b] = \begin{bmatrix} (a, b) \\ P(U) \end{bmatrix}$$

The moment generating function is given by

$$E[e^{tX}] = \sum_i e^{tx_i} P(x_i) = \frac{1}{(b-a)} \int_a^b due^{tu} = \frac{e^{bt} - e^{at}}{t(b-a)}$$

The mean is given by

$$E[U] = \frac{1}{(b-a)} \int_a^b duu = \frac{1}{2(b-a)}(b^2 - a^2) = \frac{b+a}{2}$$

The second moment is given by

$$E[U^2] = \frac{1}{(b-a)} \int_a^b duu^2 = \frac{1}{3(b-a)}(b^3 - a^3) = \frac{a^2 + b^2 + ab}{3}$$

Hence, the variance is given by

$$\sigma^2 = E[U^2] - E^2[U] = \frac{(b-a)^2}{12}$$

The linear congruential method, developed by David Knuth, yields a pseudo-random variable that behaves like $U[0, 1]$ for a sequence of the size of 2^{64}.

8.4.2 Exponential Random Variable

The exponential random variable takes values on the positive real axis. It is represented by

$$X = \begin{bmatrix} (0, \infty) \\ P(X) \end{bmatrix}$$

Its probability density function is given by

$$P(x) = \alpha \exp(-\alpha x) \ ; \ \ x \in [0, +\infty]$$

and as expected satisfies the requirement of probability density

$$\int_0^\infty dx P(x) = 1 \ ; \ \ P(x) > 0$$

The moment generating function is given by

$$Z(t) = \alpha \int_0^\infty dx e^{tx} e^{-\alpha x} = \frac{\alpha}{\alpha - t} \ ; \ \ t < \alpha$$

The mean and variance are given by

$$E[X] = \frac{1}{\alpha} \ ; \ \ \sigma^2 = E[X^2] - E^2[X] = \frac{1}{\alpha^2}.$$

8.4.3 Normal (Gaussian) Random Variable

The Normal, or Gaussian, random variable is one of the most important random variable, having a wide range of applications in both the natural and social sciences. It is a random variable X that is denoted by $X = N(\mu, \sigma)$; its probability density is given by

$$P(x) = \frac{1}{\sqrt{2\pi\sigma^2}} \exp\left\{ -\frac{1}{2\sigma^2}(x - \mu)^2 \right\} \tag{8.4.1}$$

Hence, it can be represented as follows

$$X = \begin{bmatrix} (-\infty, +\infty) \\ P(X) \end{bmatrix}$$

The moment generating function, from Eqs. 5.6.1 and 8.4.1, is given by

$$Z[J] = E_x[e^{Jx}] = \int_{-\infty}^{+\infty} dx e^{Jx} P(x)$$

$$= \frac{1}{\sqrt{2\pi\sigma^2}} \int_{-\infty}^{+\infty} dx e^{-\frac{1}{2\sigma^2}(x-\mu)^2 + Jx} = \exp\left\{J\mu + \frac{\sigma^2}{2} J^2\right\} \tag{8.4.2}$$

From Eq. 8.4.2

$$\frac{\partial \ln Z[0]}{\partial J} = E[X] \equiv \int_{-\infty}^{+\infty} x P(x) = \mu \ : \text{mean}$$

$$\frac{\partial^2 \ln Z[0]}{\partial J^2} = E[(X-\mu)^2] \equiv \int_{-\infty}^{+\infty} (x-\mu)^2 P(x) = \sigma^2 \ \ : \text{variance}$$

Any normal random variables is equivalent to the $N(0, 1)$ random variable via the following linear transformation

$$X = N(\mu, \sigma)\ ;\ Z = N(0, 1)$$
$$\Rightarrow X = \mu + \sigma Z \tag{8.4.3}$$

To prove Eq. 8.4.3, it is shown that the generating function of the two representations of X are equal. Consider

$$E_x[e^{tx}] = E_z[e^{t(\mu+\sigma z)}] = \frac{1}{\sqrt{2\pi}} \int_{-\infty}^{+\infty} dz e^{-\frac{1}{2}z^2} e^{t(\mu+\sigma z)}$$

$$= \exp\left\{t\mu + \frac{\sigma^2}{2} t^2\right\} \tag{8.4.4}$$

Equation 8.4.4 yields the expected given in Eq. 8.4.2, thus verifying Eq. 8.4.3.

All the moments of the random variable $Z = N(0, 1)$ can be determined by the generating function given by Eq. 8.4.2 and is given by

$$E[Z^n] = \frac{d^n}{dJ^n} Z[J]|_{J=0}$$

The cumulative density for the normal random variable $N(x)$ is defined by

$$F(x) = \text{Prob}(-\infty \leq z \leq x) = \frac{1}{\sqrt{2\pi}} \int_{-\infty}^{x} dz e^{-\frac{1}{2}z^2} = N(x) \qquad (8.4.5)$$

and hence

$$\text{Prob}(a \leq z \leq b) = \frac{1}{\sqrt{2\pi}} \int_{a}^{b} dz e^{-\frac{1}{2}z^2} = N(b) - N(a)$$

A sum of normal random variables is also another normal random variable

$$Z_1 = N(\mu_1, \sigma_1) \ ; \ Z_2 = N(\mu_2, \sigma_2) \ ; \ ...Z_n = N(\mu_n, \sigma_n)$$
$$\Rightarrow \ Z = \sum_{i=1}^{n} Z_i = N(\mu, \sigma) \ \Rightarrow \ \mu = \sum_{i=1}^{n} \mu_i \ ; \ \sigma^2 = \sum_{i=1}^{n} \sigma_i^2 \qquad (8.4.6)$$

Result above can be proven using the generating function given in Eq. 8.4.2.

8.5 Problems

1 A coin is tossed three times. Find the probability that there will appear the following.
(a) Three heads, (b) exactly two heads, (c) exactly one head and (d) no head.
2 A fair coin $p = 1/2, \ q = 1/2$ is tossed 6 times. This is a Binomial experiment with $N = 6$.
(a) What is the expected number of heads? (b) Find the probability that exactly 2 heads occur. (c) At least 4 heads occur. (d) At least 1 head occurs.
3 A marksman shoots at a target and hits the target with $p = 1/4$. He fires 100 times. Find the expected number of hits μ and the standard deviation σ.
4 Show explicitly using the formula for the Binomial coefficients that

$$\sum_{k=0}^{4} \binom{4}{k} = 2^4$$

5 Consider the following rate of returns (in %) with the probability of the returns (in %). Compute the mean, variance and standard deviation, the skewness and kurtosis of the returns.

Rate of returns(%)	−20	−15	−10	−5	0	5	10	15	20
Probability(%)	12	8	5	11	5	14	15	12	18

6 Consider the random variable with the probability density given by

$$P(x) = \begin{cases} \frac{1}{3} \; x \in [1, 4] \\ 0 \; x < 1 \; or \; x > 4 \end{cases} \tag{8.5.1}$$

Find the moment generating function. Find u such that $P(x < u) = 0.8$.

7 Suppose for every minute, on the average, there 2 are telephone calls to a switchboard; what is the probability that, in a minute, there will be 3 calls?

8 Consider the $Z = N(0, 1)$ normal random variable. Find the following probabilities for Z.
(a) $P(-0.5 \leq Z \leq 1.1)$, (b) $P(0.2 \leq Z \leq 1.4)$, (c) $P(-0.38 \leq Z \leq 1.72)$ and (d) $P(-1.5 \leq Z \leq -0.7)$.

9 The Binomial and Poisson distributions are given by

$$P(N, k) = \binom{N}{k} p^k q^{N-k} \quad ; \quad f(k, \lambda) = e^{-\lambda} \frac{\lambda^k}{k!}$$

Show that for $p << 1$ and $N >> 1$

$$P(N, k) \approx f(k, \lambda)$$

10 Consider

$$Z_1 = N(\mu_1, \sigma_1) \; ; \; Z_2 = N(\mu_2, \sigma_2) \; ; \; ...Z_n = N(\mu_n, \sigma_n)$$

Prove that

$$Z = \sum_{i=1}^{n} Z_i = N(\mu, \sigma) \text{ where } \mu = \sum_{i=1}^{n} \mu_i \quad ; \quad \sigma^2 = \sum_{i=1}^{n} \sigma_i^2$$

Chapter 9
Option Pricing and Binomial Model

Abstract The binomial option pricing is developed from first principles. The underlying security is assumed to evolve in discrete steps of time. An option is shown, using the principle of no arbitrage, to be equivalent to a dynamic portfolio composed of the underlying security and risk-free cash. The option price is shown to be equivalent to a random evolution of the option price provided the underlying security has a martingale evolution. In a simple and transparent discrete-time model, the underpinning of option theory is seen to be based on the binomial probability distribution. The binomial option pricing model, which is based on the binomial random variable, shows the utility of probability theory in option pricing.

9.1 Introduction

Option theory, with an emphasis on the payoff of options, has been briefly discussed in Sect. 1.12. The key results are reviewed for completeness. Consider an underlying security S. The price of a European *call option* on $S(t)$ is denoted by $C(t) = C(t, S(t))$; the owner of the instrument has the right but not an obligation (an option) to buy the security at some future time $T > t$ for the strike price of K.

At time $t = T$, when the option matures the value of the call option $C(t, S(t))$ is equal to the payoff function $g(S)$, shown in Fig. 9.1a, and is given by

$$\begin{aligned} C(T, S(T)) &= \begin{cases} S(T) - K, & S(T) > K \\ 0, & S(T) < K \end{cases} \\ &= g(S) = [S(T) - K]_+ \end{aligned} \tag{9.1.1}$$

The price of a European *put option* on $S(t)$ is denoted by $P = C(t, S(t))$, and gives the owner of the instrument the option to sell the security at some future time $T > t$ for the strike price of K.

Similar to the call option, at time $t = T$, when the option matures the value of the call option $P(t, S(t))$ is equal to the payoff function $h(S)$, shown in Fig. 9.1b, and is given by

B. E. Baaquie, *Mathematical Methods and Quantum Mathematics for Economics and Finance*, https://doi.org/10.1007/978-981-15-6611-0_9

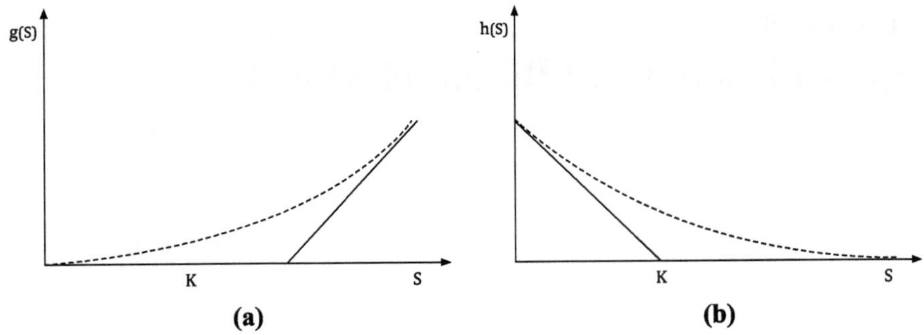

Fig. 9.1 Payoff function for call and put options; dashed line is possible values of the option before maturity. Published with permission of © Belal E. Baaquie 2020. All Rights Reserved

$$
\begin{aligned}
P(T, S(T)) &= \begin{cases} K - S(T), & S(T) < K \\ 0, & S(T) > K \end{cases} \\
&= h(S) = [K - S(T)]_+
\end{aligned}
\tag{9.1.2}
$$

The present day price of the option depends on the future value of the security; clearly, the price of the option C will be determined by how the security $S(t)$ evolves to its future value of $S(T)$.

9.2 Pricing Options: Dynamic Replication

The price of the option $C(t)$ for $t < T$ can be obtained by creating a portfolio at time t that has the same payoff as the option on maturity at time T. The dynamic portfolio is a set of static trading strategies that is extended to a set of self-financing dynamic strategies.

Self-financing dynamic strategy does not require any infusion or withdrawal of money after the initial purchase of assets at time t: Any purchase of new assets must be financed by the sale of old ones.

The principle of no arbitrage states that the two portfolios with the identical cash flows must be identical. Hence, the price of the portfolio at time t must be the price of the option.

9.2.1 Single Step Binomial Model

To simplify the model, consider an option defined for discrete time. Let the remaining time to maturity of the option be $\tau = T - t$, and let there be N steps; the size of the discrete time step is

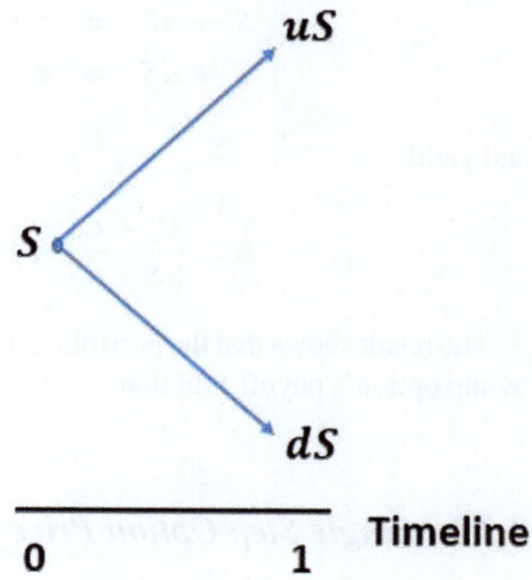

Fig. 9.2 Single step evolution of security S. Published with permission of © Belal E. Baaquie 2020. All Rights Reserved

$$\epsilon = \frac{T}{N}$$

Suppose the stock price at time t is S. Since the evolution of a stock price is random, let the stock increase to uS with some probability (to be specified later) and decrease to dS with some probability.

$$S(1) = \begin{cases} uS \ ; \ u > 1 \\ dS \ ; \ d < 1 \end{cases}$$

and is shown in Fig. 9.2. The payoff function is defined by

$$g(S(1)) = \begin{cases} C_u \ ; \ S(1) = uS \\ C_d \ ; \ S(1) = dS \end{cases}$$

9.3 Dynamic Portfolio

The portfolio's initial value is

$$\pi(0) = B + \Delta S(0) \ ; \ S(0) = S$$

The portfolio's value after one step is set equal to the payoff function. The two future values of the portfolio are set equal to the two values of the payoff of the option. Hence, for risk-free spot interest rate r, we have

$$S \to uS \;\Rightarrow\; \pi_u(1) = e^{r\epsilon}B + uS\Delta = C_u$$
$$S \to dS \;\Rightarrow\; \pi_d(1) = e^{r\epsilon}B + dS\Delta = C_d$$

and yields

$$\Delta = \frac{C_u - C_d}{uS - dS} \;;\quad B = e^{-r\epsilon}(C_d - dS\Delta) \tag{9.3.1}$$

The result shows that the portfolio consists of cash B and stock Δ both determined by the option's payoff function.

9.3.1 *Single Step Option Price*

Since the portfolio has the same cash flow as the payoff of the option, the principle of no arbitrage requires that the price of the call option be given by

$$C = \pi(0) = B + \Delta S$$

and hence, from Eq. 9.3.1

$$C = e^{-r\epsilon}(C_d - dS\Delta) + \frac{C_u - C_d}{u - d} \tag{9.3.2}$$

The derivation shows that an option is equivalent to a portfolio consists of risk-free cash B and a risky asset S. This is the key insight of Black and Scholes that allows for the pricing of options.
Example Consider

$$\Delta = 0.5 \;;\quad B = -42.86 \;;\quad S = 100$$

The price of the call option today is given by

$$C = -42.86 + 0.5 \times 100 = \$7.14$$

9.4 Martingale: Risk-Neutral Valuation

There has been no utilization of ideas from probability theory to determine the future values of the security S; the binomial model does not depend on the probability of the stock going up or down. The binomial depends only upon the initial portfolio and payoff of the option.

The concept of probability is now introduced as this provides a more powerful approach to option pricing.

Suppose the stock price at time t is S and the evolution of a stock price is random. Suppose the stock increases to uS with probability p and decreases to dS with probability $q = 1 - p$, with $p + q = 1$. Hence

$$S \to S(1) = \begin{cases} uS \ ; & \text{probability p} \\ dS \ ; & \text{probability } q = 1 - p \end{cases}$$

Note the one-step probability is determined by the Bernoulli random variable, given in Eq. 8.3.1

$$X = \begin{bmatrix} uS & dS \\ p & q \end{bmatrix} \ ; \ q = 1 - p \tag{9.4.1}$$

In the risk-neutral imaginary world, the change of the security follows a *martingale process*: this implies that the discounted *expectation value* of the future value of the security is equal to its present value. The martingale condition yields the *risk-free rate of return*. Hence

$$S = S(0) = e^{-r\epsilon}[p \cdot uS + q \cdot dS] \ : \ p + q = 1 \ ; \ p, q > 0 \tag{9.4.2}$$

Equation 9.4.2 is one of the simplest and most transparent statement of the martingale condition. In one step, it shows that the discounted value of the future expected value of the security is equal to its present value. For a security following a martingale evolution valuation is very simple: one takes the expected values and discounts at the risk-free rate.

A risk-neutral evolution of the underlying security ensures that the price of the option will be free from arbitrage opportunities. Note that we do **not** assume that investors are risk-neutral. Rather, risk-neutral pricing is an interpretation of our equations and is a very convenient method of computing option prices. One of the main results of mathematical finance is that in a complete market there always exists a unique risk-neutral probability

The risk-neutral martingale evolution of S

$$S = S(0) = e^{-r\epsilon}[p \cdot uS + (1 - p) \cdot dS] \ : \ p + q = 1 \ ; \ p, q > 0$$

yields the solution for p

$$p = \frac{e^{r\epsilon} - d}{u - d} \tag{9.4.3}$$

Note that

$$0 \le p \le 1 \ \Rightarrow \ u > e^{r\epsilon} > d$$

Since all contingent claims are given by the discounted probability of the future values of the payoff, the option price for a single time step is given by discounted value of the expectation of the payoff function, based on the martingale evolution. Hence

$$C = C(0) = e^{-r\epsilon}[p \cdot C_u + q \cdot C_d]$$

It is verified that the risk-neutral valuation yields the result obtained by the dynamic portfolio replication. From Eq. 9.3.2

$$\begin{aligned} C &= B + \frac{C_u - C_d}{u - d} \\ &= e^{-r\epsilon}\left(C_d - d\frac{C_u - C_d}{u - d}\right) + \frac{C_u - C_d}{u - d} \\ &= e^{-r\epsilon}\left[C_d + \frac{e^{r\epsilon} - d}{u - d}(C_u - C_d)\right] \\ &= e^{-r\epsilon}[(1 - p) \cdot C_d + p \cdot C_u] \end{aligned}$$

and we have the expected result.

9.5 N-Steps Binomial Option Price

Consider the case for N-steps. After N time steps, the possible final values of the stock price, as shown in Fig. 9.3, are the following

$$S_{Final} : u^N S, u^{N-1} d S, \ldots, u^{N-k} d^k S, \ldots, d^N S$$

The probability for the value $u^{N-k} d^k S$ is given by the binomial distribution

$$p^{N-k} q^k \binom{N}{k} \ : \ \binom{N}{k} = \frac{N!}{(N-k)!k!} \ ; \ k! \equiv 1 \cdot 2 \cdot 3 \ldots k$$

Recall $\binom{N}{k}$ is the number of ways of choosing u – after N steps – $N - k$ times. The option price, for N steps, is the discounted *expectation value* of the payoff. Hence

$$C = e^{-rN\epsilon} E[S_{Final} - K]_+$$

To evaluate the expectation value, using the probability for the final value that is given by the Binomial probability distribution yields

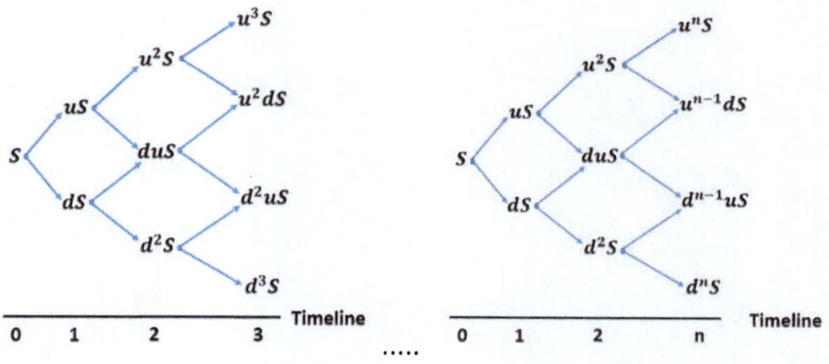

Fig. 9.3 3-step and N-step binomial tree for stock price. Published with permission of © Belal E. Baaquie 2020. All Rights Reserved

$$C(S) = e^{-rN\epsilon} \sum_{k=0}^{N} \binom{N}{k} p^{N-k} q^k [u^{N-k} d^k S - K]_+ \tag{9.5.1}$$

$$S_{Final} = u^{N-k} d^k S \ ; \ p = \frac{e^{r\epsilon} - d}{u - d} \ ; \ p + q = 1$$

The parameters u, d are related to the volatility of the security S. In the limit of continuous time, the Binomial pricing yields the continuous time Black-Scholes option price. It is shown in Eq. 11.11.4 that

$$u = e^{\sigma\sqrt{\epsilon}} \ ; \ q = e^{-\sigma\sqrt{\epsilon}}$$

where σ is the Black-Scholes volatility. Hence, from Eq. 9.4.3 and as given in Eq. 11.11.5, the value of u yields the following, to leading order in ϵ

$$p \simeq \frac{1}{2} + \frac{r\epsilon - \frac{1}{2}\sigma^2\epsilon}{2\sigma\sqrt{\epsilon}} \ ; \ q \simeq \frac{1}{2} - \frac{r\epsilon - \frac{1}{2}\sigma^2\epsilon}{2\sigma\sqrt{\epsilon}}$$

9.6 $N = 2$ Option Price: Binomial Tree

The purely algebraic derivation given in Eq. 9.5.1 for the option price has an underlying Binomial tree. Consider the two-step Binomial tree. The option price can be obtained using the binomial tree, and is shown for the $N = 2$. In all cases, one starts at the final time where the value of the payoff is defined. **One always recurses backwards using various techniques in forming the binomial tree**. The risk-neutral valuation method is used.

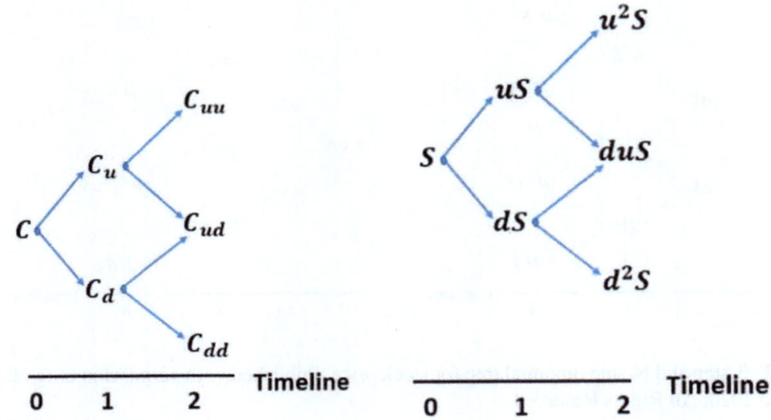

Fig. 9.4 Two-step binomial tree for option and stock price. Published with permission of © Belal E. Baaquie 2020. All Rights Reserved

Recall for one step one has the risk-neutral procedure for the call option at an earlier time C given in terms of the value at the next step by (Fig. 9.4)

$$C = C(0) = e^{-r\epsilon}[pC_u + qC_d]$$

Suppose we start at the second-step; the payoff function has the following values

$$C_{uu} = [u^2S - K]_+ \ ; \ C_{dd} = [d^2S - K]_+ \ ; \ C_{ud} = [udS - K]_+ = C_{du}$$

After one step, using risk-neutral valuation yields

$$C_u = e^{-r\epsilon}[pC_{uu} + qC_{ud}] \ ; \ C_d = e^{-r\epsilon}[pC_{du} + qC_{dd}]$$

The second step yields

$$\begin{aligned}
C &= e^{-r\epsilon}[pC_u + qC_d] \\
&= e^{-2r\epsilon}\Big[p(pC_{uu} + qC_{ud}) + q(pC_{du} + qC_{dd})\Big] \\
&= (e^{-r\epsilon})^2\Big\{p^2[u^2S - K]_+ + 2pq[udS - K]_+ + q^2[d^2S - K]_+\Big\} \\
&= (e^{-r\epsilon})^2\Big\{\binom{2}{0}p^2[u^2S - K]_+ + \binom{2}{1}pq[udS - K]_+ + \\
&\quad \binom{2}{2}q^2[d^2S - K]_+\Big\}
\end{aligned}$$

The answer is what is obtained from Eq. 9.5.1 by setting $N = 2$.

9.7 Binomial Option Price: Put-Call Parity

Equations 9.1.1 and 9.1.2 for the payoff function for the call and put options, respectively, yield the following identity

$$[S_{Final} - K]_+ - [K - S_{Final}]_+ = S_{Final} - K$$

Using $E[A + B] = E[A] + E[B]$, call and put option price yield

$$C - P = e^{-r\tau}E\Big([S_{Final} - K]_+ - [K - S_{Final}]_+\Big) \ ; \ p = \frac{e^{r\epsilon} - d}{u - d}$$

$$\Rightarrow C - P = e^{-r\tau}\sum_{k=0}^{N}\binom{N}{k}p^{N-k}q^k\Big(u^{N-k}d^kS - K\Big)$$

$$= e^{-r\tau}\Big[(pu + qd)^N S - (p + q)^N K\Big]$$

since, from the Binomial theorem

$$\sum_{k=0}^{N}\binom{N}{k}a^{N-k}b^k = (a + b)^N$$

The definition of p, for $\tau = N\epsilon$, yields

$$p = \frac{e^{r\epsilon} - d}{u - d} \ ; \ q = \frac{u - e^{r\epsilon}}{u - d} \ ; \ [e^{r\epsilon}]^N = e^{r\tau}$$

$$\Rightarrow pu + qd = \frac{1}{u - d}\Big[(e^{r\epsilon} - d)u + (u - e^{r\epsilon})d\Big] = e^{r\epsilon}$$

Hence, using the identity $p + q = 1$ yields

$$C - P = e^{-r\tau}\Big[(pu + qd)^N S - (p + q)^N K\Big] = e^{-r\tau}\left[e^{rN\epsilon}S - K\right]$$

$$= S - e^{-r\tau}K \qquad : \text{Put-Call Parity}$$

Note S has a martingale evolution, since $S_{Final} = u^{N-k}d^kS$ yields

$$e^{-r\tau}E[S_{Final}] = e^{-r\tau}\sum_{k=0}^{N}\binom{N}{k}p^{N-k}q^k\Big(u^{N-k}d^kS\Big)$$

$$= e^{-r\tau}(pu + qd)^N S = S$$

9.8 Summary

The discrete and finite model for option pricing is an intuitive and mathematically tractable path to the study of option theory. The idea of replicating portfolio does not need any ideas from probability theory, but rather is based on the idea that to avoid arbitrage, two instruments having the same cash flow must have the same price. The idea that an option is a portfolio of a risk free instrument (cash) and a risky instrument (security) plays a great role in clarifying the nature of an option.

Risk-neutral valuation introduces the idea of probability in the pricing of options. Risk-neutral valuation was realized in the Binomial model by the martingale condition. The martingale condition plays an essential role in the pricing of a wide variety of instruments and, in particular, is required for obtaining an option price free from arbitrage opportunities.

The Binomial distribution is central to the results obtained, and each step being determined by a Bernoulli random variable.

The continuum limit of time for the binomial model is discussed in detail in Sect. 11.11.

9.9 Problems

1. Consider $S = \$100, u = 1.1, d = 0.9, (1 + r) = 1.05, K = 100$. The payoff function for a put-option is given by $[100 - S]_+$. Show that the put option for a one-period Binomial tree is given by \$2.38.
2. Solve for the put option above using the risk-neutral valuation based on the Binomial probability distribution.
3. A stock price is currently at \$40. It is known that at the end of the month it will be either \$42 or \$38. The risk-free rate is 8% per annum with continuous compounding. What is the value of a one-month European call option with a strike price of \$38?
4. Consider $S = \$100, u = 1.1, d = 0.9, (1 + r) = 1.05, K = 100$. The call-option payoff function is given by $[S - 100]_+$. Show that the price of a call option for a two-period Binomial tree is given by \$10.72. [Hint. One needs to start at N = 2 and work backwards, solving for one-period replicating portfolios. At each node there is a different replicating portfolio.]
5. Solve for the call option in the problem above using the risk-neutral valuation based on the Binomial probability distribution.
6. A stock price is at present \$100. Over the next two three month periods the stock price is expected to go up by 10% or down by 10%. The risk-free rate is 8% per annum with continuous compounding. What is the value of a six-month European call option with strike price \$100?
7. Verify that the price of the call and put option for the problem above obeys the put-call parity condition.

Chapter 10
Probability Distribution Functions

Abstract The axioms of probability theory are discussed. The concept of joint probability distribution leads to the idea of the marginal probability and conditional probability. Independent random variables are defined and a detailed discussion is given of the central limit theorem, together with the limit for special cases. Conditional probability is illustrated by examples of discrete and continuous random variables.

10.1 Introduction

Let $P(X)$ be the probability density of a random variable X defined on an interval $[a, b]$. The probability that the random variable X will take values between say L and U is given by

$$P(L \leq X \leq U) = \int_L^U dx P(x)$$

The probability density $P(X)$ is characterized by its moments. The Gaussian normal random variable is fully specified by its mean and variance; however, in general for an arbitrary probability density, *all* its moments are required to fully characterize it. The cubic and quartic moments, called skewness and kurtosis, are usually used to see how far the probability density is from the Gaussian random variable.

The mean and variance of $P(X)$ is the following

$$\mu = E[X] \ ; \ \sigma^2 = E[(X - \mu)^2]$$

Skewness and kurtosis are defined as follows

$$\text{Skewness} = E\left[\left(\frac{X - \mu}{\sigma}\right)^3\right] \ ; \ \text{Kurtosis} = E\left[\left(\frac{X - \mu}{\sigma}\right)^4\right] - 3$$

B. E. Baaquie, *Mathematical Methods and Quantum Mathematics for Economics and Finance*, https://doi.org/10.1007/978-981-15-6611-0_10

Kurtosis is defined to make the kurtosis of the Gaussian random variable equal to zero.

10.1.1 Cumulative Distribution Function

The cumulative distribution function $F(X)$ of a probability distribution function $P(x)$ is given by

$$F(x) = P(a \leq X \leq x) = \int_a^x dy P(y)$$

Note $F(x) \in [0, 1]$ since

$$F(a) = 0 \ ; \ F(b) = 1 \ \Rightarrow \ F(x) \in [0, 1]$$

The cumulative density is a mapping from $P(X)$ to $F(X) \in [0, 1]$. This is a mapping of the random variable X to the uniform random variable U.

$$U = F(X) \ \Rightarrow \ X = F^{-1}(U) \tag{10.1.1}$$

The above result shows that *all random variables* can be mapped into the uniform random variable $U(0, 1)$.

Consider the exponential random variable. Its cumulative density is given by

$$F(x) = \alpha \int_0^x dy e^{-\alpha y} = 1 - e^{-\alpha x} \ \Rightarrow X = -\frac{1}{\alpha} \ln(1 - U)$$

The sample values of $U(0, 1)$ are numerically generated by linear congruential or other methods. The formula in Eq. 10.1.1 can be used to generate the sample values of any other random variable, such as the exponential random variable given above.

10.2 Axioms of Probability Theory

Probability theory underpins the description and understanding of random phenomena and is briefly reviewed.

Consider a large collection of black and white billiard balls in a container, as in Fig. 10.1. Suppose there are N_w white and N_b black billiard balls, with $N = N_w + N_b$. It is assumed each ball has *equal likelihood* of being picked. The probability of picking white and black billiard balls is given by

$$P(w) = \lim_{N \to \infty} \frac{N_w}{N} \ ; \ P(b) = \lim_{N \to \infty} \frac{N_b}{N}$$

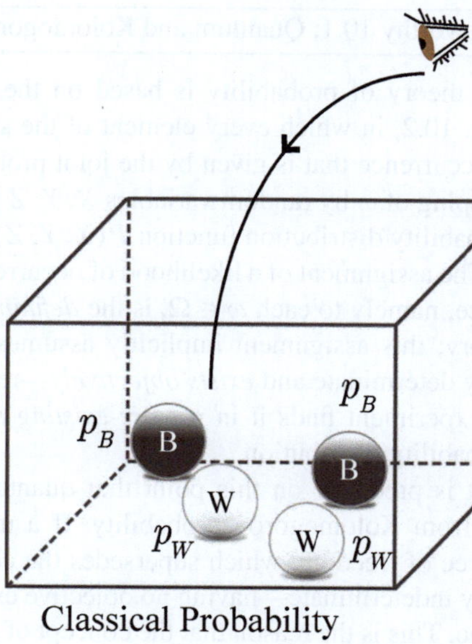

Fig. 10.1 Classical random events: the number of black and white balls inside a closed box *exist objectively*, independent of being observed or not. Furthermore, there exists a probability p_B, p_W that is *intrinsic* to each possible outcome and is assigned to *each* black and white ball. Published with permission of © Belal E. Baaquie 2020. All Rights Reserved

Each ball *objectively exists* in some determinate state before being picked and the *uncertainty in the knowledge* of which billiard is picked is attributed to our *ignorance* of the exact configuration of a very large collection of billiard balls. The billiard balls themselves are in no sense random; rather it is our knowledge that is incomplete leading to the description of the billiard balls using concepts based on probability.

Probability is based on the concept of a random variable, which takes a range of values, and which exists in an objective state regardless of whether it is sampled or not. A unique probability, called the joint probability density, is assigned to a collection of random variables and predicts how frequently will a collection of specific values appear when the random variables are sampled.

Following Kolomogorov, probability theory is defined by the following postulates.

- A collection of all possible allowed random sample values labeled by ω, which forms a *sample space* Ω.
- A *joint probability density* function $P(\omega)$ that determines the probability for the simultaneous occurrence for these random events and provides an exhaustive and complete description of the random system.

The events can be enumerated by *random variables*, say $\vec{X} = (X, Y, Z, \ldots)$, that map the random events ω of the sample space Ω to real numbers, namely

$$\begin{aligned}
&\vec{X} : \Omega \ \rightarrow \ \Re^N \\
&X, Y, Z : \omega \rightarrow \Re \otimes \Re \otimes \Re \ ; \ \omega \in \Omega \\
&P(X, Y, Z) \ : \ \text{joint probability density}
\end{aligned}$$

Noteworthy 10.1: Quantum and Kolomogorov probability

The theory of probability is based on the Kolgomorov axioms—as discussed in Sect. 10.2, in which every element of the sample space Ω is assigned a likelihood of occurrence that is given by the joint probability distribution function $P(\omega)$. The mapping of ω by random variables X, Y, Z to the real numbers is given by the joint probability distribution function $P(X, Y, Z)$.

The assignment of a likelihood of occurrence $P(\omega)$ to *each element* of the sample space, namely to each $\omega \in \Omega$, is the *defining property* of Kolomogorov probability theory; this assignment implicitly assumes that *each element* ω of Ω is intrinsically determinate and *exists objectively*—regardless of being observed or not—and an experiment finds it in it's *pre-existing state* with a likelihood specified by the probability distribution.

It is precisely on this point that quantum probability is fundamentally different from Kolomogorov probability. If a measurement is not made, the quantum degree of freedom, which supersedes the concept of the random variable, is inherently indeterminate—having no objective existence, having no intrinsic determinate value. This is the reason that the concept of the quantum degree of freedom replaces the concept of the random variable in the description of quantum phenomenon.

10.3 Joint Probability Density

Two or more random variables can have a joint density function. The outcomes of the sampling the two random variables can influence each other, depending on whether they are correlated or not. The joint probability density gives rise to the concept of independent and correlated random variables, as well as the marginal and conditional probability densities. A discussion in some detail is given of these aspects of a joint probability density in later Sections.

Consider a set S that contains events A and B, with union $A \cup B$ and intersection $A \cap B$—shown in Fig. 10.2. Consider n trials that fall into the following mutually exclusive categories.

- n_1 = number of outcomes for which A occurs but not B.
- n_2 = number of outcomes for which B occurs but not A.
- n_3 = number of outcomes for which both A and B occur.
- n_4 = number of outcomes for which neither A nor B occur.

Since the four events are mutually exclusive, we have

$$n_1 + n_2 + n_3 + n_4 = n$$

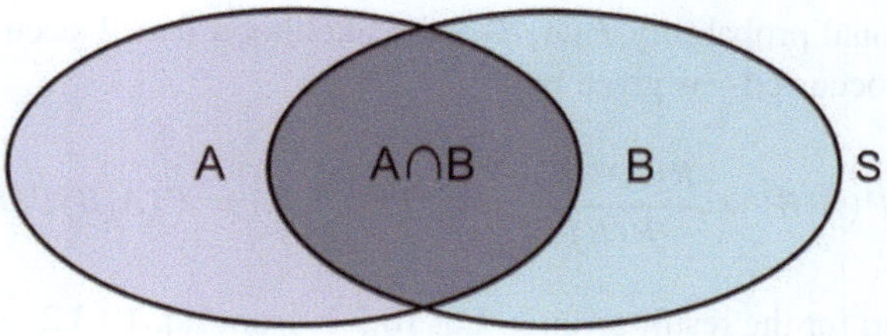

Fig. 10.2 Sets A and B and intersection $A \cap B$ are contained in a larger set S. The set $A \cup B$ is the shaded area, with the elements in the intersection being counted only once. Published with permission of © Belal E. Baaquie 2020. All Rights Reserved

Let $P(A)$ and $P(B)$ be the probability that events A and B occur; $P(A+B) = P(A \cup B)$ is the joint probability density that events A or B occurs; and $P(AB) = P(A \cap B)$ is the probability that events A, B occur simultaneously. Probabilities $P(A)$ and $P(B)$ are given by

$$P(A) = \frac{n_1 + n_3}{n} \; ; \; P(B) = \frac{n_2 + n_3}{n}$$

The probability that both occur is the joint probability denoted by $P(AB)$ and given by

$$P(AB) = P(A \cap B) = \frac{n_3}{n}$$

The probability of either A or B occur is given by

$$P(A+B) = P(A \cup B) = \frac{n_1 + n_2 + n_3}{n} \tag{10.3.1}$$

Define *conditional probability* that A occurs, given that B has occurred, by

$$P(A|B) = \frac{n_3}{n} \Big/ \frac{n_2 + n_3}{n} = \frac{n_3}{n_2 + n_3}$$

and similarly

$$P(B|A) = \frac{n_3}{n_1 + n_3}$$

Two important general lessons can be obtained from this example.

- The probability of either A or B occur is given by

$$P(A+B) = P(A \cup B) = P(A) + P(B) - P(A \cap B) \tag{10.3.2}$$

- The conditional probability $P(A|B)$—the likelihood that A occurs given that B has already occurred—is given by

$$P(A|B) = \frac{P(A \cap B)}{P(B)} \quad \Rightarrow \quad P(A \cap B) = P(A|B)P(B) \tag{10.3.3}$$

In particular, for the result given in Eq. 10.3.1, from Eq. 10.3.2

$$P(A + B) = \frac{n_1 + n_3}{n} + \frac{n_2 + n_3}{n} - \frac{n_3}{n} = \frac{n_1 + n_2 + n_3}{n}$$

As one can see from the case above, subtraction by the probability $P(A \cap B)$ compensates for the over-counting in calculating the probability $P(A + B)$.

An example of Eq. 10.3.2 is the likelihood of drawing two aces from two different fair decks, say event A and B. Then $P(A) = 1/13 = P(B)$; note that $P(AB)$ is the probability of drawing two aces and hence $P(AB) = 1/13 \times 1/13$. Equation 10.3.2 yields

$$P(A + B) = \frac{1}{13} + \frac{1}{13} - \frac{1}{169} = \frac{25}{169}$$

Furthermore, from Eq. 10.3.3

$$P(A|B)P(B) = P(B|A)P(A) \tag{10.3.4}$$

A similar relation can be written for events A and C. Dividing the two conditional probabilities and using Eq. 10.3.4 yields Bayes theorem given by

$$\frac{P(B|A)}{P(C|A)} = \frac{P(B)P(A|B)}{P(C)P(A|C)} \quad : \text{ Bayes theorem}$$

An **example** of Eq. 10.3.3 is the probability of consecutively drawing two hearts from a fair deck. Let B be the probability of drawing a heart and is given by $P(B) = P(1 \text{ Heart}) = 1/4$. Let A be the probability of drawing a heart, given that B has occurred. If B has occurred,then there is now 51 cards in the deck and one less card of hearts; hence, $P(A|B) = P(1\text{Heart}|1\text{Heart}) = 12/51$ and Eq. 10.3.3 yields

$$\begin{aligned} P(A \cap B) &= P(2\text{Heart}) = P(A|B)P(B) \\ &= P(1 \text{ Heart}|1\text{Heart})P(1\text{Heart}) = \frac{12}{51} \cdot \frac{1}{4} = \frac{1}{17} \end{aligned}$$

If events A and B are *mutually exclusive*, then $P(A \cap B) = 0$ and

$$P(A + B) = P(A \cup B) = P(A) + P(B)$$

Events A and B being *independent* is defined by

$$P(A \cap B) = P(A)P(B)$$

Independence means that the occurrence of A is not affected by B and visa versa. Hence, it follows that

$$P(A|B) = P(A) \text{ and } P(B|A) = P(B)$$

Example. Consider tossing a fair coin 3 times Let A be the event that H appears on the first throw and event B that T appears on the second throw. The probability of A and B is the following.

$$P(A) = P[HHT, HTH, HTT, HHH] = \frac{4}{8} = \frac{1}{2} = P(B)$$

$$P(A \cap B) = P[HTH, HTT] = \frac{2}{8} = \frac{1}{4} = P(A)P(B)$$

The joint probability density function obeys all the laws of probability. Consider random variables X, Y. Their joint probability density is given by

$$P(X, Y) \geq 0 \ ; \ \int_{-\infty}^{\infty} dxdyP(x, y) = 1$$

In other words, $P(X = x, Y = y)$ yields the probability for the *simultaneous occurrence* of the specific values x, y of the random variables X, Y. Let $A = [a, b]$ and $B = [u, v]$; then

$$P(A) = \int_a^b dx \int_{-\infty}^{\infty} dyP(x, y) \ ; \ P(B) = \int_{-\infty}^{\infty} dx \int_u^v dyP(x, y)$$

Furthermore

$$P(A \cap B) = \int_a^b dx \int_u^v dyP(x, y)$$

Consider a function H that depends on the random variables X, Y; its (average) expectation value is given by

$$E[H] = \int dxdyH(x, y)P(x, y)$$

If the random variables are *independent*, the joint probability density function factorizes and yields

$$P(x, y) = P_1(x)P_2(y)$$

For multiple random variables one can form the *marginal probability density*. The probability that random variables are observed having random values of X, regardless of the value of Y, is given by the *marginal density* for two random variables, namely

$$P(X) = \int_{-\infty}^{\infty} dy y P(X, y) \ ; \quad \int_{-\infty}^{\infty} dx P(x) = 1 \tag{10.3.5}$$

Consider the case of two securities; the probability of finding a security with value x, given that a second security has already been observed to have a value y, is given by the *conditional probability*

$$P(X|Y) = \frac{P(X, Y)}{P(Y)} = \frac{P(X, Y)}{\int_{-\infty}^{\infty} dx P(x, Y)}$$

$$\int_{-\infty}^{\infty} dx P(x|Y) = 1 \ ; \quad \int_{-\infty}^{\infty} dy P(X|y) \neq 1 \tag{10.3.6}$$

$P(X|Y)$ is a probability distribution for X, given Y, and not vise versa. Note that the normalization given in Eq. 10.3.6 holds for any Y, but not vise versa.

10.4 Independent Random Variables

In general, the probability density function for independent random variables, is factorized and given by

$$P(X_1, X_2, \ldots, X_N) = P_1(X_1)P_2(X_2)\ldots P_N(X_N)$$

All the correlation functions factorize

$$E[X_i X_j \ldots X_N] = E[X_i] \ldots E[X_N]$$

The criterion that the random variables are correlated is given by $E[X_i X_j] - E[X_i]E[X_j] \neq 0$. For independent random variables, as expected.

$$E[X_i X_j] - E[X_i]E[X_j] = 0 \ ; \quad i \neq j$$

Consider the Bernoulli random variable given by

$$X = \begin{bmatrix} 0 & 1 \\ p & q \end{bmatrix} \ ; \quad q = 1 - p \tag{10.4.1}$$

The probability distrubtion function can be written as

$$P(X) = (p)^{1-x}(1-p)^x \ ; \ x = 0, 1$$

Consider two Bernoulli random variables X_1, X_2 that are determined by the p_1, p_2. The joint probability density and sample space is given by

$$P(X_1, X_2) \ : \ \Omega = \{00, 01, 10, 11\}$$

The random variables then has the following representation

$$(X_1, X_2) = \begin{bmatrix} 00 & 01 & 10 & 11 \\ p_1p_2 & p_1q_2 & q_1p_1 & q_1q_2 \end{bmatrix} \ ; \ q_1 = 1 - p_1, q_2 = 1 - p_2$$

We see that the two Bernoulli random variables are independent since they are decorrelated. The joint probability density factorizes since, for $X_1 = X, X_2 = Y$

$$P(X, Y) = P_1(x)P_2(y) = (p_1)^{1-x}(1-p_1)^x(p_2)^{1-y}(1-p_2)^y \qquad (10.4.2)$$

The moment generating function factorizes for the independent Bernoulli random variables. From Eq. 10.7.1

$$\begin{aligned} Z(t, s) &= E[e^{tX_1+sX_2}] = \sum_{i,j} e^{tx_{1i}+sx_{2j}} P(x_{1i}, x_{2j}) \\ &= \Big(\sum_i e^{tx_{1i}+} P_1(x_{1i})\Big)\Big(\sum_j e^{sx_{2j}} P_2x_{2j})\Big) \\ &= (p_1 + e^t q_1)(p_2 + e^s q_2) \end{aligned} \qquad (10.4.3)$$

The mean is given by

$$E[X_1] = \frac{\partial}{\partial t} \ln Z(0, 0) = q_1 \ ; \ E[X_2] = \frac{\partial}{\partial s} \ln Z(0, 0) = q_2$$

The new quantity for a joint probability density is the correlation of two random variable, given by

$$\begin{aligned} E[X_1X_2] - E[X_1]E[X_2] &= \frac{\partial^2}{\partial t \partial s} \ln Z(0, 0) = 0 \\ \Rightarrow \ E[X_1X_2] &= E[X_1]E[X_2] \end{aligned}$$

Consider the case of N independent and identical Bernoulli random variables X_n. All the random variables are identical to the random variable given in Eq. 10.4.1 with the probability density function $P(X)$. Since the random variables are independent and identical, the probability density function is given by

$$P = P_1(X_1)P_2(X_2)\dots P_N(X_N) = \prod_{i=1}^{N} P(X_i)$$

To connect to the earlier notation, let $H = 1$ and tail $T = 0$. The sum of the N-random variables given by

$$K = \sum_{n=1}^{N} X_n \;\Rightarrow K \in [0, N] \;:\; X_n = \begin{bmatrix} 0 & 1 \\ p & q \end{bmatrix} \;;\; q = 1 - p$$

In other words, K is a random integer taking values between 0 and N. The moment generating function is given by

$$E[e^{tK}] = E\Big[\exp\Big\{t\sum_{n=1}^{N} X_n\Big\}\Big] = \prod_{n=1}^{N} E[e^{tX_n}] = \prod_{n=1}^{N}\Big[\sum_{x_n=0}^{1} e^{tx_n}P(x_n)\Big]$$
$$= \prod_{n=1}^{N}(p + e^t q) = (p + e^t q)^N \tag{10.4.4}$$

Equation 10.4.4 is the moment generating function of the binomial random variable given in Eq. 8.3.4. Hence, from the generating function, we conclude that K is a Binomial random variable with

$$B(N, k) = \frac{N!}{k!(N-k)!}q^k(1-q)^{N-k} = \binom{N}{k}q^k(1-q)^{N-k}$$

Consider two independent Gaussian random variables $X = N(\mu_x, \sigma_x)$, $Y = N(\mu_y, \sigma_y)$; their joint probability density function is given by

$$P(x, y) = \frac{1}{2\pi\sigma_x\sigma_y}\exp\Big\{-\frac{1}{2}\Big(\frac{1}{\sigma_x^2}(x-\mu_x)^2 + \frac{1}{\sigma_y^2}(y-\mu_y)^2\Big)\Big\}$$

To simplify the notation, define two $N(0, 1)$ normal random variables

$$u = \frac{1}{\sigma_x}(x - \mu_x) \;;\; v = \frac{1}{\sigma_y}(y - \mu_y) \tag{10.4.5}$$

We have

$$P(x, y)dxdy = P(u, v)dudv$$

with

$$P(u, v) = \frac{1}{2\pi}\exp\Big\{-\frac{1}{2}\Big(u^2 + v^2\Big)\Big\} \tag{10.4.6}$$

10.5 Central Limit Theorem: Law of Large Numbers

Consider N independent and identical random variables X_i , $i = 1, \ldots, N$ with probability density given by

$$P(X_1, X_2, \ldots, X_N) = P(X_1)PX_2) \ldots P(X_N)$$

Consider the average of the random variables

$$S = \frac{1}{N}\sum_{i=1}^{N} X_i$$

The moment generating function of Z is given by

$$E[\exp\{tS\}] = \prod_{i=1}^{N} \int dx_i P(x_i) e^{\frac{t}{N}x_i} = \left(\int dx P(x) e^{\frac{t}{N}x}\right)^N \tag{10.5.1}$$

The expectation value yields

$$\int dx P(x) e^{\frac{t}{N}x} = \int dx P(x)\left[1 + \frac{t}{N}x + \frac{t^2}{2N^2}x^2 + O(1/N^3)\right] \tag{10.5.2}$$

$$= 1 + \frac{t}{N}\mu + \frac{t^2}{2N^2}E[X^2] + O(1/N^3) = \exp\left\{\frac{t}{N}\mu + \frac{t^2}{2N^2}\sigma^2 + O(1/N^3)\right\}$$

since

$$\ln(1 + x) \simeq x - \frac{1}{2}x^2 + \cdots \; ; \; \sigma^2 = E[X^2] - (E[X])^2$$

Hence, from Eqs. 10.5.5 and 10.5.2, moment generating function is given by

$$E[\exp\{tS\}] = \left(\exp\left\{\frac{t}{N}\mu + \frac{t^2}{2N^2}\sigma^2 + O(1/N^3)\right\}\right)^N$$

$$= \exp\left\{t\mu + \frac{1}{2}t^2\frac{\sigma^2}{N}\right\} + O(1/N^2) \tag{10.5.3}$$

The moment generating function of Z is the same as that of a Gaussian random variable given in Eq. 8.4.2, with the following mean and volatility

$$S = N\left(\mu, \frac{\sigma}{\sqrt{N}}\right) \tag{10.5.4}$$

Equation 10.5.4 is a general result and holds for all random variables. From Eq. 8.4.3, the normal random variable $N(\mu, \sigma)$ can be mapped to $N(0, 1)$ by

$$S = \mu + \frac{\sigma}{\sqrt{N}} Z = \mu + vZ \ : \ v = \frac{\sigma}{\sqrt{N}} \ ; \ Z = N(0, 1)$$

In particular, from Eq. 8.4.2

$$P(a < S < b) = P\left(\frac{a-\mu}{v} \leq z \leq \frac{b-\mu}{v}\right) \ ; \ Z = N(0, 1) \quad (10.5.5)$$

Form the sum from identical and independent random variables X_i

$$K = \sum_{i=1}^{N} X_i \ : \ E[X_i] = \mu \ ; \ E[X_i^2] - (E[X_i])^2 = \sigma^2$$

The moment generating function of $Z[K]$, similar to Eq. 10.5.5, is given by

$$\begin{aligned} E[\exp\{tK\}] &= \prod_{i=1}^{N} \int dx_i P(x_i) e^{tx_i} = \left(\int dx P(x) e^{tx}\right)^N \\ &= \left(\int dx P(x)(1 + tx + \frac{1}{2}t^2x^2 + \cdots)\right)^N \\ &= \left(1 + t\mu + \frac{1}{2}t^2(\sigma^2 + \mu^2) + \cdots\right)^N \\ &\simeq \exp\left\{tN\mu + \frac{1}{2}t^2\sigma^2 N + \cdots\right\} \end{aligned} \quad (10.5.6)$$

Another proof is given on Eq. 10.5.6. Note that

$$\begin{aligned} E[K] = N\mu \ ; \quad E[K^2] &= \sum_{ij} E[X_i X_j] = \sum_{i \neq j} E[X_i X_j] + \sum_i E[X_i^2] \\ &= (N^2 - N)\mu^2 + N(\sigma^2 + \mu^2) = N\sigma^2 + N^2\mu^2 \end{aligned}$$

Hence

$$\begin{aligned} E[\exp\{tK\}] &= [1 + tE[K] + \frac{1}{2}t^2 E[K^2] + \cdots] \\ &= 1 + tN\mu + \frac{1}{2}t^2(N\sigma^2 + N^2\mu^2) + \cdots \\ &\simeq \exp\left\{tN\mu + \frac{1}{2}t^2\sigma^2 N + \cdots\right\} \end{aligned} \quad (10.5.7)$$

Equation 10.5.6 yields the central limit theorem that

$$\begin{aligned} & K = N(\mu N, \sqrt{N}\sigma) \ ; \ K = N\mu + \sqrt{N}\sigma Z \ ; \ Z = N(0, 1) \\ & \Rightarrow P(a < K < b) = P\left(\frac{a - \mu N}{\sqrt{N}\sigma} \leq z \leq \frac{b - \mu N}{\sqrt{N}\sigma}\right) \end{aligned} \quad (10.5.8)$$

One of the far reaching consequences of Eq. 10.5.4 is in statistics. Suppose there is a random variable X with an unknown mean with a value μ. To determine the mean of X, suppose the random variable is sampled N times and with sample values x_i. The central limit theorem then states that, to a confidence level of 66%, the estimate of the mean μ_{Est} obtained from the samples x_i of the random variable is given by

$$\mu_{\text{Est}} = \frac{1}{N}\sum_{i=1}^{N} x_i = \mu \pm \frac{\sigma}{\sqrt{N}} \tag{10.5.9}$$

Similarly, for Monte Carlo simulations, one can generate N sample values x_i of random variable X; then Eq. 10.5.9 states the estimate of μ is given by

$$\frac{1}{N}\sum_{i=1}^{N} x_i = \mu \pm \frac{\sigma}{\sqrt{N}}$$

The accuracy of the simulation is given by $1/\sqrt{N}$; hence, for example, to increase the accuracy by one decimal position, the sample size must be increased by 100.

10.5.1 Binomial Random Variable

Consider the Bernoulli random variable given by

$$X = \begin{bmatrix} 0 & 1 \\ p & q \end{bmatrix} \; ; \; p = 1 - q \; ; \; \mu = q \; ; \; \sigma^2 = pq \tag{10.5.10}$$

The moment generating function of the sum of identical Bernoulli random variables is given by

$$K = \sum_{i=1}^{N} X_i$$

which, as shown in Eq. 10.4.4, is identical to a Binomial random variable with probability distribution given by $P(N, p)$. Hence, from Eq. 10.4.4 and using Eq. 10.5.10[1]

[1] The generating function needs to be evaluated only in the neighborhood of $t = 0$.

$$E[e^{tK}] = E\left[\exp\left\{t\sum_{n=1}^{N} X_n\right\}\right] = (p + e^t q)^N$$
$$= \left(p + q + tq + \frac{t^2}{2}q^2 + \cdots\right)^N \simeq \exp\left\{tNq + \frac{t^2}{2}N(q - q^2)\right\}$$
$$= \exp\left\{tm + \frac{t^2}{2}v^2\right\} \ ; \ m = Nq \ ; \ v^2 = Npq \tag{10.5.11}$$

For large N, the Binomial random variables with probability distribution given by $B(k, N)$, from Eq. 10.5.11, is equivalent to the Normal random variable $N(m, v)$. Hence

$$B(k, N) = \binom{N}{k} p^{N-k} q^k \ ; \ p + q = 1$$
$$\lim_{N\to\infty} B(k, N) \to N(m, v) \ : \ m = Nq \ ; \ v^2 = Npq \tag{10.5.12}$$

The result given for the general case in Eq. 10.5.5 has been obtained in Eq. 10.5.11 for the special case of the Bernoulli random variables.

Example. A fair coin is tossed 100 times; hence $p = q = 1/2$. Hence

$$K_h = \text{Number of H in 100 throws}$$

Estimate the probability that the number of heads lies S_h lies between 40 and 60. The expected number of heads and the standard deviation for the number of heads are given by

$$\mu = nq = 100 \times 1/2 = 50 \ ; \ \sigma\sqrt{N} = \sqrt{Npq} = \sqrt{100 \cdot 1/2 \cdot 1/2} = 5$$

Since $n = 100$ is reasonably large, the result can be approximately given by the normal distribution

$$K = N(qN, \sigma\sqrt{n}) = N(50, 5)$$

Hence, from Eq. 10.5.8

$$P(40 \le K_h \le 60) \approx P\left(\frac{40 - 50}{5} \le z \le \frac{60 - 50}{5}\right)$$
$$= \frac{1}{\sqrt{2\pi}} \int_{-2}^{+2} dz \exp\left\{-\frac{1}{2}z^2\right\} \ ; \ Z = N(0, 1)$$
$$= 2\left[N(2) - \frac{1}{2}\right] = 2 \times 0.9972 - 1 = 0.9944 \tag{10.5.13}$$

The actual value is 0.9648, to four decimal places. Note that the probability has been computed based on the deviation away from the mean being equal to one standard deviation.

If one computes the probability that the number of successes is between 35 and 65, then this amounts to the result for the mean deviating from by three standard deviations. The estimate is the area under the standard normal curve between −3 and 3 is given 0.9972. The actual answer in this case, to four places, is 0.9982.

10.5.2 Limit of Binomial Distribution

The limit for the Binomial distribution for large N can be obtained directly from the Binomial probability distribution. For this, when n is large, one needs Stirlings approximation to $n!$ given by

$$n! \approx \sqrt{2\pi n} n^n e^{-n}\left(1+\frac{1}{12n}+\frac{1}{288n^2}+\cdots\right)=\sqrt{2\pi n} n^n e^{-n}\left(1+O\left(\frac{1}{n}\right)\right)$$

Recall from the Binomial distribution, the probability for k heads in N throws is given by

$$\begin{aligned}
B(k,N) &= \frac{N!}{k!(N-k)!}p^{N-k}q^k \ : \ m = Nq \ ; \ v^2 = Npq \\
&\simeq \frac{\sqrt{2\pi N}N^N e^{-N}}{\sqrt{2\pi k}k^k e^{-k}\sqrt{2\pi(N-k)}(N-k)^{N-k}e^{-N+k}}p^{N-k}q^k\left(1+O\left(\frac{1}{N}\right)\right) \\
&= \left(\frac{Np}{N-k}\right)^{N-k}\left(\frac{Nq}{k}\right)^k\sqrt{\frac{N}{2\pi k(N-k)}}
\end{aligned} \quad (10.5.14)$$

For large N, it is expected that the deviation of k from the mean m is going to be small. Hence, define a new variable ξ by

$$\xi \equiv k-m = k-Nq \ \Rightarrow \ N-k = N-\xi-Nq = Np-\xi$$

$$\Rightarrow \ \ln\left(\frac{Np}{N-k}\right) = \ln\left(\frac{Np}{Np-\xi}\right) = -\ln\left(1-\frac{\xi}{Np}\right)$$

Similarly

$$k = Nq+\xi \ \Rightarrow \ \ln\left(\frac{Nq}{k}\right) = \ln\left(\frac{Nq}{Nq+\xi}\right) = -\ln\left(1+\frac{\xi}{Nq}\right)$$

Using the Taylor expansion of $\ln(1+x) \simeq x - x^2/2 + \cdots$ yields

$$\ln\Big[\Big(\frac{Np}{N-k}\Big)^{N-k}\Big(\frac{Nq}{k}\Big)^{k}\Big] = (N-k)\ln\Big(\frac{Np}{N-k}\Big) + k\ln\Big(\frac{Nq}{k}\Big)$$
$$= -(Np-\xi)\ln\Big(1-\frac{\xi}{Np}\Big) - (Nq+\xi)\ln\Big(1+\frac{\xi}{Nq}\Big)$$
$$\simeq -(Np-\xi)\Big[-\frac{\xi}{Np} - \frac{1}{2}\Big(\frac{\xi}{Np}\Big)^2\Big] - (Nq+\xi)\Big[\frac{\xi}{Nq} - \frac{1}{2}\Big(\frac{\xi}{Nq}\Big)^2\Big]$$
$$= Np\Big[\frac{1}{2}\Big(\frac{\xi}{Np}\Big)^2\Big] + Nq\Big[\frac{1}{2}\Big(\frac{\xi}{Nq}\Big)^2\Big] + \xi\Big[-\frac{\xi}{Np}\Big] - \xi\Big[\frac{\xi}{Nq}\Big] + O\Big(\frac{\xi^3}{N^2}\Big)$$
$$= \frac{1}{2}\frac{\xi^2}{N}\Big[\frac{1}{p}+\frac{1}{q}\Big] - \frac{\xi^2}{N}\Big[\frac{1}{p}+\frac{1}{q}\Big] = -\frac{1}{2}\frac{\xi^2}{Npq} \quad (10.5.15)$$

Furthermore

$$\frac{\sqrt{N}}{\sqrt{2\pi k(N-k)}} = \frac{\sqrt{N}}{\sqrt{2\pi(Nq+\xi)(Np-\xi)}} \simeq \frac{1}{\sqrt{2\pi Nqp}} + O(\frac{1}{N^{3/2}}) \quad (10.5.16)$$

Collecting the results, Eqs. 10.5.14, 10.5.15 and 10.5.16 yields the result

$$B(k,N) \simeq \sqrt{\frac{1}{2\pi Nqp}}\exp\Big\{-\frac{1}{2}\frac{\xi^2}{Npq}\Big\}$$
$$\Rightarrow B(k,N) = \sqrt{\frac{1}{2\pi Npq}}\exp\Big\{-\frac{1}{2}\frac{(k-Nq)^2}{Npq}\Big\} \quad (10.5.17)$$

Equation 10.5.17 shows that we have recovered the central limit theorem, obtained earlier in Eq. 10.5.12, that $B(k,N) \to N(Nq, \sqrt{Npq})$ for $N \to \infty$.

10.6 Correlated Random Variables

Correlation is not the same as causation. One random variable, say U does not cause the other random variable V to take any particular value. Instead, both the random variables U, V are equally important, with their correlation being due to the joint probability density $P(u, v)$.

The probability density function for random variables that are correlated, and hence not independent, cannot be factorized. Hence

$$P(X_1, X_2, \ldots, X_N) \neq P_1(X_1)P_2(X_2)\ldots P_N(X_N)$$

10.6.1 Bernoulli Random Variables

For simplicity consider two Bernoulli random variables X, Y that are identical. Choose the following probability density function for the random variables

$$(X, Y) = \begin{bmatrix} 00 & 01 & 10 & 11 \\ cp^2 & cpq & cqp & c(q^2 + \rho^2) \end{bmatrix} \; ; \; p + q = 1 \qquad (10.6.1)$$

It is not necessary to choose $p + q = 1$; the choice is made for simplicity. Note all the probabilities are positive. The scale is fixed by the requirement

$$1 = \sum_{ij} P(x_i, y_j) = c(p^2 + 2pq + q^2 + \rho^2) = c(1 + \rho^2) \;\Rightarrow\; c = \frac{1}{1 + \rho^2}$$

Repeating the calculation given in Eq. 10.4.3 for the moment generating function yields

$$\begin{aligned} Z(t, s) &= E[e^{tX+sY}] = \sum_{i,j} e^{tx_i + sy_j} P(x_i, y_j) \\ &= c\{p^2 + e^s pq + e^t qp + e^{t+s}(q^2 + \rho^2)\} \end{aligned} \qquad (10.6.2)$$

The mean is given by

$$E[X] = c(pq + q^2 + \rho^2) = c(q + \rho^2) = E[Y]$$

The variance is

$$\begin{aligned} \sigma_y^2 = \sigma_x^2 &= E[X^2] - E^2[X] = c(pq + q^2 + \rho^2) - c^2(q + \rho^2)^2 \\ &= pc^2(q + \rho^2) > 0 \end{aligned} \qquad (10.6.3)$$

For the random variables to be correlated, we need to have $E[XY] - E[X]E[Y] \neq 0$. Consider the cross-correlator given by

$$E[XY] - E[X]E[Y] = c(q^2 + \rho^2) - c^2(q + \rho^2)^2 = pc^2(q^2 + \rho^2) \neq 0$$

and hence the Bernoulli random variables are correlated. The covariance is given

$$cov(XY) = \frac{1}{\sigma_x \sigma_y}(E[XY] - E[X]E[Y]) = \frac{q^2 + \rho^2}{q + \rho^2} \neq 0$$

10.6.2 Gaussian Random Variables

Consider two correlated Gaussian random variables given by

$$P(x,y) = \frac{1}{2\pi\sigma_x\sigma_y\sqrt{1-\rho^2}} \times \exp\left\{-\frac{1}{2(1-\rho^2)}\Big(\frac{1}{\sigma_x^2}(x-\mu_x)^2 + \frac{1}{\sigma_y^2}(y-\mu_y)^2 - 2\frac{\rho}{\sigma_x\sigma_y}(x-\mu_x)(y-\mu_y)\Big)\right\} \tag{10.6.4}$$

To simplify the notation, define as in Eq. 10.4.5 two $N(0,1)$ normal random variables

$$u = \frac{1}{\sigma_x}(x-\mu_x) \ ; \ v = \frac{1}{\sigma_y}(y-\mu_y)$$

We have

$$P(x,y)dxdy = P(u,v)dudv$$

and from Eq. 10.4.6

$$P(u,v) = \frac{1}{2\pi\sqrt{1-\rho^2}} \exp\left\{-\frac{1}{2(1-\rho^2)}\Big(u^2+v^2-2\rho uv\Big)\right\}$$

In matrix notation

$$\begin{aligned} P(u,v) &= \frac{1}{2\pi\sqrt{1-\rho^2}} \exp\left\{-\frac{1}{2(1-\rho^2)}\begin{bmatrix} u & v\end{bmatrix}\begin{bmatrix} 1 & -\rho \\ -\rho & 1\end{bmatrix}\begin{bmatrix} u \\ v\end{bmatrix}\right\} \\ &= \frac{1}{2\pi\sqrt{1-\rho^2}} \exp\left\{-\frac{1}{2}\begin{bmatrix} u & v\end{bmatrix} M \begin{bmatrix} u \\ v\end{bmatrix}\right\} \end{aligned}$$

where

$$M = \frac{1}{(1-\rho^2)}\begin{bmatrix} 1 & -\rho \\ -\rho & 1\end{bmatrix} \ ; \ M^{-1} = \begin{bmatrix} 1 & \rho \\ \rho & 1\end{bmatrix} \ ; \ \det(M) = \frac{1}{(1-\rho^2)}$$

From Eq. 5.7.5, for $N = 2$

$$\int dudvP(u,v) = \frac{1}{2\pi\sqrt{1-\rho^2}} \frac{(2\pi)^{N/2}}{\sqrt{\det M}} = 1$$

The mean and variance of U, V are given by

$$E[U] = 0 = E[V] \ ; \ \sigma_u^2 = E[U^2] = 1 = E[V^2] = \sigma_v^2 \tag{10.6.5}$$

From the moment generating function and Eq. 5.7.6, the covariance (cross-correlation) function is given by

$$cov(U, V) = \frac{E[UV] - E[U]E[V]}{\sigma_u \sigma_v} = E[UV] = \int dudvP(u, v)uv = M_{12}^{-1} = \rho$$

The parameter ρ is a measure of the correlation of the two Gaussian random variables. The occurrence of one of the random variable is influenced by the other random variable.

Noteworthy 10.2: $\rho = 1$ Limit

Equation 10.4.6 is the joint probability density for correlated Gaussian random variables U, V

$$P(u, v) = \frac{1}{2\pi\sqrt{1 - \rho^2}} \exp\left\{-\frac{1}{2(1 - \rho^2)}\left(u^2 + v^2 - 2\rho uv\right)\right\} \tag{10.6.6}$$

The exponent can be rewritten in the following manner

$$\frac{1}{2(1 - \rho^2)}(u^2 + v^2 - 2\rho uv) = \frac{1}{2(1 - \rho^2)}(u - \rho v)^2 + \frac{1}{2}v^2$$

and yields

$$P(u, v) = \frac{1}{2\pi\sqrt{1 - \rho^2}} \exp\left\{-\frac{1}{2(1 - \rho^2)}(u - \rho v)^2 - \frac{1}{2}v^2\right\}$$

Note that $\rho = \epsilon(\rho)|\rho|$, where $\epsilon(\rho)$ is the sign ρ. The limit $|\rho| \to 1$ is given by

$$\lim_{|\rho| \to 1} \frac{1}{\sqrt{2\pi(1 - \rho^2)}} \exp\left\{-\frac{1}{2(1 - \rho^2)}(u - \rho v)^2\right\} = \delta(u - \epsilon(\rho)v)$$

Hence

$$\lim_{|\rho| \to 1} P(u, v) = \frac{1}{\sqrt{2\pi}} e^{-\frac{1}{2}v^2} \delta(u - \epsilon(\rho)v)$$

The Dirac delta-function $\delta(u - \epsilon(\rho)v)$ in $P(u, v)$ is a reflection of the fact that for $|\rho| = 1$, random variables U, V are exactly correlated (or anticorrelated). For any function $f(U, V)$ of the random variables, its expectation value is given by

$$\lim_{|\rho| \to 1} E[f(U, V)] = \lim_{|\rho| \to 1} \int dudvP(u, v) f(u, v) = \int dv \frac{1}{\sqrt{2\pi}} e^{-\frac{1}{2}v^2} f(\epsilon(\rho)v, v)$$

10.7 Marginal Probability Density

For two or more random variables, one may choose to observe the outcomes of only one of the random variables, say X, and make no observation of the other random variables. One will then obtain the marginal probability for the outcomes of only X. The marginal density is obtained by summing the joint probability density over all the random variables that are not being observed.

Consider the Bernoulli joint probability density for two independent random variables X, Y given in Eq. 10.7.1

$$P(x, y) = P_1(x)P_2(y) = (p_1)^{1-x}(1 - p_1)^x(p_2)^{1-y}(1 - p_2)^y \qquad (10.7.1)$$

The marginal probability density of X, the sum over the random variable is taken and yields

$$P(x) = \sum_{y=0}^{1} P(x, y) = (p_1)^{1-x}(1 - p_1)^x \sum_{y=0}^{1}(p_2)^{1-y}(1 - p_2)^y = (p_1)^{1-x}(1 - p_1)^x$$

As expected, the marginal probability for independent random variables reproduces the density for the random variable X.

In general, for independent random variables, since the probability density factorizes, the marginal probability density is equal to the probability density of the random variable being observed.

For correlated random variables, the random variables are coupled and yields nontrivial results. Consider the correlated Bernoulli random variables with probabilities given by Eq. 10.6.1. To evaluate the marginal probability the joint probability is written as ($c = 1/(1 + \rho^2)$)

$$P(x, y) = c\Big(p^{2-x-y}q^{x+y} + (1 + \rho^2)^{xy} - 1\Big) \ ; \ x, y = 0, 1 \qquad (10.7.2)$$

The marginal probability is given by

$$\begin{aligned} P(x) &= \sum_{y=0}^{1} P(x, y) = c\Big(p^{2-x}q^x + p^{1-x}q^{1+x} + (1 + \rho^2)^x - 1\Big) \\ &= cp^{1-x}(q + \rho^2)^x \ ; \ x = 0, 1 \ ; \ p + q = 1 \end{aligned} \qquad (10.7.3)$$

It can seen that $\sum_x P(x) = 1$. Re-writing the marginal probability given in Eq. 10.7.3 in terms of new parameters $\tilde{p}$ yields the marginal Bernoulli random variable

$$\begin{aligned} &P(x) = \tilde{p}^{1-x}\tilde{q}^x \ : \ \tilde{p} + \tilde{q} = 1 \\ &X = \begin{bmatrix} 0 & 1 \\ \tilde{p} & \tilde{q} \end{bmatrix} \ ; \ \tilde{p} = cp \ ; \ \tilde{q} = c(q + \rho^2) = \frac{q + \rho^2}{1 + \rho^2} \end{aligned} \qquad (10.7.4)$$

Hence, the marginal probability is also a Bernoulli random variable. The marginal probability has new probability $\tilde{p}$ that incorporates the effect of the correlation encoded in ρ.

Consider the correlated Gaussian random variables U, V given by Eq. 10.4.6. The marginal density for U is given by

$$P(u) = \int dv P(u, v) = \frac{1}{2\pi\sqrt{1-\rho^2}} \int dv e^{-\frac{1}{2(1-\rho^2)}\left(u^2+v^2-2\rho uv\right)}$$
$$= \frac{1}{2\pi\sqrt{1-\rho^2}} \int dv e^{-\frac{1}{2(1-\rho^2)}\left((1-\rho^2)u^2+(v-\rho)^2\right)} = \frac{1}{\sqrt{2\pi}} e^{-\frac{1}{2}u^2} \tag{10.7.5}$$

Similar to the Bernoulli random variable, the marginal probability density of the Gaussian random variable is also a Gaussian random variable. The marginal probability density for U has no memory of its correlation with random variable V, unlike the Bernoulli case given in Eq. 10.7.4.

10.8 Conditional Expectation Value

For joint probability density, a value of one of the random variables is observed, and one can then ask about the likelihood of the other random variables, given that one of the outcomes is known. In other words, what is the expected value of the other random variables, given the value of one of the variables. Let the probability density for the conditional expectation be denoted by $P(Y|X)$. It follows that

$$P(Y|X) = \frac{P(X, Y)}{P(X)} \quad \Rightarrow \quad P(Y|X)P(X) = P(X|Y)P(Y) \tag{10.8.1}$$

where $P(X, Y)$ is the joint probability density of both X, Y occurring and $P(X)$ is the probability of the occurrence of the given value of X.

For the case of X, Y being independent random variables, $P(X, Y) = P_1(X)P_2(Y)$ and hence

$$P(Y|X) = P_2(Y) \quad : \quad \text{Independent random variables}$$

10.8.1 Bernoulli Random Variables

For the correlated case, consider the Bernoulli random variables. To evaluate the conditional probability, one needs the joint probability, given in Eq. 10.7.2

$$P(x, y) = c\Big(p^{2-x-y}q^{x+y} + (1+\rho^2)^{xy} - 1\Big) \; ; \; x, y = 0, 1 \; ; \; c = \frac{1}{1+\rho^2}$$

Suppose we want the probability of the occurrence of the different values of Y given that $X = 1$. From Eq. 10.7.2 the probability that $X = 1$ is given by

$$P(x=1,y) = \sum_{y=0}^{1} P(1,y) = c\sum_{y=0}^{1}\left(p^{1-y}q^{1+y} + (1+\rho^2)^y - 1\right)$$
$$= c(pq + q^2 + \rho^2) = c(q+\rho^2) = \frac{q+\rho^2}{1+\rho^2}$$

Hence

$$P(Y|1) = \frac{P(1,Y)}{P(1)} = \frac{p^{1-y}q^{1+y} + (1+\rho^2)^y - 1}{q+\rho^2} = \frac{(pq)^{1-y}(q^2+\rho^2)^y}{q+\rho^2}$$

For the general case of $P(Y|X)$, Eq. 10.7.2 yields

$$P(Y|X) = \frac{P(X,Y)}{P(X)} = \frac{p^{2-x-y}q^{x+y} + (1+\rho^2)^{xy} - 1}{p^{2-x}q^x + p^{1-x}q^{1+x} + (1+\rho^2)^x - 1}; \quad \sum_y P(Y|X) = 1$$

10.8.2 Binomial Random Variables

Consider two independent Binomial random variables specified by $X_1 = (N_1, p)$ and $X_2 = (N_2, p)$. We calculate the conditional probability density for X_1, given the condition that $X_1 + X_2 = K$. The general result given in Eq. 10.7.2 yields

$$P(X_1|X_1+X_2=K) = \frac{P(X_1 \cap X_2)}{P(X_1+X_2=K)}$$
$$= \frac{P(X_1)P(X_2)}{P(X_1+X_2=K)} = \frac{P(X_1)P(K-X_1)}{P(X_1+X_2=K)} \tag{10.8.2}$$

where the probability for the Binomial random variable, for $q = 1 - p$, is given by

$$P(X) = P(N,m;p) = \binom{N}{m} p^m q^{N-m}$$

The denominator is given by

$$P(X_1+X_2=K) = \sum_{m=0}^{N_1}\sum_{n=0}^{N_2}\binom{N_1}{m}p^m q^{N_1-m}\binom{N_2}{n}p^n q^{N_2-n}\delta_{m+n-K}$$

Recall from Eq. 2.6.3

$$\delta_{n+m-K} \equiv \begin{cases} 1 \; n+m=K \\ 0 \; n+m \neq K \end{cases} : m, n = 1, \ldots, N$$

$$\Rightarrow \delta_{n+m-K} = \int_0^{2\pi} \frac{d\theta}{2\pi} e^{i(m+n-K)\theta} \tag{10.8.3}$$

Hence, the denominator is given by

$$\begin{aligned} &P(X_1 + X_2 = K) \\ &= \int_0^{2\pi} \frac{d\theta}{2\pi} e^{-iK\theta} \sum_{m=0}^{N_1} \sum_{n=0}^{N_2} e^{i(m+n)\theta} \binom{N_1}{m} p^m q^{N_1-m} \binom{N_2}{n} p^n q^{N_2-n} \\ &= \int_0^{2\pi} \frac{d\theta}{2\pi} e^{-iK\theta} (e^{i\theta} p + q)^{N_1+N_2} \\ &= \int_0^{2\pi} \frac{d\theta}{2\pi} e^{-iK\theta} \sum_{m=0}^{N_1+N_2} \binom{N_1+N_2}{m} e^{im\theta} p^m q^{N_1+N_2-m} \\ &= \binom{N_1+N_2}{K} p^K q^{N_1+N_2-K} \end{aligned} \tag{10.8.4}$$

The conditional probability, from Eqs. 10.8.4 and 10.8.4, is given by

$$\begin{aligned} &P(X_1 = m | X_1 + X_2 = K) = \frac{P(X_1) P(K - X_1)}{P(X_1 + X_2 = K)} \\ &= \binom{N_1}{m} p^m q^{N_2-m} \binom{N_2}{K-m} p^{K-m} q^{N_2-K+m} \Big/ \Big[\binom{N_1+N_2}{K} p^K q^{N_1+N_2-K} \Big] \\ &= \binom{N_1}{m} \binom{N_2}{K-m} \Big/ \binom{N_1+N_2}{K} \end{aligned}$$

which is independent of p, q. Hence, one obtains the result that

$$P(X_1 = m | X_1 + X_2 = K) = \binom{N_1}{m} \binom{N_2}{K-m} \Big/ \binom{N_1+N_2}{K} \tag{10.8.5}$$

Equation 10.8.5 is the probability density for the *hypergeometric* distribution. It is the probability of obtaining m blue balls from a sample of size K drawn from an urn that contains N_1 blue balls and N_2 red balls.

10.8.3 Poisson Random Variables

Consider two independent Poisson random variables N, M with

$$P(n, m) = \frac{\lambda_1^n \lambda_2^m}{n!m!} e^{-\lambda_1 - \lambda_2}$$

What is expectation value of N, given that $N + M = K$? From Eq. 10.8.1,

$$P(N|N + M = K) = \frac{P(N, K - N)}{P(N + M = K)}$$

The conditional expectation value of N is given by

$$E[N|N + M = K] = \sum_{n=0}^{\infty} nP(n, K - n|n + m = K) = \sum_{n=0}^{\infty} n \frac{P(n, K - n)}{P(n + M = K)}$$

Using Eq. 10.8.3 yields that the probability density for $P(n + m = K)$ is given by

$$\begin{aligned}
P(n + m = K) &= e^{-\lambda_1 - \lambda_2} \sum_{m,n=0}^{\infty} \delta_{n+m-K} \frac{\lambda_1^n \lambda_2^m}{n!m!} \\
&= e^{-\lambda_1 - \lambda_2} \int_0^{2\pi} \frac{d\theta}{2\pi} e^{-iK\theta} \sum_{m,n=0}^{\infty} e^{i\theta(m+n)} \frac{\lambda_1^n \lambda_2^m}{n!m!} \\
&= e^{-\lambda_1 - \lambda_2} \int_0^{2\pi} \frac{d\theta}{2\pi} e^{-iK\theta} \exp\{e^{i\theta}(\lambda_1 + \lambda_2)\} \\
&= e^{-\lambda_1 - \lambda_2} \frac{1}{K!} (\lambda_1 + \lambda_2)^K
\end{aligned}$$

Hence

$$P(n, m|N + M = K) = \frac{\lambda_1^n \lambda_2^{K-n}}{n!(K - n)!} \frac{K!}{(\lambda_1 + \lambda_2)^K}$$

Using $(K - n)! = \infty$ for $n > K$ yields the final answer

$$P(n|N + M = K) = \frac{\lambda_1^n \lambda_2^{K-n}}{n!(K - n)!} \frac{K!}{(\lambda_1 + \lambda_2)^K} \quad ; \quad n = 0, 1, \ldots, K$$

This yields the conditional expectation

$$E[N|N+M=K] = \sum_{n=0}^{K} nP(n, K-n|n+m=K)$$
$$= \frac{1}{(\lambda_1+\lambda_2)^K} \sum_{n=0}^{K} n \frac{K!}{n!(K-n)!} \lambda_1^n \lambda_2^{K-n}$$

Hence

$$E[N|N+M=K] = \frac{\lambda_1}{(\lambda_1+\lambda_2)^K} \frac{\partial}{\partial \lambda_1} (\lambda_1+\lambda_2)^K = K \frac{\lambda_1}{\lambda_1+\lambda_2}$$

In contrast, *without* the condition, the expectation value is given by

$$E[N] = \lambda_1$$

10.8.4 Gaussian Random Variables

The probability density for the correlated Gaussian random variables, from Eq. 10.4.6, is given by

$$P(u, v) = \frac{1}{2\pi\sqrt{1-\rho^2}} \exp\left\{-\frac{1}{2(1-\rho^2)}\left(u^2+v^2-2\rho uv\right)\right\}$$

and from Eq. 10.7.5

$$P(u) = \int dv P(u, v) = \frac{1}{\sqrt{2\pi}} \exp\left\{-\frac{1}{2}u^2\right\}$$

Hence, the conditional probability density for V, given that U is fixed at some value, is

$$P(v|u) = \frac{P(u, v)}{\int dv P(u, v)} = \frac{P(u, v)}{P(u)} = \mathcal{N} e^F$$

where

$$F = -\frac{1}{2(1-\rho^2)}\left(u^2+v^2-2\rho uv\right) + \frac{1}{2}u^2 = -\frac{1}{2(1-\rho^2)}(v-\rho u)^2$$

Hence

$$P(v|u) = \frac{1}{\sqrt{2\pi(1-\rho^2)}} \exp\left\{-\frac{1}{2(1-\rho^2)}(v-\rho u)^2\right\}$$

Note that $P(v|u)$ is a probability density function for V since

$$\int dv P(v|u) = 1$$

In contrast, $P(v|u)$ is *not* a probability density function for U since

$$\int du P(v|u) = \frac{1}{\rho} \neq 1$$

The expectation value of V, conditional on U being at some fixed value, is

$$E[V|u] = \int dv v P(v|u) = \rho u$$

Note the unconditioned expectation value of V is given by $E[V] = 0$, as in Eq. 10.6.5. Hence forming the conditional expectation changes the probability of outcomes for the random variable V.

10.9 Problems

1. Let $X_1, \ldots, X_N$ be identical exponential random variables with probability distribution given by
$$P(x) = a \exp\{-ax\} \ ; \ a > 0$$
Define K by
$$K = \sum_{i=1}^{N} X_i$$
Find the mean and variance for K. Based on the central limit theorem, for large N, write-down the Normal distribution that the random variable K satisfies. Directly calculate the generating function
$$Z[j] = E[\exp\{jK\}] \ ; \ j < a$$
for K using the exponential probability distribution and verify the result that, for large N, K has the probability distribution given by the central limit theorem.
2. A fair dice is rolled 420 times. Show that the probability that the sum of the rolls lies between 1400 and 1550 is 0.9670.
3. Consider the limit of the Binomial distribution such that $N \to \infty$ and $q \to 0$ such that $Nq = a$ is fixed. In this limit, show that the Binomial probability distribution converges to the Poisson distribution given by
$$B(k, N) \to P(k) = \frac{a^k}{k!} e^{-a}$$

4. Consider tossing a fair coin 3 times Let A be the event that H appears on the first throw and event B that H appears on the second throw. Find $P(A|B)$.
5. Consider the joint distribution of two Bernoulli random variables

$$P(x, y) = c\Big(p^{2-x-y}q^{x+y} + (1+\rho^2)^{xy} - 1\Big) \; ; \; x, y = 0, 1 \; ; \; c = \frac{1}{1+\rho^2}$$

   Find $E[X]$, $E[Y]$ and the conditional probability $P(X|Y)$.
6. For two correlated Gaussian random variables given in Sect. 10.8.4, find the generating function $Z[j] = E[\exp\{jV\}]$ and hence find the conditional variance

$$E_u[V^2] - E_u^2[V]$$

Chapter 11
Stochastic Processes and Black–Scholes Equation

Abstract The value of the security changes constantly and over very short time scales: FX transactions can be measured in microseconds, or millionths of a second. The actual values of each FX trade does indeed exist, but it becomes near to impossible to describe its motion in complete detail. Financial instruments, in particular, options are modeled based on considering the underlying security to be following a stochastic process. The Black–Scholes equation, one of the cornerstones of option pricing, is studied in great detail.

11.1 Introduction

The study of random variables in Chaps. 8–10 was confined to a few random variables, and of a large number of identical and independent random variables in analyzing the central limit theorem. A stochastic process is an infinite collection of random variables, indexed by a real variable that is usually the time variable. A stochastic process, also called white noise—exemplified by the time-evolution of a security—has an independent (usually taken to be identical but not necessary) random variable for every instant of time. The independent random variable for white noise is the Gaussian random variable.[1]

The continuous index for the Gaussian random variables leads to new features in the calculus of white noise. Taylor expansion leads to new terms that go to zero for ordinary functions, but yield a new term for functions of white noise. A careful analysis of the functions of white noise lead to results encoded in Ito calculus. The analysis of white noise shows that both differentiation and integration have extra features that don't exist for ordinary functions.

One of the main application of white noise in finance is in the study of option theory, and the stochastic process underlying the Black–Scholes option price is studied in detail. The Black–Scholes equation is derived using the concept of hedging, which entails creating a portfolio with a rate of return that is free of white noise. The Binomial option price is shown to converge to the Black–Scholes case, leading to the conclusion that the Binomial tree is a discrete version of a stochastic process.

---

[1] 'Pink' noise and 'brown' noise can also be defined but will not be considered.

B. E. Baaquie, *Mathematical Methods and Quantum Mathematics for Economics and Finance*, https://doi.org/10.1007/978-981-15-6611-0_11

11.2 Stochastic Differential Equation

Consider the time evolution of security, shown for a typical financial instrument in Figure 11.1. To encode the lack of knowledge of the future trajectory of a security, randomness is introduced in the description of the motion of the security: it is assumed that the security is taking many possible trajectories with varying likelihood. A statistical knowledge of the security replaces an exact knowledge of its future evolution.

At every instant t, it is assumed that the value of the security is random, and is being driven by a random variable. A stochastic process is a collection of random variables labeled by a continuous index that can be thought of as time t. Figure 11.2 shows that there is an independent (Gaussian) random variable $R(t)$ for each instant of time t.

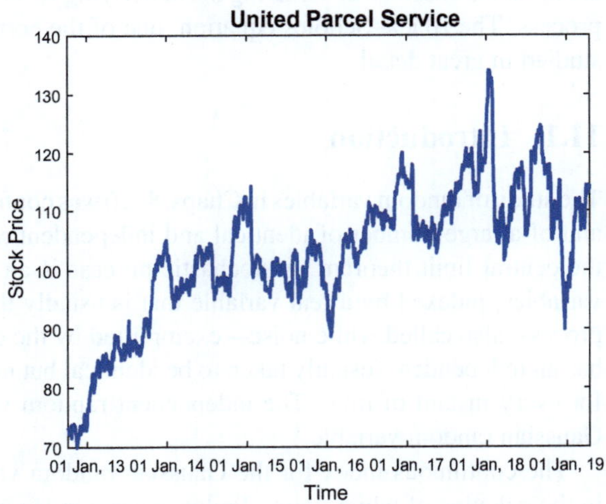

Fig. 11.1 The time evolution of a security's price. Published with permission of © Belal E. Baaquie 2020. All Rights Reserved

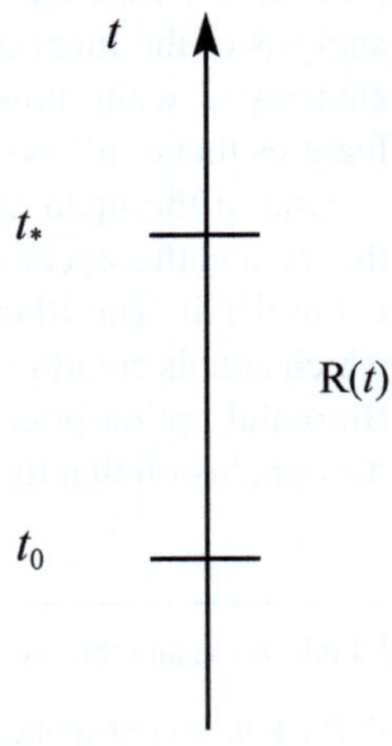

Fig. 11.2 One random variable $R(t)$ for each instant of time. Published with permission of © Belal E. Baaquie 2020. All Rights Reserved

In theoretical finance the stock price $S(t)$ is modeled as a continuous *stochastic process* that defines the following *stochastic differential equation* given by

$$\frac{dS(t)}{dt} = \phi S(t) + \sigma S(t)R(t) \tag{11.2.1}$$

where ϕ is the expected return on the security S, σ is its volatility, and R is a Gaussian white noise with zero mean defined by the following correlations

$$E[R(t)] = 0 \ ; \ E[R(t)R(t')] = \delta(t - t')$$

The Dirac delta function $\delta(t)$ is discussed in Sect. 11.3 below; a more detailed discussion is given in Sect. 13.11. From the properties of the delta function it follows that

$$E[R(t)R(t')] = 0 \ : \ t \neq t'$$

The equation above shows that the correlation of white noise for two instants is zero and hence white noise $R(t)$ is an *independent* (Gaussian) random variable for each t.

σ is a measure of the randomness in the evolution of the stock price. For the special case of $\sigma = 0$, the stock price has deterministic evolution, with its future value given by

$$S(t) = e^{\phi t} S(0) \ : \ \text{deterministic} \tag{11.2.2}$$

It is the randomness arising from white noise $R(t)$ that makes the stock price $S(t)$ a stochastic process.

11.3 Gaussian White Noise and Delta Function

Randomness in Eq. 11.2.1 arises from white noise and the properties of white noise are analyzed. To mathematically describe white noise $R(t)$, one needs to introduce what is called the Dirac delta function $\delta(t)$, which is defined as follow

$$\delta(t) = \begin{cases} 0 & t \neq 0 \\ \infty & t = 0 \end{cases}$$

The delta function $\delta(t)$ has the further property

$$\int dt f(t)\delta(t - \xi) = f(\xi)$$

Intuitively, the $\delta(t)$ is inverse of dt and can be thought of as $1/dt$.

The correlation of Gaussian white noise $R(t)$ is defined by

$$E[R(t)] = 0 \ ; \ \ E[R(t)R(t')] = \delta(t - t') \tag{11.3.1}$$

The probability distribution for white noise can be explicitly written by discretizing $t = n\epsilon$ with $n = 1, 2 \ldots$ and $R(t) \to R_n$. For discrete time, the Dirac delta function has the representation

$$\delta(t - t') \to \frac{1}{\epsilon}\delta_{n-n'}$$

From Eq. 11.3.1

$$E[R_n] = 0 \ ; \ \ E[R_n R_{n'}] = \frac{1}{\epsilon}\delta_{n-n'} \ \ \Rightarrow R_n^2 = \frac{1}{\epsilon} + \text{random}$$

Note that R_n is a Gaussian random variable with zero mean and $1/\sqrt{\epsilon}$ variance, and is given by $N(0, 1/\sqrt{\epsilon})$. The probability distribution function of white noise is hence given by

$$P(R_n) = \sqrt{\frac{\epsilon}{2\pi}} \exp\left\{-\frac{\epsilon}{2}R_n^2\right\} \tag{11.3.2}$$

The following result is essential in deriving the rules of Ito calculus

$$R_n^2 = \frac{1}{\epsilon} + \ \ \text{random terms of } 0(1) \tag{11.3.3}$$

Although R_n is a random variable having no fixed value, what $R_n^2 \simeq 1/\epsilon$ means is that in any correlation function where R_n^2 appears, one can replace it by $1/\epsilon$.

Noteworthy 11.1: White Noise

To prove result stated in Eq. (11.3.3), it will be shown that, to leading order in $1/\epsilon$, the generating function of R_n^2 can be derived by considering R_n^2 to be deterministic. All the moments of R_n^2 can be determined from its generating function, namely

$$E\big[(R_n^2)^k\big] = \frac{d^k}{dt^k} E\big[e^{tR_n^2}\big]\Big|_{t=0}$$

Note one needs to evaluate the generating function $E\big[e^{tR_n^2}\big]$ only in the limit of $t \to 0$. Hence, for ϵ small but fixed and for $t \to 0$

$$\lim_{t\to 0} E\big[e^{tR_n^2}\big] = \int_{-\infty}^{+\infty} dR_n e^{tR_n^2}\sqrt{\frac{\epsilon}{2\pi}}e^{-\frac{\epsilon}{2}R_n^2}$$
$$= \frac{1}{\sqrt{1-\frac{2t}{\epsilon}}} \sim \exp\left(\frac{t}{\epsilon}\right) + O(1) \tag{11.3.4}$$

The results given in Eq. 11.3.4 is valid for $t \to 0$, which is the limit required for computing $E\big[(R_n^2)^k\big]$. The limit of $\epsilon \to 0$ is taken only after taking the limit of $t \to 0$.

The probability distribution function $P(R_n^2)$ for R_n^2, which gives the above generating function, is given by

$$P(R_n^2) = \delta\left(R_n^2 - \frac{1}{\epsilon}\right)$$
$$\Rightarrow E\big[e^{tR_n^2}\big] = \int_{-\infty}^{+\infty} dR_n e^{tR_n^2}\delta\left(R_n^2 - \frac{1}{\epsilon}\right) = \exp\left(\frac{t}{\epsilon}\right)$$

In other words, the quantity R_n^2, to leading order in ϵ is not a random variable, but is instead fixed at the value of $1/\epsilon$. This singularity of white noise is the basis of Ito calculus.

11.3.1 Integrals of White Noise

Consider the following integral of white noise

$$I = \int_t^T dt' R(t') \sim \epsilon \sum_{n=0}^{M} R_n \ ; \ M = \left[\frac{T-t}{\epsilon}\right]$$

where ϵ is an infinitesimal. For Gaussian white noise

$$R_n = N\left(0, \frac{1}{\sqrt{\epsilon}}\right) \quad \Rightarrow \epsilon R_n = N(0, \sqrt{\epsilon})$$

The integral of white noise is a sum of normal random variables and hence, from Eq.(8.4.6) and above, is also a Gaussian random variable given by

$$I \sim N(0, \sqrt{\epsilon M}) \to N(0, \sqrt{T-t}) \tag{11.3.5}$$

In general, for

$$Z = \int_t^T dt' a(t') R(t') \Rightarrow Z = N(0, \sigma^2) \; ; \; \sigma^2 = \int_t^T dt' a^2(t')$$

Due to the (Dirac delta function) singularity in white noise $R(t)$ there is an ambiguity in discretizing white noise for nonlinear integrands. Consider the following integral of white noise

$$I = \int_0^T dt R(t) f(R(t); t) = \int_0^T dW(t) f(W(t); t) \; ; \; dW(t) = R(t)dt$$

Using

$$dW(t) = R(t)dt \Rightarrow W(t) = \int_0^t d\xi R(\xi)$$

The correlation function of $W(t)$, for $t > t'$, is given by

$$E[W(t)W(t')] = \int_0^t d\xi \int_0^{t'} d\xi' E[R(\xi)R(\xi')] = \int_0^t d\xi \int_0^{t'} d\xi' \delta(\xi - \xi')$$
$$\Rightarrow E[W(t)W(t')] = \int_0^{t'} d\xi = t' \; ; \; E[W^2(t)] = t \tag{11.3.6}$$

where Eq 11.3.6 has been obtained using Eq. 11.6.12. Hence, in general

$$E[W(t)W(t')] = t'\theta(t - t') + t\theta(t' - t) = \min(t, t') \tag{11.3.7}$$

and where the theta function is defined in Eq. 13.8.23.

Let $t_n = n\epsilon$ and with $\tau = N\epsilon$; $W(t) \to W_n$; There are two different and *inequivalent* ways of discretizing the integral, namely the Ito and Strantovich discretizations, which are given by the following

$$E[W_n W_{n'}] = \big[\min(n, n')\big]\epsilon \; ; \; E[W_n^2] = n\epsilon \; ; \; dW(t) = W_{n+1} - W_n$$
$$I_I = \sum_{n=0}^{N-1} f(W_n; t_n)(W_{n+1} - W_n) \tag{11.3.8}$$
$$I_S = \sum_{n=0}^{N-1} f\left(\frac{W_{n+1} + W_n}{2}; \frac{t_{n+1} + t_n}{2}\right)(W_{n+1} - W_n)$$

The expectation value of the stochastic integrals for the case of $f = W(t)$ is used to illustrate the ambiguity in the discretization of the stochastic integral. From Eq. 11.3.8

$$E[I_I] = \sum_{n=0}^{N-1} E[W_n(W_{n+1} - W_n)] = \sum_{n=0}^{N-1}(n\epsilon - n\epsilon) = 0 \tag{11.3.9}$$

In contrast

$$E[I_S] = \frac{1}{2}\sum_{n=0}^{N-1} E[(W_{n+1} + W_n)(W_{n+1} - W_n)]$$
$$= \frac{1}{2}\sum_{n=0}^{N-1} E[W_{n+1}^2 - W_n^2] = \frac{1}{2}\sum_{n=0}^{N-1}\{(n+1)\epsilon - n\epsilon\} = \frac{T}{2} \qquad (11.3.10)$$

Hence, Eqs. 11.3.9 and 11.3.10 show the two methods for discretizing the stochastic integral are inequivalent.

Both the Ito and Strantovich discretizations are used extensively. It can be shown that Ito discretization is the correct one for mathematical finance since it is consistent with the requirement the market is free from arbitrage opportunities; the Strantovich discretization requires a *future* value of $f(W)$ and which in turn is in conflict with the requirement of no arbitrage. The Strantovich discretization is used in the study of the Fokker–Planck equation.

11.4 Ito Calculus

The application of stochastic calculus to finance is widely used, and a brief discussion is given to relate Ito calculus to the properties of white noise. Due to the singular nature of white noise $R(t)$, functions of white noise such as the security $S(t)$ and the option $C(t)$ have new features. In particular, the infinitesimal behavior of such functions, as seen in their Taylor expansions, acquire new terms.

Let f be some arbitrary function of white noise $R(t)$. From the definition of a derivative

$$\frac{df}{dt} = \lim_{\epsilon\to 0}\frac{1}{\epsilon}\{f(t+\epsilon, S(t+\epsilon)) - f(t, S(t))\}$$
$$= \lim_{\epsilon\to 0}\frac{1}{\epsilon}\left\{f\left(t+\epsilon, S(t) + \epsilon\frac{dS}{dt} + O(\epsilon^2)\right) - f(t, S(t))\right\} \qquad (11.4.1)$$

Taylor expansion yields

$$f\left(t+\epsilon, S(t) + \epsilon\frac{dS}{dt} + O(\epsilon^2)\right) = f(t, S(t)) + \epsilon\frac{\partial f}{\partial t} + \epsilon\frac{dS}{dt}\frac{\partial f}{\partial S} + \frac{\epsilon^2}{2}\left[\frac{dS}{dt}\right]^2\frac{\partial^2 f}{\partial S^2} + \cdots$$

The neglected terms in Taylors expansion are of order $\epsilon^{1/2}$ for smooth functions, and go to zero as $\epsilon \to 0$. Hence

$$\frac{df}{dt} = \frac{\partial f}{\partial t} + \frac{dS}{dt}\frac{\partial f}{\partial S} + \frac{\epsilon}{2}\left[\frac{dS}{dt}\right]^2\frac{\partial^2 f}{\partial S^2} + 0(\epsilon^{1/2}) \qquad (11.4.2)$$

However, due to the singular nature of white noise

$$\left[\frac{dS}{dt}\right]^2 = \sigma^2 S^2 R^2 + 0(1) = \frac{1}{\epsilon}\sigma^2 S^2 + 0(1) \qquad (11.4.3)$$

Hence, from Eqs. (11.2.1), (11.4.2) and (11.4.3), for $\epsilon \to 0$

$$\begin{aligned}\frac{df}{dt} &= \frac{\partial f}{\partial t} + \frac{\partial f}{\partial S}\frac{dS}{dt} + \frac{\sigma^2}{2}S^2\frac{\partial^2 f}{\partial S^2} \\ &= \frac{\partial f}{\partial t} + \phi S\frac{\partial f}{\partial S} + \sigma S\frac{\partial f}{\partial S}R + \frac{1}{2}\sigma^2 S^2\frac{\partial^2 f}{\partial S^2}\end{aligned} \qquad (11.4.4)$$

Suppose $g(t, R(t)) \equiv g_t$ is another function of the white noise $R(t)$. The abbreviated notation $\delta g_t \equiv g_{t+\epsilon} - g_t$ yields

$$\begin{aligned}\frac{d(fg)}{dt} &= \lim_{\epsilon\to 0}\frac{1}{\epsilon}[f_{t+\epsilon}g_{t+\epsilon} - f_t g_t] \\ &= \lim_{\epsilon\to 0}\frac{1}{\epsilon}[\delta f_t g_t + f_t\delta g_t + \delta f_t\delta g_t]\end{aligned}$$

Usually the last term $\delta f_t\delta g_t$ is of order ϵ^2 and goes to zero. However, due to the singular nature of white noise

$$\frac{d(fg)}{dt} = \frac{df}{dt}g + f\frac{dg}{dt} + \frac{df}{\sqrt{dt}}\frac{dg}{\sqrt{dt}} \quad : \text{ Ito's Chain Rule} \qquad (11.4.5)$$

Equation (11.4.5), in terms of infinitesimals, is given by

$$d(fg) = dfg + fdg + dfdg$$

Since Eq. (11.4.4) is of central importance for the theory of security derivatives a derivation is given based on Ito-calculus. Rewrite Eq. (11.2.1) in terms differentials as

$$dS = \phi S dt + \sigma S dz \ ; \ dz = R dt \qquad (11.4.6)$$

where dz is a Wiener process. Since from Eq. (11.3.3)

$$R^2(t) = \frac{1}{dt}$$

we have

$$(dz)^2 = R_t^2(dt)^2 = dt + 0(dt^{3/2}) \ \Rightarrow \ (dS)^2 = \sigma^2 S^2 dz^2 + 0(dt^{3/2}) = \sigma^2 S^2 dt$$

From the equations for dS and $(dS)^2$ given above and $S(t+dt) = S + dS$ yields

$$\begin{aligned} df &= f(t+dt, S+dS) - f(t, S) \\ &= \frac{\partial f}{\partial t}dt + \frac{\partial f}{\partial S}dS + \frac{1}{2}\frac{\partial^2 f}{\partial S^2}(dS)^2 + 0(dt^{3/2}) \\ &= \left(\frac{\partial f}{\partial t} + \phi S\frac{\partial f}{\partial S} + \frac{1}{2}\sigma^2 S^2\frac{\partial^2 f}{\partial S^2}\right)dt + \sigma S\frac{\partial f}{\partial S}dz \end{aligned}$$

and Eq. (11.4.4) is recovered using $dz/dt = R$.

11.5 Lognormal Stock Price

Recall from Eq. 11.2.1, that the stock price can be modeled as a stochastic process driven by white noise $R(t)$ as follows

$$\frac{dS(t)}{dt} = \phi S(t) + \sigma S(t)R(t) \tag{11.5.1}$$

where ϕ is the expected return on the security S, σ is its volatility, and R is a Gaussian white noise with zero mean.

The stochastic differential equation Eq. (11.5.1) is integrated using the results from stochastic calculus. Consider the change of variable

$$S(t) = S(0)e^{x(t)} \Rightarrow x(t) = \ln[S(t)/S(0)]$$

Hence, using Eq. (11.4.2)

$$\begin{aligned} \frac{dx}{dt} &= \frac{1}{\epsilon}[x(t+\epsilon) - x(t)] = \frac{1}{\epsilon}[\ln S(t+\epsilon) - \ln S(t)] \\ &= \frac{1}{\epsilon}\left[\ln\left(S + \epsilon\frac{dS}{dt}\right) - \ln S(t)\right] = \frac{1}{\epsilon}\left[\ln S\left(1 + \epsilon S^{-1}\frac{dS}{dt}\right) - \ln S\right] \\ &= \frac{1}{\epsilon}\ln\left(1 + \epsilon S^{-1}\frac{dS}{dt}\right) = S^{-1}\frac{dS}{dt} - \frac{1}{2}\epsilon S^{-2}\left(\frac{dS}{dt}\right)^2 + O(\epsilon) \\ \Rightarrow \frac{dx}{dt} &= \phi + \sigma R(t) - \frac{\sigma^2}{2} \end{aligned}$$

Hence

$$\frac{dx}{dt} = \phi - \frac{\sigma^2}{2} + \sigma R(t) \tag{11.5.2}$$

$$\Rightarrow x(T) = x(t) + \left(\phi - \frac{\sigma^2}{2}\right)(T-t) + \sigma\int_t^T dt' R(t') \tag{11.5.3}$$

To obtain the generating function for $x(T)$, discretize time t into $t \to n\epsilon$ with $N = (T-t)/\epsilon$, and which yields

$$x(T) = x(t) + \left(\phi - \frac{\sigma^2}{2}\right)(T-t) + \sigma\epsilon\sum_{n=0}^{N} R_n$$

To compute the generating function, from Eq. 13.11.1, consider

$$E\left[\exp\left\{j\sigma\epsilon\sum_{n=0}^{N} R_n\right\}\right] = \int DRe^{j\sigma\epsilon\sum_{n=0}^{N} R_n} P[R]$$
$$= \prod_{n=0}^{N}\left(\sqrt{\frac{\epsilon}{2\pi}}\int_{-\infty}^{+\infty} dR_n \exp\left\{j\sigma\epsilon R_n - \frac{\epsilon}{2}R_n^2\right\}\right)$$
$$= \exp\left\{\frac{1}{2}j^2\sigma^2\epsilon\sum_{n=0}^{N}\right\} = \exp\left\{\frac{1}{2}j^2\sigma^2(T-t)\right\}$$

Hence, the generating function for $x(T)$ is given by

$$E[\exp\{jx(T)\}] = \exp\left\{j(x(t) + \left(\phi - \frac{\sigma^2}{2}\right)(T-t) + \frac{1}{2}j^2\sigma^2(T-t)\right\} \quad (11.5.4)$$

The moment generating function for $x(T)$ is that of a Gaussian random variable with mean $(x(t) + (\phi - \frac{\sigma^2}{2})(T-t)$ and variance $\sigma\sqrt{T-t}$. Hence, for the initial value of $x(t)$ evolving to $x(T)$ is given by

$$x(T) = x(t) + \left(\phi - \frac{\sigma^2}{2}\right)(T-t) + (\sigma\sqrt{T-t})Z \ : \ Z = N(0,1)$$
$$\Rightarrow P(x(T); x(t)) = \frac{1}{\sqrt{2\pi\sigma^2(T-t)}} \times$$
$$\exp\left\{-\frac{1}{2\sigma^2(T-t)}\left[x(T) - x(t) - \left(\phi - \frac{\sigma^2}{2}\right)(T-t)\right]^2\right\} \quad (11.5.5)$$

and

$$S(T) = \exp\{x(T)\} \ ; \ S(t) = \exp\{x(t)\} \quad (11.5.6)$$

The result obtained in Eq. 11.5.6 can be obtained directly by noting that the random variable $\int_0^t dt' R(t')$ is a sum of normal random variables and is shown in Eq. 11.3.5 to be equal to a normal $N(0, \sqrt{T-t})$ random variable.

The stock price evolves randomly from its initial value of $S(t)$ at time t to a whole range of possible values $S(T)$ at time T. Since the random variable $x(T)$ is a normal (Gaussian) random variable, the security $S(t)$ is a lognormal random variable.

11.5.1 Geometric Mean of Stock Price

The probability distribution of the (path dependent) geometric mean of the stock price can be exactly evaluated. For $\tau = T - t$ and $m = x(t) + (\phi - \frac{\sigma^2}{2})\tau$, Eq. (11.5.3) yields,

$$\begin{aligned} S_{\text{GM}} &= e^G \\ G &\equiv \frac{1}{\tau}\int_t^T dt' x(t') = m + \frac{\sigma}{\tau}\int_t^T dt' \int_t^{t'} dt'' R(t'') \\ &= m + \frac{\sigma}{\tau}\int_t^T dt'(T - t')R(t') \end{aligned}$$

From Eq. (11.3.5) the integral of white noise is a Gaussian random variable, which is completely specified by its means and variance. Hence, using $E[G] = m$ and $E[R(t)R(t')] = \delta(t - t')$ yields

$$\begin{aligned} E[(G-m)^2] &= \left(\frac{\sigma}{\tau}\right)^2 \int_t^T dt'(T-t') \int_t^T dt''(T-t'')E[R(t')R(t'')] \\ &= \left(\frac{\sigma}{\tau}\right)^2 \int_t^T dt'(T-t') \int_t^T dt''(T-t'')\delta(t'-t'') \\ &= \left(\frac{\sigma}{\tau}\right)^2 \int_t^T dt'(T-t')^2 = \frac{\sigma^2\tau}{3} \end{aligned}$$

Hence

$$G = N\left(m, \sigma\sqrt{\frac{\tau}{3}}\right) \tag{11.5.7}$$

The geometric mean of the stock price is lognormal with the same mean as the stock price, but with its volatility being one-third of the stock price's volatility.

11.6 Linear Langevin Equation

Consider the price of a security that undergoes a random evolution due to its constant interaction with the market, with buyers and sellers influencing the stock price to move in opposite directions. One can think of the stock price evolving randomly due to its interaction with a background medium, which is the market composed of many traders.

Since one is modeling the effect of the market on the stock price, one needs to incorporate the 'collisions' that the security has with the market. Let the security be given by $S = e^x$; collisions will change the velocity of the security, with velocity v

being defined by

$$S(t) = e^{x(t)} \ ; \ v = \frac{dx}{dt}$$

One can use an equation, proposed historically by Langevin, to model the effect of market that is responsible for the random motion of the security. A *random force* is introduced that changes the velocity of the security, and yields a *stochastic differential equation* given by

$$m\frac{dv}{dt} + \gamma v + \Phi(v) = \sqrt{2A}R(t) \tag{11.6.1}$$

The random force $\sqrt{2A}R(t)$ is given by Gaussian white noise $R(t)$.

The introduction of a random force $\sqrt{2A}R(t)$ makes the trajectories of the security random. For every choice of a random sample drawn for white noise $R(t)$, the trajectory of the stock price changes.

Consider the case with no external potential, namely $\Phi(v) = \phi$, with ϕ taken to be a constant. One can rewrite the Langevin equation given in Eq. 11.6.1 as follows

$$\frac{dv}{dt} + \frac{\gamma v}{m} + \phi = \frac{\sqrt{2A}}{m}R \tag{11.6.2}$$

From a mathematical point of view, Eq. 11.6.2 is a generalization of the stochastic differential equation given by Eq. 11.2.1 since Eq. 11.6.2 has a new term $\gamma v/m$. Redefine v by $v - m\phi/\gamma$. This yields

$$\frac{dv}{dt} + \frac{\gamma v}{m} = \frac{\sqrt{2A}}{m}R \tag{11.6.3}$$

R is white noise, with probability density given in Eq. 13.11.2 by the following

$$P[R] = \frac{1}{Z}\exp\left(-\frac{1}{2}\int R^2(t)dt\right)$$

Integrating the first order stochastic differential given in Eq. 11.6.3 yields for $v(t)$

$$\frac{dv}{dt} + \frac{\gamma}{m}v = \frac{\sqrt{2A}}{m}R = e^{-\frac{\gamma}{m}t}\frac{d}{dt}\left(e^{\frac{\gamma}{m}tv}\right)$$

Hence

$$v = v_0 e^{-\gamma t/m} + \sqrt{2A}\frac{e^{-\frac{\gamma t}{m}}}{m}\int_0^t dt' e^{\frac{\gamma}{m}t'}R(t') \tag{11.6.4}$$

Define

$$u = v - v_0 e^{-\gamma t/m} \tag{11.6.5}$$

such that $E[u(t)] = 0$; hence

$$\begin{aligned} E[u^2(t)] &= \frac{\sqrt{2A}}{m^2} \int_0^t d\eta d\tau e^{-\frac{\gamma}{m}(t-\eta)} e^{-\frac{\gamma}{m}(t-\tau)} E[R(\eta)R(\tau)] \\ &= \frac{2A}{m^2} \int_0^t d\eta e^{-2\frac{\gamma}{m}(t-\eta)} = \frac{A}{m\gamma}(1 - e^{-2\gamma t/m}) \\ \Rightarrow \quad \lim_{t\to\infty} E[u^2(t)] &= E[v^2(t)] = \frac{A}{m\gamma} \end{aligned}$$

The velocity $v(t)$ of the logarithm of a security is a random variable, and hence is determined by a probability distribution function $P[v; t]$. Solving a stochastic differential equation, unlike an ordinary differential equation, consists of determining the probability distribution $P[v; t; v_0]$ for the occurrence of different values for $v(t)$—given the initial velocity v_0. For a stochastic process, the solution consists of specifying the *likelihood* of different trajectories, rather than a unique trajectory.

Recall from Eq. 11.6.4 that the Langevin equation yields

$$v = v_0 \exp\left\{-\frac{\gamma t}{m}\right\} + \int_0^t dt' \rho(t, t') R(t') \tag{11.6.6}$$

The linear sum of Gaussian random variables is also a Gaussian random variable. Hence all we need to do is to determine the mean and variance of $v(t)$ to determine $P[v; t]$. Recall from Eq. 11.6.5 that $u = v - v_0 \exp\{-\gamma t/m\}$ and hence

$$\begin{aligned} E[v] &= v_0 \exp\left\{-\frac{\gamma t}{m}\right\} \\ \Rightarrow \sigma_v^2 &= \sigma_u^2 = E[u^2(t)] = \frac{A}{m\gamma}\left(1 - \exp\{-2\frac{\gamma t}{m}\}\right) \end{aligned} \tag{11.6.7}$$

Hence

$$\begin{aligned} P[v; t; v_0] &= \frac{1}{\sqrt{2\pi\sigma_v^2}} \exp\left\{-\frac{1}{2\sigma_v^2}\left(v - v_0 \exp\left\{-\frac{\gamma t}{m}\right\}\right)^2\right\} \\ &= \sqrt{\frac{m\gamma}{2\pi A\left(1 - \exp\left\{-2\frac{\gamma t}{m}\right\}\right)}} \exp\left\{-\frac{m\gamma}{2A} \cdot \frac{\left(v - v_0 \exp\left\{-\frac{\gamma t}{m}\right\}\right)^2}{\left(1 - \exp\left\{-\frac{\gamma t}{m}\right\}\right)}\right\} \end{aligned} \tag{11.6.8}$$

Note, as expected

$$\lim_{t\to 0} P[v;t;v_0] \to \delta(v-v_0)$$

$$\lim_{t\to\infty} P[v;t;v_0] \to \sqrt{\frac{m\gamma}{2\pi A}}\exp\left\{-\frac{m\gamma}{2A}v^2\right\} \qquad (11.6.9)$$

11.6.1 Security's Random Paths

From Eq. 11.6.3, for $\beta = \gamma/m$, the Langevin equation is given by

$$\frac{dv}{dt} = -\beta v + \frac{\sqrt{2A}}{m}R$$

The random path $x(t) = \ln S(t)$, as determined by the Langevin equation, is given by the following

$$\frac{dx(t)}{dt} = v(t) \ ; \ x(t) = x_0 + \int_0^t d\xi v(\xi) \qquad (11.6.10)$$

$$\Rightarrow E[x(t)] = x_0 + \int_0^t d\xi E[v(\xi)]$$

$$= x_0 + v_0 \int_0^t d\xi e^{-\beta\xi} = x_0 + \frac{v_0}{\beta}(1-e^{-\beta t}) \qquad (11.6.11)$$

Let $x_0 = -v_0/\beta$ then

$$E[x(t)] = -\frac{v_0}{\beta}e^{-\beta t} \to 0 \quad \text{as} \quad t\to\infty$$

Let

$$y(t) = x(t) + \frac{v_0}{\beta}e^{-\beta t} \ ; \ y(0) = 0$$

The change from $x(t)$ to $y(t)$ is made to remove un-necessary terms arising from the boundary conditions on $x(t)$ and $v(t)$. Defining $\dot{y} = dy/dt$ yields

$$\dot{y} = \frac{dy}{dt} = v - e^{-\beta t}v_0 \ \Rightarrow \ \dot{y} = \sqrt{2A}\frac{e^{-\beta t}}{m}\int_0^t d\xi e^{\beta\xi}R(\xi)$$

The unequal time correlation function is given by

$$E[\dot{y}(t)\dot{y}(t')] = \frac{2A}{m^2}\int_0^t d\xi \int_0^{t'} d\eta e^{-\beta(t-\xi)}e^{-\beta(t'-\eta)}E[R(\xi)R(\eta)]$$

$$= e^{-\beta(t+t')}\frac{2A}{m^2}\int_0^t d\xi \int_0^{t'} d\eta e^{\beta\xi}e^{\beta\eta}\delta(\xi-\eta)$$

Note

$$\int_0^{t'} d\eta\delta(\xi-\eta)=\begin{cases}0 & \xi>t'\\ 1 & \xi\leqslant t'\end{cases}=\theta(t'-\xi) \tag{11.6.12}$$

Let $t \geq t'$; then

$$E[\dot{y}(t)\dot{y}(t')]=\frac{A}{m^2}e^{-\beta(t+t')}\int_0^{t'}d\xi e^{2\beta\xi}=\frac{A}{2\beta m^2}e^{-\beta(t+t')}[-1+e^{2\beta t'}]$$

Due to time ordering expressed in $t \geq t'$, the correlator is given by

$$\begin{aligned}E[y(t)y(t')]&=\int_0^t d\tau\int_0^{t'}d\tau' E[\dot{y}(t)\dot{y}(t')]\\&=\frac{A}{2\beta m^2}\int_0^t d\tau\int_0^{t'}d\tau'\left[-e^{-\beta(\tau+\tau')}+e^{-\beta|\tau-\tau'|}\right]\end{aligned}$$

For $t=t'$

$$E[y^2(t)]=\frac{A}{2\beta m^2}\int_0^t d\tau d\tau'\left[-e^{-\beta(\tau+\tau')}+e^{-\beta|\tau-\tau'|}\right] \tag{11.6.13}$$

Carrying out the integrating in the equation above yields

$$E[y^2(t)]=\frac{2A}{\beta^2}t+\frac{Am}{\beta^3}\left[-3+4e^{-\frac{\gamma}{m}t}-e^{-\frac{2\gamma}{m}t}\right] \tag{11.6.14}$$

In the limit of $t\to\infty$ we have that $y(t)\to x(t)$ and yields

$$\sqrt{E[x^2(t)]}=\sqrt{\frac{2A}{\gamma^2}}\cdot\sqrt{t} \tag{11.6.15}$$

where $\sqrt{t}$ is the characteristic signal of a stochastic process undergoing a Gaussian random walk.

For a uniforming moving security $x\sim vt$, as given in Eq. 11.5.1. In contrast, for a randomly evolving security $x\sim\sqrt{t}$, which is much slower than a freely moving security—due to the frequent random interactions that the security has with the market.

It is an empirical question whether the Langevin equation correctly models the market's effect on the security, and needs to be studied using data from a liquid market, such as the FX market.

11.7 Black–Scholes Equation; Hedged Portfolio

The price of the call option was derived in Chap. 9 using the binomial distribution. Time was discretized and the finite N step call option was derived exactly.

For most applications, the option is defined for continuous time and the Black–Scholes equation determines the option price for continuous time. The option price has to primarily obey the condition of no arbitrage.

Black and Scholes made the fundamental observation that if one could *perfectly hedge* an option, then one could price it as well. The reason being that a perfectly hedged portfolio has no uncertainty, and hence has a *risk-free rate* of return—given by the spot rate r. In order to form a perfectly hedged portfolio, the time evolution of the option has to be analyzed.

To form a hedged portfolio, the instantaneously change of the portfolio must be *independent* of the white noise R. Such a portfolio is perfectly hedged since it has no randomness.

Consider the portfolio

$$\Pi = C - \frac{\partial C}{\partial S} S \tag{11.7.1}$$

Π is a portfolio in which an investor holds an option C and short sells $\partial C/\partial S$ amount of security S. Hence, from Eqs. (11.4.4) and (11.2.1)[2]

$$\begin{aligned}
\frac{d\Pi}{dt} &= \frac{dC}{dt} - \frac{\partial C}{\partial S}\frac{dS}{dt} \\
&= \frac{\partial C}{\partial t} + \frac{dS}{dt}\frac{\partial C}{\partial S} + \frac{1}{2}\sigma^2 S^2 \frac{\partial^2 C}{\partial S^2} - \frac{\partial C}{\partial S}\frac{dS}{dt} \\
&= \frac{\partial C}{\partial t} + \frac{1}{2}\sigma^2 S^2 \frac{\partial^2 C}{\partial S^2}
\end{aligned} \tag{11.7.2}$$

At time t, since $S(t)$ is known, the price $C(t, S(t))$ is deterministic; $\partial C/\partial t$ is also deterministic since it does not entail changing the value of the stochastic stock price $S(t)$. Hence—from Eq. 11.7.2 above—the change in the value of the portfolio Π is *deterministic*.

The random term coming from dS/dt has been removed due to the choice of the portfolio, and $d\Pi/dt$ is consequently free from risk that comes from the stochastic nature of the security. This technique of canceling the random fluctuations of one security (in this case of C) by another security (in this case S) is a key feature of *hedging*.

Since the rate of (change) return on Π is deterministic, it must equal the risk-free return given by the short-term risk-free interest rate r, since otherwise one could arbitrage. Hence, based on the absence of arbitrage opportunities the price of an option has, from Eqs. 11.7.1 and 11.7.2, the following evolution

[2]The term $(\partial^2 C/\partial S\partial t)S$ can be shown to be negligible.

$$\frac{d\Pi}{dt} = r\Pi$$

$$\frac{\partial C}{\partial t} + \frac{1}{2}\sigma^2 S^2 \frac{\partial^2 C}{\partial S^2} = r\Big[C - \frac{\partial C}{\partial S} S\Big] \tag{11.7.3}$$

which yields the famous Black–Scholes equation

$$\frac{\partial C}{\partial t} + rS\frac{\partial C}{\partial S} + \frac{1}{2}\sigma^2 S^2 \frac{\partial^2 C}{\partial S^2} = rC \tag{11.7.4}$$

The parameter ϕ of Eq. (11.2.1) has dropped out of Eq. (11.7.4) showing that a risk-neutral portfolio Π is independent of the investor's expectation as reflected in ϕ; or equivalently, the pricing of the security **derivative** is based on a risk-free process that is independent of the investor's risk preferences.

Consider the change of variable

$$S = e^x \ ; \ -\infty \leq x \leq \infty$$

We have the following

$$\frac{\partial}{\partial S} = S^{-1}\frac{\partial}{\partial x} \ ; \ \frac{\partial^2}{\partial S^2} = -S^{-2}\frac{\partial}{\partial x} + S^{-2}\frac{\partial^2}{\partial x^2}$$

From Eq. 11.7.5 the Black–Scholes equation is given by

$$\frac{\partial C}{\partial t} = -\frac{1}{2}\sigma^2 \frac{\partial^2 C}{\partial x^2} + \left(\frac{1}{2}\sigma^2 - r\right)\frac{\partial C}{\partial x} + rC \tag{11.7.5}$$

The Black–Scholes framework for the pricing of options hinges on the concept of a risk-less, hedged portfolio—something that can never be achieved in practice. Many generalizations of the Black–Scholes have been made.

11.8 Assumptions in the Black–Scholes

The following assumptions were made in the derivation of the Black–Scholes equation.

- The portfolio satisfies the no arbitrage condition.
- To form the hedged portfolio Π the stock has to be infinitely divisible, and that short selling of the stock is possible.[3]
- The spot rate r is constant [this can be generalized to a stochastic spot rate].

[3] Short selling entails taking a share on loan and returning it when the option matures—by buying it from the market.

- The portfolio Π can be re-balanced continuously.
- There is no transaction cost.

The conditions above are not fully met in the financial markets. In particular, transactions costs are significant. In spite of this, the market uses the Black–Scholes option pricing as the industry standard, and forms the basis for the pricing of more complex options.

11.9 Martingale: Black–Scholes Model

There are several ways of obtaining the solution of the Black–Scholes equation. One can directly solve the Black–Scholes equation as a partial differential equation; a solution will be given in Sect. 14.2 using techniques of functional analysis and of Fourier transform.

The hedging of an instrument in effect implies that for the hedged portfolio, there exists a risk-neutral evolution of the security $S(t)$—also called a risk-free evolution. Since this principle is used extensively in pricing of options, the price of a European call option is solved using the technique of risk-neutral valuation, as was the case for the Binomial pricing model in Chap. 9.

The concept of a martingale measure is a mathematical formulation of a risk-free evolution. An elegant and simple solution is obtained by using the principle of risk-neutral valuation. For

$$S(t) = e^{x} \;\; ; \;\; S(T) = e^{x'}$$

recall from Eq. 11.5.5, with $\tau = T - t$, the probability distribution of the security is given by

$$P(x, x') = \frac{1}{\sqrt{2\pi\sigma^2\tau}} \exp\Big\{ -\frac{1}{2\sigma^2\tau}\big[x' - x - \tau(\phi - \frac{\sigma^2}{2})\big]^2\Big\} \qquad (11.9.1)$$

Equation 11.9.1 yields the following *conditional probability* distribution

$$P(x, x') = \text{Prob of x}'\text{occuring, given x} \;\; ; \;\; \int_{-\infty}^{+\infty} dx' P(x, x') = 1$$

The martingale condition states the following: *given* that the stock value today is $S(t)$, the discounted expectation value of the future value of the stock price $S(T)$ is equal to its present day value $S(t)$. Hence, the risk-neutral martingale probability distribution satisfies the martingale condition

$$S(t) = e^{x} = E[e^{-r\tau} S(T)|S(t)] = e^{-r\tau} \int_{-\infty}^{+\infty} dx' e^{x'} P(x, x') \qquad (11.9.2)$$

From Eq. 11.9.1, the right hand side of Eq. 11.9.2 yields

$$\int_{-\infty}^{+\infty} dx' e^{x'} P(x, x') = \frac{e^{x+\left(\phi-\frac{\sigma^2}{2}\right)\tau}}{\sqrt{2\pi\sigma^2\tau}} \int_{-\infty}^{+\infty} dx' \exp\left\{-\frac{1}{2\sigma^2\tau}x'^2 + x'\right\}$$
$$= \exp\{x + \phi\tau\} \tag{11.9.3}$$

Hence, from Eqs. 11.9.2 and 11.9.3

$$1 = e^{-r\tau} e^{\phi\tau} \quad \Rightarrow \quad \phi = r \quad : \quad \text{risk-neutral condition}$$

Once the martingale condition is satisfied, one can use the conditional probability distribution function of the stock price to calculate the payoff's present discounted value.

11.10 Black–Scholes Option Price

The martingale probability distribution is given by ($\tau = T - t$)

$$P_m(x, x') = \frac{1}{\sqrt{2\pi\sigma^2\tau}} \exp\left\{-\frac{1}{2\sigma^2\tau}\left[x' - x - \tau\left(r - \frac{\sigma^2}{2}\right)\right]^2\right\} \tag{11.10.1}$$

ϕ in Eq. 11.2.1, which is the expected return on the security S, has been replaced by the risk-neutral spot rate r to obtain the probability distribution function $P_m(x, x')$, in accordance with the principle of risk-neutral valuation. The stock price evolves randomly from its given value of $S(t)$ at time t to a whole range of possible values $S(T)$ at time T.

The principle of risk-neutral valuation implies that the present value of the European call option is the expected final value $E[\max(S - K, 0)]$ determined by the risk-neutral probability distribution $P_m(x, x')$, discounted by the risk-free interest rate. In short, the present day price of the option is the discounted value of the payoff function, using the martingale measure. Hence

$$C(x; K; T - t) = e^{-r\tau} E[\max(S(T) - K, 0)]$$
$$= e^{-r\tau} \int_{-\infty}^{+\infty} dx' (e^{x'} - K)_+ P_m(x, x') \tag{11.10.2}$$

The interpretation of Eq. 11.10.2 is the following: the call option price is the average value of the discounted payoff—taken over all the final values of the stock price $e^{x'}$—weighted by the *conditional probability* $P_m(x, x')$ that the value of the security at the maturity of the option is $e^{x'}$, given that its initial value is e^x.

An explicit derivation is given of the call option price has been given in Sect. 5.6.1, and is briefly summarized for completeness. Let $x_0 = x + \tau(r - \frac{\sigma^2}{2})$. The option price, from Eqs. 11.10.1 and 11.10.2, is given by

$$\begin{aligned}
C(\widetilde{x}; K; T-t) &= e^{-r\tau} \int_{-\infty}^{+\infty} \frac{dx'}{\sqrt{2\pi\tau\sigma^2}} (e^{x'} - K)_+ e^{-\frac{1}{2\tau\sigma^2}(x'-x_0)^2} \\
&= e^{-r\tau} \int_{-\infty}^{+\infty} \frac{dx'}{\sqrt{2\pi\tau\sigma^2}} (e^{x'+x_0} - K)_+ e^{-\frac{1}{2\tau\sigma^2}x'^2} \\
&= e^{-r\tau} \int_{\ln K - x_0}^{+\infty} \frac{dx'}{\sqrt{2\pi\tau\sigma^2}} (e^{x'+x_0} - K) e^{-\frac{1}{2\tau\sigma^2}x'^2} \\
&= S \int_{\ln K - x_0}^{+\infty} \frac{dx'}{\sqrt{2\pi\tau\sigma^2}} e^{-\frac{1}{2\tau\sigma^2}[x'+\tau\sigma^2]^2} - e^{-r\tau} K N(d_2)
\end{aligned}$$

Hence, the Black–Scholes option price is given by

$$C(x; K; T-t) = SN(d_1) - e^{-r\tau} K N(d_2) \tag{11.10.3}$$

The cumulative distribution for the normal random variable $N(x)$, from Eq.(8.4.5), is defined by

$$N(x) = \frac{1}{\sqrt{2\pi}} \int_{-\infty}^{x} e^{-\frac{1}{2}z^2} dz \ ; \ \tau = T - t \ ; \ S = e^x \tag{11.10.4}$$

and

$$d_1 = \frac{1}{\sigma\sqrt{\tau}} \left[\ln\left(\frac{S}{K}\right) + \tau \left(r + \frac{\sigma^2}{2} \right) \right] \ ; \ d_2 = d_1 - \sigma\sqrt{\tau} \tag{11.10.5}$$

The European put option is defined by

$$\begin{aligned}
P(x; K; T-t) &= e^{-r\tau} E[\max(K - S, 0)] \\
&= e^{-r\tau} \int_{-\infty}^{+\infty} dx' (K - e^{x'})_+ P_m(x, x')
\end{aligned} \tag{11.10.6}$$

A derivation similar to the one for the call option given in Eq. 11.10.3 yields the result

$$P(x; K; T-t) = e^{-r\tau} K N(-d_2) - SN(-d_1) \tag{11.10.7}$$

11.10.1 Put-Call Parity

Put-call parity in Black-Schloes follows from the identity

$$[S-K]_+ - [K-S]_+ = S-K \tag{11.10.8}$$

Taking the discounted expectation value of the left hand side of Eq. 11.10.8, using the risk neutral probability distribution given in Eq. 11.10.1, yields

$$e^{-r\tau}E\Big([S-K]_+ - [K-S]_+\Big) = C(S) - P(S) \tag{11.10.9}$$

where C, P are the call and put option respectively. Taking the expectation value of the right hand side of Eq. 11.10.8, and using the martingale condition for S given in Eq. 11.9.2, yields

$$e^{-r\tau}E[S-K] = S - e^{-r\tau}K \tag{11.10.10}$$

Hence, collecting the results above yields the put-call parity relation

$$C(S) - P(S) = S - e^{-r\tau}K \tag{11.10.11}$$

Put-call parity is a result that is model independent, the result of the absence of arbitrage opportunities for the pricing of options. Two requirements of the model is that it obeys the martingale condition and that the probability for the future values of the underlying be given by a conditional probability. Both these conditions are met by the Binomial model, as in Sect. 9.7, and for the Black–Scholes.

There are models for option pricing that do not fulfill these conditions [11], but nevertheless yield option prices that are empirically within the arbitrage bound for option pricing. The arbitrage bound for option pricing refers to the put-call parity being violated within the bounds set by transaction costs and the interest rates for long and short positions.

11.11 Black–Scholes Limit of the Binomial Model

The Binomial option pricing model is defined for a security that evolves on a finite size time lattice with spacing given by ϵ; the option matures after taking N steps, with the remaining time to maturity given by $\tau = N\epsilon$. The continuum limit is taken with $N \to \infty$; $\epsilon = \tau/N \to 0$—holding τ fixed.

The price of the call option in the Binomial model is given, from Eq. 9.5.1, by the following

$$C(S) = e^{-rN\epsilon} \sum_{k=0}^{N} \binom{N}{k} p^{N-k} q^k \Big[u^{N-k} d^k S - K \Big]_+ \tag{11.11.1}$$

$$S_{Final} = u^{N-k} d^k S \ ; \ p = \frac{e^{r\epsilon} - d}{u - d} \ ; \ p + q = 1$$

The Binomial probability distribution function is given by

$$B(p, N) = \binom{N}{k} p^{N-k} q^k$$

The mean and variance of the Binomial random variable is given by

$$m = E[k] = Nq \ ; \ v^2 = E[k^2] - (E[k])^2 = Npq$$

The law of large number yields—as derived in Eq. 10.5.12—that, as $N \to \infty$, the Binomial probability converges to the normal random variable. Hence[4]

$$\lim_{N \to \infty} B(p, N) \to N(m, v) = \frac{1}{\sqrt{2\pi Npq}} \exp\left\{ -\frac{1}{2Npq} (k - Nq)^2 \right\} \tag{11.11.2}$$

and from Eq. 11.11.8 the option price is given by

$$C(S) = \frac{e^{-rN\epsilon}}{\sqrt{2\pi Npq}} \sum_{k=0}^{N} \exp\left\{ -\frac{1}{2Npq} (k - Nq)^2 \right\} \Big[u^{N-k} d^k S - K \Big]_+ \tag{11.11.3}$$

To obtain the Black–Scholes limit, introduce parameter σ by choosing

$$u = e^{\sigma\sqrt{\epsilon}} \ ; \ d = e^{-\sigma\sqrt{\epsilon}} \tag{11.11.4}$$

Hence, to leading order in ϵ

$$p \simeq \frac{1}{2} + \frac{r\epsilon - \frac{1}{2}\sigma^2\epsilon}{2\sigma\sqrt{\epsilon}} \ ; \ q \simeq \frac{1}{2} - \frac{r\epsilon - \frac{1}{2}\sigma^2\epsilon}{2\sigma\sqrt{\epsilon}} \tag{11.11.5}$$

Let the initial value of the security be given by $S = e^{x_0}$. The final (random) value of the security is given by

$$u^{N-k} d^k S = \exp\{(N - 2k)\sigma\sqrt{\epsilon} + x_0\} \tag{11.11.6}$$

Let N be an even number and define

---

[4] A detailed proof is given in Sect. 10.5.15.

$$k = \ell + \frac{N}{2} \ : \ 0 \le k \le N \ \Rightarrow \ -\frac{N}{2} \le \ell \le \frac{N}{2} \tag{11.11.7}$$

Hence

$$u^{N-k} d^k S = \exp\{-2\sigma\sqrt{\epsilon}\ell + x_0\} \tag{11.11.8}$$

and

$$\begin{aligned} k - Nq &= \ell + \frac{N}{2} - N\Big[\frac{1}{2} - \frac{r\epsilon - \frac{1}{2}\sigma^2\epsilon}{2\sigma\sqrt{\epsilon}}\Big] \\ &= \frac{1}{2\sigma\sqrt{\epsilon}}\Big[2\sigma\sqrt{\epsilon}\ell + N\epsilon(r - \frac{1}{2}\sigma^2)\Big] \end{aligned} \tag{11.11.9}$$

due to a nontrivial cancellation. Note that from Eqs. 11.11.3 and 11.11.9

$$\begin{aligned} Npq(2\sigma\sqrt{\epsilon})^2 &= N\frac{1}{4}\left[1 - \frac{\epsilon}{\sigma^2}\left(r - \frac{1}{2}\sigma^2\right)^2\right](4\sigma^2\epsilon) \\ &= N\sigma^2\epsilon + O(\epsilon^2) = \sigma^2\tau \end{aligned} \tag{11.11.10}$$

Hence, from Eqs. 11.11.3, 11.11.8 and 11.11.9

$$\begin{aligned} C(x_0) &= e^{-r\tau}\frac{(2\sigma\sqrt{\epsilon})}{\sqrt{2\pi Npq(2\sigma\sqrt{\epsilon})^2}} \\ &\times \sum_{\ell=-\frac{N}{2}}^{\frac{N}{2}} \exp\Big\{-\frac{1}{2Npq(2\sigma\sqrt{\epsilon})^2}\Big(2\sigma\sqrt{\epsilon}\ell + N\epsilon(r - \frac{1}{2}\sigma^2)\Big)^2\Big\} \\ &\times \Big[\exp\{-2\sigma\sqrt{\epsilon}\ell + x_0\} - K\Big]_+ \end{aligned} \tag{11.11.11}$$

Defining a continuous variable x, and dx from Eq. 5.1.2, yields

$$x = 2\sigma\sqrt{\epsilon}\ell = \frac{2\sigma\epsilon\ell}{\sqrt{\epsilon}} \ \Rightarrow \ dx = x(\ell+1) - x(\ell) = 2\sigma\sqrt{\epsilon}$$

Recall that

$$\frac{N}{2} \le \ell \le \frac{N}{2}$$

Hence, as $\epsilon \to 0$, the limits on ℓ yield the following limits for x

$$\frac{\sigma\epsilon N}{\sqrt{\epsilon}} = \frac{\sigma\tau}{\sqrt{\epsilon}} \ \Rightarrow \ -\frac{\sigma\tau}{\sqrt{\epsilon}} \le x \le \frac{\sigma\tau}{\sqrt{\epsilon}} \ \Rightarrow \ -\infty \le x \le +\infty$$

Note that

$$\lim_{N\to\infty}(2\sigma\sqrt{\epsilon})\sum_{\ell=-\frac{N}{2}}^{\frac{N}{2}} \to \int_{-\infty}^{+\infty} dx$$

Equations 11.11.10 and 11.11.11 yield

$$\begin{aligned} C(x_0) &= \frac{e^{-r\tau}}{\sqrt{2\pi\sigma^2\tau}}\int_{-\infty}^{+\infty} dx\left\{-\frac{1}{2\sigma^2\tau}\left(x+\tau(r-\frac{1}{2}\sigma^2)\right)^2\right\}\left[e^{-x+x_0}-K\right]_+ \\ &= \frac{e^{-r\tau}}{\sqrt{2\pi\sigma^2\tau}}\int_{-\infty}^{+\infty} dx\exp\left\{-\frac{1}{2\sigma^2\tau}\left(x-x_0-\tau(r-\frac{1}{2}\sigma^2)\right)^2\right\}\left[e^{x}-K\right]_+ \end{aligned} \quad (11.11.12)$$

The continuum limit of the Binomial option pricing yields the Black–Scholes pricing equation. The parameter σ introduced in Eq. 11.11.4, can be seen from the Black–Scholes analysis to be the volatility of the stock price S. In particular, the result obtained in Eq. 11.11.12 is the result obtained earlier for the Black–Scholes equation and given in Eq. 11.10.1.

The continuum limit of the Binomial option price to the Black–Scholes case shows that the Binomial tree is a discrete version of a stochastic process: the discrete paths from the initial security to the value at the payoff function are a discrete version of the random paths of a stochastic process. The Binomial tree converges to paths generated by white noise due to the central limit theorem.

11.12 Problems

1. Let a stock price be given by

$$dS = 0.7Sdt + 0.11Sdz$$

Suppose $S(0) = \$100$. Find $S(t)$.

2. Consider two geometric Brownian motion for stock prices S_1, S_2 given by

$$S_1(t) = S(0)e^{\mu_1 t+(\sigma\sqrt{t})Z_1} \;\; ; \;\; S(t) = S_2(0)e^{\mu_2 t+(2\sigma\sqrt{t})Z_2}$$

where $Z_1, Z_2 = N(0, 1)$ are Gaussian random variables.

- Show that

$$P\Big(S_1(t) > (S_2(t))^2\Big) = N\left(\frac{(\mu_1-\mu_2)\sqrt{t}}{\sigma}\right)$$

- $$E[S_1(t) - \{S_2(t)\}^2] = S(0)[e^{\mu_1 t + \frac{\sigma^2 t}{2}} - e^{\mu_2 t + 2\sigma^2 t}]$$

- The variance of the ratio is given by

$$E\left[\left\{\frac{S_1(t)}{S_2(t)}\right\}^2\right] - E^2\left[\frac{S_1(t)}{S_2(t)}\right] = e^{2(\mu_1 - \mu_2)t + \sigma^2 t}\left[e^{\sigma^2 t} - 1\right]$$

3. Show that the function of stock price S given by

$$V(S,t) = a(t)S(t) + b(t)$$

satisfies the Black–Scholes equation if and only if

$$a(t) = a \ ; \ b(t) = be^{rt}$$

where a, b are constants.

4. Consider a stock whose price evolves as in the Black–Scholes equation with drift of $\mu = 10\%$/year and volatility of $\sigma = 40\%$/year. The current price of the stock is $S = 16$€.

   - The risk free rate is $r = 4\%$/year. Find the price of the call and put option for strike price $K = 18$ Euros and maturity of $T = 1$ year.
   - Verify that the prices obey Put-Call parity.

5. Define the Δ for an option F by $\Delta = \partial F/\partial S$. Show that for the call and put options

$$\Delta_C = \frac{\partial C}{\partial S} = N(d_+) \ ; \ \Delta_P = \frac{\partial P}{\partial S} = N(d_+) - 1$$

6. An option on a security matures in future time T; the volatility of the security is σ and the risk-free interest rate is r. For stock price given by $S(t)$, with $t < T$, what is the *price of the options* having the following payoff functions?

   - Bull spread. Let $K_1 < K_2$. Buy a call option with strike price K_2 and sell a call option with strike price K_1.
   - Bear spread. Buy a call option with strike price K_1 and sell a call option with strike price K_2.
   - Straddle. Sell a call and put option with same maturity and same strike price.
   - Strangle. Let $K_1 < K_2$. Sell a put option with strike price K_1 and sell a call option with strike price K_2.

Chapter 12
Stochastic Processes and Merton Equation

Abstract Corporate finance, among other major branches of finance, deals with raising of capital by either selling equity or issuing debt. A major advance in the analysis of corporate debt was made by Merton, in which he modeled all contingent claims on a firm as a branch of option theory. In particular, Merton's model of risky coupon bonds are options on the valuation of a firm, with the model taking into account firms that do and don't default.

12.1 Introduction

Merton, in his seminal paper, proposed a model of risky coupon bonds defined to be an option on the issuing firm's valuation [29, 36]. Merton's model is the industry standard for pricing corporate bonds, and has also been applied for calculating the default probability, as well as the recovery rates, of defaulting corporate bonds.

Merton's theory of corporate debt is based on the mathematics of stochastic processes and is an important application of the techniques developed in the study of stochastic processes.

In this Chapter various aspects of Merton's formulation are discussed, with focus on zero coupon bonds due to their simplicity. Important features of a zero coupon corporate bond, including the spread of the rate of return of a risky bond above the risk free rate, the likelihood of default and the rate of recovery are studied in detail using Merton's formulation.

Merton's theory of the firm, including the pricing of coupon bonds, is discussed later using quantum mathematics in Chaps. 14 and 15.

12.2 Firm's Stochastic Differential Equation

The value of a firm V is equal to the sum of its total debt $\mathcal{B}$ plus value of equity E minus cash. Hence

$$V = \mathcal{B} + E - \text{Cash}$$

B. E. Baaquie, *Mathematical Methods and Quantum Mathematics for Economics and Finance*, https://doi.org/10.1007/978-981-15-6611-0_12

Since cash does not form a big component of V for most listed firms, it will be ignored and the following will be taken to always hold

$$V = \mathcal{B} + E$$

In Merton's model of the firm, its valuation follows a stochastic process given by the following differential equation (R is white noise)

$$\frac{1}{V}\frac{dV}{dt} = \alpha + \upsilon + \sigma R \ : \ \upsilon = \upsilon(V) \tag{12.2.1}$$

α is the instantaneous rate of return per unit time of V. $\upsilon = \upsilon(V)$ is a function of V and represents the exogenous change per unit time in the value of the firm. A *positive* υ stands for net new in-flow of capital from investors of the amount $V\upsilon$; a *negative* υ is due to payouts of the amount $V\upsilon$ to its shareholders or due to paying off other liabilities, such as coupon payments to bond holders and so on. In particular, for the case of coupon payments by the firm to bondholders

$$\text{Rate of coupon payment} = |\upsilon(V)|V \ : \ \$\ \text{value/time} \tag{12.2.2}$$

12.3 Contingent Claims on Firm

Let C be any contingent claim on V and hence $C = C(V)$ is a function of the risky asset V; it is *assumed* by Merton that C also follows a stochastic process given by the stochastic differential equation

$$\frac{1}{C}\frac{dC}{dt} = \beta + q + \gamma Q \ : \ q = q(C) \tag{12.3.1}$$

where Q is another Gaussian white noise. β is the instantaneous rate of return per unit time of C. For $q > 0$, $-Cq$ is the coupon payment, per unit time, and which reduces the value of the contingent claim. Using Eqs. 11.4.4 and 12.2.1 yield

$$\begin{aligned}\frac{dC}{dt} &= \frac{\partial C}{\partial t} + \frac{\sigma^2}{2}V^2\frac{\partial^2 C}{\partial V^2} + \frac{\partial C}{\partial V}\frac{dV}{dt} \\ &= \frac{\partial C}{\partial t} + \frac{1}{2}\sigma^2 V^2 \frac{\partial^2 C}{\partial V^2} + (\alpha + \upsilon)V\frac{\partial C}{\partial V} + \sigma V \frac{\partial C}{\partial V}R \end{aligned} \tag{12.3.2}$$

Comparing Eqs. 12.3.1 and 12.3.2 yields

$$\beta + q = \frac{1}{C}\left[\frac{\partial C}{\partial t} + \frac{1}{2}\sigma^2 V^2 \frac{\partial^2 C}{\partial V^2} + (\alpha V + \upsilon)\frac{\partial C}{\partial V}\right] \tag{12.3.3}$$

$$\gamma C = \sigma V \frac{\partial C}{\partial V} \ ; \ Q = R \tag{12.3.4}$$

Equation 12.3.4 relates the volatility of the contingent claim to the valuation and volatility of the issuing firm, and is useful in computing the probability of default, and rate of recovery, of risky bonds.

12.4 No Arbitrage Portfolio

To ensure the contingent claim C is free from arbitrage opportunities, suppose that C, V are used to create a three-security portfolio, the third security being cash with a risk free return given by r. The net investment in the portfolio is zero; and hence

$$p_1 + p_2 + p_3 = 0$$

Form a portfolio of three securities, consisting of p_1 invested in V, p_2 invested in C and $p_3 = -p_1 - p_2$ invested in cash.

The inflow of investment means that the intrinsic change in the value of the firm, given by V, grows at a slower pace because of the increase caused exogenously by the inflow of υ. Hence, in the presence of υ the instantaneous rate of return for the asset is

$$\frac{1}{V}\frac{dV}{dt} - \upsilon$$

Similarly, due to the inflow of investment q, the intrinsic change in the value of the contingent claim C grows at a slower pace and hence the instantaneous rate of return on claim C is

$$\frac{1}{C}\frac{dC}{dt} - q$$

The instantaneous rate of return on p, from Eqs. 12.3.1 and 12.2.1, is given by $(p_3 = -p_1 - p_2)$

$$\begin{aligned}\frac{dp}{dt} &= p_1\left[\frac{1}{V}\frac{dV}{dt} - \upsilon\right] + p_2\left[\frac{1}{C}\frac{dC}{dt} - q\right] + p_3 r \\ &= p_1(\alpha + \sigma R) + p_1(\beta + \gamma R) + p_3 r \\ &= p_1(\alpha - r) + p_2(\beta - r) + (p_1\sigma + p_2\gamma)R \end{aligned} \tag{12.4.1}$$

Since the portfolio has an aggregate value of zero, to be free from arbitrage the rate of return must be deterministic and zero. Hence, to have $dp/dt = 0$, the coefficients of the stochastic term as well as the deterministic term must *both* be zero. Hence, from the no arbitrage condition requires

$$\frac{dp}{dt} = 0 \Rightarrow p_1(\alpha - r) + p_2(\beta - r) = 0 \text{ and } (p_1\sigma + p_2\gamma) = 0$$

$$\Rightarrow \qquad -\frac{p_1}{p_2} = \frac{\beta - r}{\alpha - r} = \frac{\gamma}{\sigma} = \frac{V}{C}\frac{\partial C}{\partial V}$$

Replacing the value of β using Eq. 12.3.3 yields

$$(\alpha - r)\frac{V}{C}\frac{\partial C}{\partial V} = \frac{1}{C}\left[\frac{\partial C}{\partial t} + \frac{1}{2}\sigma^2 V^2 \frac{\partial^2 C}{\partial V^2} + (\alpha V + \upsilon)\frac{\partial C}{\partial V}\right] - q - r$$

Simplifying the expression above yields Merton's equation for the contingent claim given by

$$\frac{\partial C}{\partial t} = -\frac{1}{2}\sigma^2 V^2 \frac{\partial^2 C}{\partial V^2} - (r + \upsilon)V\frac{\partial C}{\partial V} + (r + q)C \tag{12.4.2}$$

α and β are the risky rates of returns on the valuation and contingent claim, respectively. They have canceled out from Eq. 12.4.2 reflecting the fact that the contingent claim is free from arbitrage opportunities—not depending on the risky returns of the underlying security or its contingent claim.

The cancellation of α and β are shows that two investors with different expectations for the company's future but who agree with the firm's risk profile, as encoded in and the firm's value and volatility, will agree on the value of the contingent claim C.

12.5 Merton Equation

Consider the change of variable

$$V = e^x \Rightarrow \frac{\partial}{\partial V} = V^{-1}\frac{\partial}{\partial x}\ ; \ \frac{\partial^2}{\partial V^2} = -V^{-2}\frac{\partial}{\partial x} + V^{-2}\frac{\partial^2}{\partial x^2}$$

Hence, Eq. 12.4.2 yields Merton's equation

$$\frac{\partial C}{\partial t} = -\frac{1}{2}\sigma^2\frac{\partial^2 C}{\partial x^2} + \left(\frac{1}{2}\sigma^2 - r - \upsilon\right)\frac{\partial C}{\partial x} + (r + q)C \tag{12.5.1}$$

No assumptions have been made on the nature of the firm V or about the contingent claim C. It is the boundary conditions on Eq. 12.4.2 that differentiates the various types of securities. In particular, the boundary conditions imposed on C determines whether C is a debt instrument issued by the firm or is an equity.

A special case of Eq. 12.4.2, with $v, q \neq 0$ will be studied later using quantum mathematics.

From Eq. 12.2.1, and similar to the derivation of Eq. 11.5.3, the change of variable yields the following for the firm's stochastic differential equation

$$V = e^x \Rightarrow \frac{dx}{dt} = \alpha - \frac{1}{2}\sigma^2 + v(x) + \sigma R \qquad (12.5.2)$$

If one sets $v = 0 = q$, Eq. 12.4.2 yields the Black–Scholes equation with constant volatility σ^2 and risk-free spot interest rate r

$$\frac{\partial C}{\partial t} = -\frac{1}{2}\sigma^2 V^2 \frac{\partial^2 C}{\partial V^2} - rV\frac{\partial C}{\partial V} + rC \ : \ \text{Black–Scholes eq.} \qquad (12.5.3)$$

12.6 Risky Corporate Coupon Bond

Consider the case of the contingent claim $C(t, x)$ on the firm to be a *coupon bond*, denoted by $\mathcal{B}(t, x)$, that has a principal of L and matures at time $T > t$. The corporate coupon bond is a risky bond, carrying a risk of default of both the coupons and of the principal. The general contingent claim $C(t, x)$ is relabeled as $\mathcal{B}(t, x)$ for greater clarity. Hence

$$C(t, x) \rightarrow \mathcal{B}(t, x)$$

and the equation for the coupon bond is written as

$$\frac{\partial \mathcal{B}(t, x)}{\partial t} = -\frac{1}{2}\sigma^2 \frac{\partial^2 \mathcal{B}(t, x)}{\partial x^2} + \left(\frac{1}{2}\sigma^2 - r - v\right)\frac{\partial \mathcal{B}(t, x)}{\partial x} + (r + q)\mathcal{B}(t, x) \qquad (12.6.1)$$

According to Merton the coupon bond is an option on the company's valuation. Suppose the company does not default on the coupon payments and the bond matures at future time T. Then on maturity, the investor will receive the face value of the bond L; however, if on maturity the value of the firm is less than L, the investor will then take charge of the firm and, by liquidating the firm, will receive the residual value V.

Hence, the payoff of the option is the value of the bond on maturity and is given by

$$\mathcal{B}(T, x) = \min\{V, L\} = \begin{cases} L; & V > L \\ V; & V < L \end{cases}$$

and shown in the Fig. 12.1.

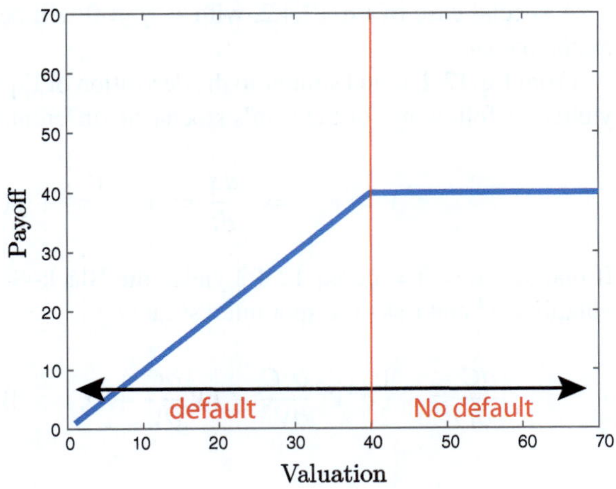

Fig. 12.1 The payoff function for a coupon bond, with the default barrier at $V = L$. Published with permission of © Belal E. Baaquie 2020. All Rights Reserved

The firm issuing the corporate coupon bond $\mathcal{B}(t, x)$ continuously pays coupons at the rate of $-\upsilon V$ to the holder of the coupon bond $\mathcal{B}(t)$. Hence, due to the inflow of new cash, the change in the value of the coupon bond in Eq. 12.6.1 is given by

$$q\mathcal{B} \rightarrow \upsilon V \tag{12.6.2}$$

The price of the coupon bond $\mathcal{B}(t, x)$ obeys a *special case* of Eq. 12.6.1 and is given by the following

$$\frac{\partial \mathcal{B}}{\partial t} = -\frac{1}{2}\sigma^2 \frac{\partial^2 \mathcal{B}}{\partial x^2} + \left(\frac{1}{2}\sigma^2 - r - \upsilon\right)\frac{\partial \mathcal{B}}{\partial x} + r\mathcal{B} + \upsilon V \tag{12.6.3}$$

with final value, shown in Fig. 12.1, given by

$$\mathcal{B}(T, x) = \min\{V, L\}$$

One can use the Modigliani–Miller theorem—which states that the value of a firm is the sum of its assets and liabilities—to simplify Eq. 12.6.3. Consider the change of variables from $\mathcal{B}(t, x) \rightarrow E(t, x)$ given by

$$\mathcal{B}(t, x) + E(t, x) = V = e^{x(t)} V_0 \tag{12.6.4}$$

where $E(t, x)$ is the *equity* of the firm. The Modigliani–Miller theorem is required for interpreting $E(t, x)$ as equity—and is used for empirically studying bond valuation.

The change of variables in Eq. 12.6.3, given in Eq. 12.6.4 and discussed in Noteworthy 12.1, yields the following

$$\frac{\partial E}{\partial t} = -\frac{1}{2}\sigma^2 \frac{\partial^2 E}{\partial x^2} + \left(\frac{1}{2}\sigma^2 - r - \upsilon\right)\frac{\partial E}{\partial x} + rE \tag{12.6.5}$$

The final value of $E(t, x)$, from Eq. 12.6.4, is given by[1]

$$E(T, x) = V - \mathcal{B}(T, x) = V - \min\{V, L\} = [V - L]_+ \tag{12.6.6}$$

Equation 12.6.5 is for an equity paying out coupons (dividends) at the rate of υV. The cancellation of the term υV in Eq. 12.6.3 makes Eq. 12.6.5 into a *homogeneous* linear partial differential equation and amenable to analysis using the Hamiltonian of quantum mathematics, as discussed in Chap. 14.

Noteworthy 12.1: Change of Variables

The change of variables given in Eq. 12.6.4

$$\mathcal{B}(t, x) + E(t, x) = V = e^{x(t)} V_0$$

yields the following

$$\frac{\partial \mathcal{B}}{\partial t} = -\frac{\partial E}{\partial t} \; ; \; \frac{\partial \mathcal{B}}{\partial x} = -\frac{\partial E}{\partial x} + V \; ; \; \frac{\partial^2 \mathcal{B}}{\partial x^2} = -\frac{\partial^2 E}{\partial x^2} + V$$

Hence, using the results above in Eq. 12.6.3 yields

$$-\frac{\partial E}{\partial t} = \frac{1}{2}\sigma^2 \left[\frac{\partial^2 E}{\partial x^2} - V\right] - \left(\frac{1}{2}\sigma^2 - r - \upsilon\right)\left[\frac{\partial E}{\partial x} - V\right] - r[E - V] + \upsilon V$$

All the terms that depend on V cancel, and equation above yields Eq. 12.6.5

$$\frac{\partial E}{\partial t} = -\frac{1}{2}\sigma^2 \frac{\partial^2 E}{\partial x^2} + \left(\frac{1}{2}\sigma^2 - r - \upsilon\right)\frac{\partial E}{\partial x} + rE$$

12.7 Zero Coupon Corporate Bond

Consider the case of the contingent claim $\mathcal{B}(T, x)$ on the firm to be a *zero coupon corporate bond*, denoted by $Z(t, x)$, that has a principal of L and matures at time $T > t$. Then $\mathcal{B}_0(T, x) = Z(t, x)L$. Note that a zero coupon corporate bond, unlike a risk-less US Treasury Bond, is a risky bond with a likelihood of default.

---

[1] The final value of equity at maturity, given by $E(T, x)$, is similar to the payoff of a call option.

For zero coupons $v = 0 = q$. From Eq. 12.6.5, the equation for the equity of a firm paying zero coupons is given by

$$\frac{\partial E}{\partial t} = -\frac{1}{2}\sigma^2\frac{\partial^2 E}{\partial x^2} + \left(\frac{1}{2}\sigma^2 - r\right)\frac{\partial E}{\partial x} + rE \tag{12.7.1}$$

The final condition on $E(t, x)$, from Eq. 12.6.4 is given

$$E(T, x) = V - \mathcal{B}_0(T, x) = V - \min\{V, L\} = [V - L]_+ \tag{12.7.2}$$

Equation 12.7.1 is the Black–Scholes equation given in Eq. 11.7.5, with the security for the Black–Scholes equation being replaced by the valuation of the firm V and the strike price of the option being replaced by the face value of the zero coupon bond L. The payoff for $E(t, x)$ given in Eq. 12.7.2 is the payoff for a call option. To emphasize that $E(t, x)$ is a Black-Schloes option price, the following notation is adopted

$$E(t, x) \equiv \mathcal{C}_{BS}(t, x) \tag{12.7.3}$$

Put-Call parity for the zero coupon bond, from Eq. 11.10.11, is given by

$$\mathcal{C}_{BS}(t, x) - \mathcal{P}_{BS}(t, x) = V - e^{-r(T-t)}L$$

Hence, using Put-Call parity yields for the risky zero coupon bond

$$\mathcal{B}_0(t, x) = V - \mathcal{C}_{BS}(t, x) = e^{-r(T-t)}L - \mathcal{P}_{BS}(t, x) \tag{12.7.4}$$

This is Merton's well known result, that for the special case of a risky zero coupon bond, the price of the bond is equivalent to holding a risk-free zero coupon bond and buying a put option; this result shows that the price of risk carried by a firm's zero coupon bond is given precisely by the price of a put option on the valuation. A generalized version of the identity given in Eq. 12.7.4 can be shown to hold for the case of a risky coupon bond [14]. In summary

$$\text{price of risky bond} = \text{price of risk-free bond} - \text{price of put option}$$

Figure 12.2 shows that the payoff function for the coupon bond given in Fig. 12.1, which for the zero coupon corporate bond is equivalent to the payoff of a risk-free bond and of a put option.

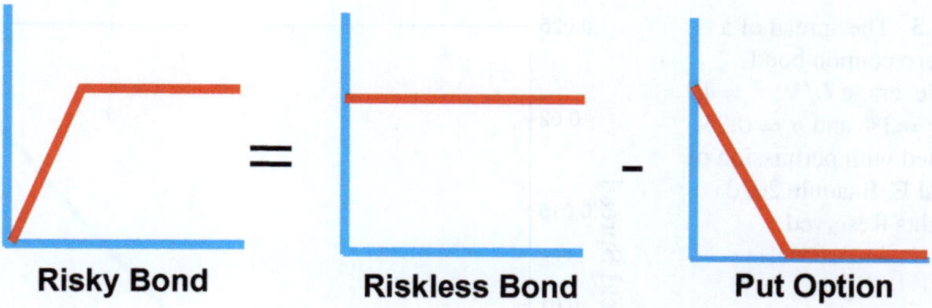

Fig. 12.2 Zero coupon bond payoff function equivalent to a portfolio of a treasury bond and a put option. Published with permission of © Belal E. Baaquie 2020. All Rights Reserved

12.8 Zero Coupon Bond Yield Spread

A measure of the risk in holding a corporate coupon bond is the spread of yield (rate of return) of a risky corporate coupon bond above the risk-free Treasury coupon bond yield. The yield spread increases as the risk increases. The yield spread of a risky bond is a function of the leverage undertaken by the firm in issuing its debt, with leverage being defined by

$$\ell = \frac{L}{V}$$

ℓ is also a measure of risk for an investor; the larger is the leverage ℓ the more risky is the bond. Merton's model for a risky zero coupon bond allows for an explicit calculation of spread of the yield as a function of ℓ.

The Put option, given in Eq. 11.10.7, can be re-written as follows. Note that

$$\begin{aligned} d_1 &= \frac{1}{\sigma\sqrt{\tau}}\big[\ln(V/L) + \tau(r + \frac{1}{2}\sigma^2)\big] \ ; \ V = V(t) \qquad (12.8.1)\\ &= \frac{1}{\sigma\sqrt{\tau}}\big[\ln(1/\ell) + \tau(r + \frac{1}{2}\sigma^2)\big] \ ; \ d_2 = d_1 - \sigma\sqrt{\tau} \end{aligned}$$

and hence the Put-option is expressed in terms of leverage as follows

$$\begin{aligned} \mathcal{P}_{BS}(t, x) &= e^{-r(\tau)\tau} L N(-d_2) - V N(-d_1)\\ &= L\big[e^{-r(\tau)\tau} N(-d_2) - \frac{1}{\ell} N(-d_1)\big] \equiv L F(\ell) \end{aligned}$$

where

$$F(\ell) = e^{-r(\tau)\tau} N(-d_2) - \frac{1}{\ell} N(-d_1) \qquad (12.8.2)$$

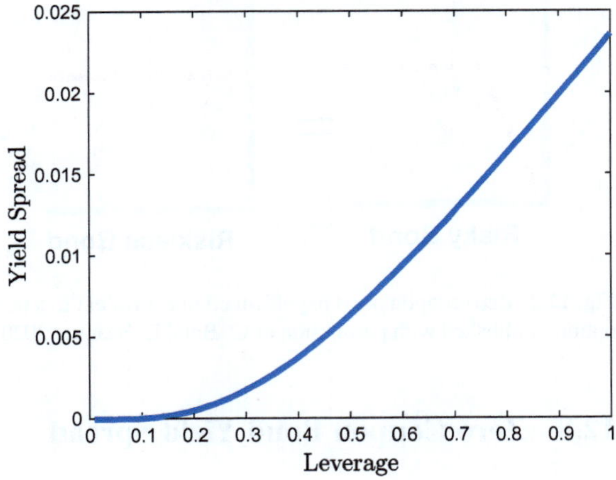

Fig. 12.3 The spread of a BBB zero coupon bond versus leverage L/V. $\tau = 10$ years, r =3% and $\sigma = 0.25$. Published with permission of © Belal E. Baaquie 2020. All Rights Reserved

In terms of the leverage ℓ, the zero coupon bond is given by

$$\mathcal{B}_0(t, x) = e^{-(T-t)r(\tau)}L - \mathcal{P}_{BS}(t, x) = L\big[e^{-r(\tau)\tau} - F(\ell)\big]$$

The zero coupon bond's yield $R(\tau)$ is given by

$$\mathcal{B}_0(t, x) \equiv e^{-\tau R(\tau)}L$$
$$\Rightarrow \quad R(\tau) = -\frac{1}{\tau}\ln\left[\frac{\mathcal{B}_0(t, x)}{L}\right] = -\frac{1}{\tau}\ln\left[e^{-r(\tau)\tau} - F(\ell)\right]$$

Hence the yield spread of the zero coupon bond, as a function of leverage ℓ, using Eq. 12.8.2, is given by

$$\text{spread}(\ell) = R(\tau) - r(\tau) = -\frac{1}{\tau}\ln([e^{-r(\tau)\tau} - F(\ell)]) - r(\tau)$$
$$= -\frac{1}{\tau}\ln\left[1 - N(-d_2) + e^{r(\tau)\tau}\frac{1}{\ell}N(-d_1)\right]$$

Figure 12.3 shows the spread for a BBB bond.

12.9 Default Probability and Leverage

The probability of default is the likelihood of the valuation having a final value $V(T)$ less than L. The Black–Scholes probability for the final stock price $V(t) = e^x$; $V(T) = e^{x'}$, for $\tau = T - t$ is given by Eq. 11.9.1.

$$P(x; x') = \frac{1}{\sqrt{2\pi\sigma^2\tau}} \exp\left\{-\frac{1}{2\sigma^2\tau}(x' - \mu)^2\right\} = P(x; x') \quad (12.9.1)$$

$$\mu = x + \tau(\alpha - \sigma^2/2) \quad (12.9.2)$$

The risk free interest rate r in Eq. 11.9.1 is replaced in μ above by α since one is considering the evolution of a risky asset with α rate of return.

The default probability is given by [25, 36]

$$\text{PD} = P(x' < L; x) = \int_{-\infty}^{\ln L} dx' P(x; x') = \int_{-\infty}^{\ln L} dx' \frac{e^{-\frac{1}{2\sigma^2\tau}(x'-\mu)^2}}{\sqrt{2\pi\sigma^2\tau}}$$

$$= \int_{-\infty}^{U} dx \frac{e^{-\frac{1}{2}x^2}}{\sqrt{2\pi}} = N(U)$$

The upper limit U is given by

$$U = \frac{\ln L - \mu}{\sqrt{\sigma^2\tau}} = -d_2 = -\frac{1}{\sigma\sqrt{\tau}}\left[\ln(V/L) + \tau\left(\alpha - \frac{1}{2}\sigma^2\right)\right]$$

Hence, the default probability is

$$PD = N(-d_2)\ ;\ d_2 = \frac{1}{\sigma\sqrt{\tau}}\left[\ln(1/\ell) + \tau\left(\alpha - \frac{1}{2}\sigma^2\right)\right] \quad (12.9.3)$$

One needs to estimate V, α and σ. There are many data providers, like Thomson-Reuter, who provide daily values for V, from which α and σ can be estimated.

Another approach, in the absence of the daily value of V, is to estimate the risk-neutral default that entails setting $\alpha = r$, and one is left with the task of determining V, σ. The price of equity $E(t, x)$ is known from the market, and Eq. 12.7.3 gives one equation. The volatility γ of the zero coupon bond $\mathcal{B}_0(t, x)$ is provided by the market. Hence, the value of V, σ can then be estimated from Eqs. 12.7.3 and 12.4.1 that are given below [25]

$$E(t, x) = \mathcal{C}_{BS}(t, x)\ ;\ \ \gamma\mathcal{B}_0(t, x) = \sigma V \frac{\partial \mathcal{B}_0(t, x)}{\partial V}\ ;\ \ V = e^x$$

To calculate default probabilities we need α, the return on the company valuation. The Sharpe ratio for all contingent claims on the assets, including coupon bonds, are the same; in particular, the Sharpe ratio for any option is the same as the Sharpe ratio for the stock.

The typical Sharpe ratio (SR) for equity is around 0.22; for a BBB-bond

$$\text{SR} = \frac{\alpha - r}{\sigma} = 0.22 : \ \ \alpha = 0.22 \times 0.25 + 0.03 = 8.5\%$$

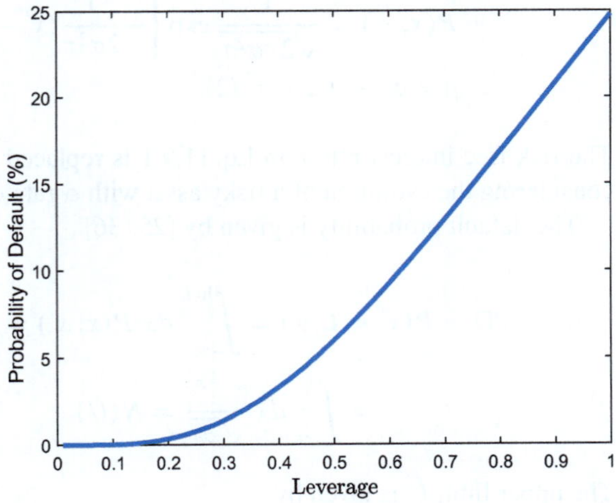

Fig. 12.4 Probability of default of a BBB zero coupon bond versus leverage. $\tau = 10$ years, α =8.5% and $\sigma = 0.25$. Published with permission of © Belal E. Baaquie 2020. All Rights Reserved

The risk-neutral default probability of the stock price is $N(-d_2)$. To get the default probability one needs to use α instead of r in the calculation of d_2. The default probability for a BBB-rated bond as a function of leverage ℓ is shown in Fig. 12.4.

The default probability is $N(-d_2)$ with

$$d_2 = \frac{1}{\sigma\sqrt{\tau}}\left[\ln(V/L) + \tau\left(\alpha - \frac{1}{2}\sigma^2\right)\right]$$

where V is enterprise value (asset value), L is the face value of debt, and α is the expected return on the assets.

For a *one-year* default probabilities this reduces to

$$d_2 = \frac{1}{\sigma}[\ln(V/L) + (\alpha - \frac{1}{2}\sigma^2)]$$

α is difficult to estimate and $\alpha - \frac{1}{2}\sigma^2$ is small in many cases and as a first approximation is sometimes ignored; then (Fig. 12.5)

$$d_2 \simeq \frac{1}{\sigma}\ln(V/L) \tag{12.9.4}$$

Default distance (DD) is defined in the market by

$$DD \equiv d_2 = \frac{1}{\sigma}\left[\ln(V/L) + \left(\alpha - \frac{1}{2}\sigma^2\right)\right]$$

Noteworthy 12.2: Example of Default Distance

Consider the following financial data of a zero coupon corporate bond.

- $V = \$130,00$; $L = \$100,000$. Hence $\ln(V/L) = \ln(1.3) = 0.262$.
- Expected rate of return $\alpha = 5.0\%$/year.
- Time to maturity $\tau = 1$ year ; volatility $\sigma = 20\%$.

The data yields

$$\ln(V/L) + \left(\alpha - \frac{1}{2}\sigma^2\right) = \ln(1.3) + 0.05 - 0.5 \times 0.04 = 0.292$$

Hence

$$DD = d_2 = 0.292/0.20 = 1.462 \Rightarrow N(-1.462) = 7.19\%$$

Hence, there is approximately 7.19% likelihood that the firm will default in one year. Note that if we ignore the term $\alpha - \sigma^2/2$, then Eq. 12.9.4 yields

$$DD \simeq \ln(1.3)/0.20 = 0.262/0.20 = 1.31 \Rightarrow N(-1.31) = 9.18\%$$

Hence, the approximation given in Eq. 12.9.4 over-estimates the likelihood of default.

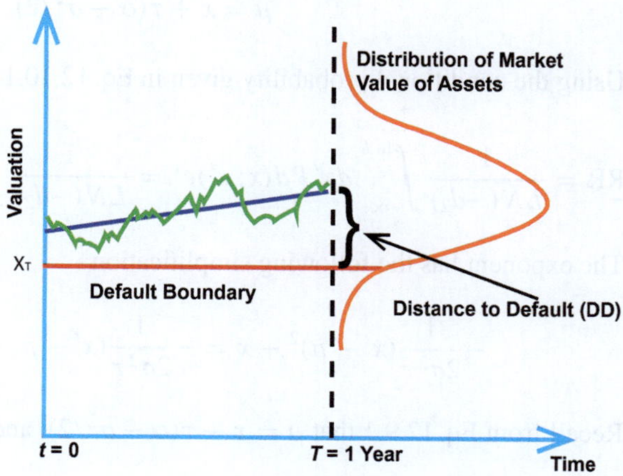

Fig. 12.5 Distance to default of a zero coupon bond. Published with permission of © Belal E. Baaquie 2020. All Rights Reserved

12.10 Recovery Rate of Defaulted Bonds

The recovery rate (RR) is given by the conditional probability that $V(T) < L$ and which leads to a default. This is given, from Eq. 12.9.1 by

$$Pd(x'|x' < \ln L) = \frac{P(x, x')}{P(x' < L, x)} = \frac{P(x, x')}{\int_{-\infty}^{\ln L} dx' P(x, x')}$$

The denominator is the default probability evaluated earlier in Eq. 12.9.3:

$$P(x' < L, x) = \int_{-\infty}^{\ln L} dx' P(x, x') = N(-d_2)$$

Hence

$$Pd(x'|x' < \ln L) = \frac{1}{N(-d_2)} P(x, x') \Rightarrow \int_{-\infty}^{\ln L} dx' Pd(x, x') = 1 \quad (12.10.1)$$

The recovery rate RR is equal to the conditional expectation value of the valuation at maturity $V(T)$ divided by the face value of the bond L, and is

$$\text{RR} = E[\frac{V(T)}{L}|V(T) < L] \ ; \ V(T) = e^{x'} = \text{valuation on maturity}$$

Let

$$\mu = x + \tau(\alpha - \sigma^2/2)$$

Using the conditional probability given in Eq. 12.10.1 yields

$$\text{RR} = \frac{1}{LN(-d_2)} \int_{-\infty}^{\ln L} dx' Pd(x; x') e^{x} = \frac{1}{LN(-d_2)} \int_{-\infty}^{\ln L} dx' \frac{e^{-\frac{1}{2\sigma^2\tau}(x'-\mu)^2 + x}}{\sqrt{2\pi\sigma^2\tau}}$$

The exponent has the following simplification

$$-\frac{1}{2\sigma^2\tau}(x' - \mu)^2 + x = -\frac{1}{2\sigma^2\tau}(x' - \mu - \sigma^2\tau)^2 + \alpha\tau + x$$

Recall from Eq. 12.9.1 that $\mu = x + \tau(\alpha - \sigma^2/2)$, and from Eq. 12.8.1

$$d_1 = \frac{1}{\sigma\sqrt{\tau}} \left[\ln(V/L) + \tau\left(\alpha + \frac{1}{2}\sigma^2\right) \right] \ ; \ V = e^x$$

which yields the result [26].

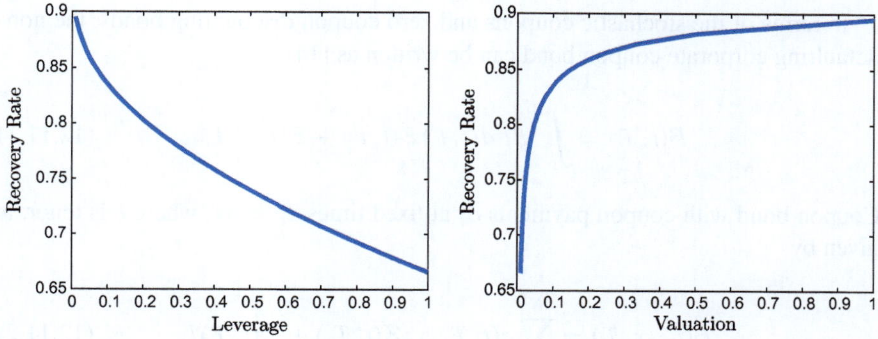

Fig. 12.6 Recovery rate as a function of leverage and valuation. Published with permission of © Belal E. Baaquie 2020. All Rights Reserved

$$\mathrm{RR} = \frac{e^{\alpha T} V}{LN(-d_2)} \int_{-\infty}^{-d_1} dx \frac{e^{-\frac{1}{2}x^2}}{\sqrt{2\pi}} = e^{\alpha T} \frac{V}{L} \frac{N(-d_1)}{N(-d_2)} \ : V < L \qquad (12.10.2)$$

For no default, $L \to \infty$ in the integral and yields

$$RR = e^{\alpha T} \frac{V}{L}$$

For the example given in Noteworthy 12.2, the recovery rate, as a function of leverage and valuation, is given in Fig. 12.6. Furthermore, from Noteworthy 12.2, $d_1 = 1.97$, $N(-d_1) = 0.024$, $\alpha = 0.05/\text{year}$ and yields the recovery rate

$$RR = \frac{V}{L}(1 + 0.05)\frac{0.024}{0.072} = 0.356\frac{V}{L} \ ; \ V < L$$

Due to the default $E[V(T)] = 0.356V$. Hence, when the bond matures in one year, the investor can expect to be paid only $0.356V$, $V < L$—instead of the face value L of the bond.

12.11 Merton's Risky Coupon Bond

In Merton's model, a firms risky coupons are paid continuously, for $\upsilon(V) < 0$, at the rate of $|\upsilon(V)|V$ and the coupon bond's price, at time t, is given by $\mathcal{B}(t, x)$. Figure 16.2b shows coupons being paid daily, reflecting the nature of a bond that pays coupons continuously.

In terms of the stochastic coupons and zero coupon discounting bonds, the non-defaulting corporate coupon bond can be written as [14]

$$\mathcal{B}(t,T) = \int_t^T dt' d(t,t')\mathcal{Z}(t,t') + \mathcal{Z}(t,T)L \tag{12.11.1}$$

Coupon bond with coupon payments c_n at fixed times $T_n = n\ell$, where ℓ is tenor, is given by

$$\mathcal{B}(t,,x,T) = \sum_{n=1}^{N} c(t,T_n,x)\mathcal{Z}(t,T_n) + \mathcal{Z}(t,T)L \tag{12.11.2}$$

Coupons $c(t,T_n,x)$ are stochastic since the future value of the coupon depends on the valuation, which is stochastic. To express the coupon bond that pays continuous coupons, one needs to sum (integrate) the coupons $d(t,t')$ over the period of one tenor ℓ.

Note that $Z(t,t) = 1 = Z(T,T)$; hence

$$\mathcal{B}(t,t) = L\mathcal{Z}(t,t) = L$$

and yields the identity

$$\mathcal{B}(t,T) = \int_t^T dt' \frac{\partial}{\partial t'}[\mathcal{B}(t,t') - \mathcal{Z}(t,t')L] + \mathcal{Z}(t,T)L \tag{12.11.3}$$

Hence, from Eqs. 12.11.1

$$d(t,t';x) = \frac{1}{\mathcal{Z}(t,t')}\frac{\partial}{\partial t'}[\mathcal{B}(t,t') - \mathcal{Z}(t,t')L] \tag{12.11.4}$$

The value of the stochastic coupon $c(t,T_n;x)$ at time t, and to be paid at future time T_n, from Eqs. 12.11.1 and 12.11.2, is given by

$$c(t,T_n;x) = \frac{1}{\mathcal{Z}(t,T_n)}\int_{T_{n-1}}^{T_n} dt' d(t,t';x)\mathcal{Z}(t,t')$$

The stochastic coupon is approximately the following

$$c(t,T_n) \approx \ell d(t,T_n)$$

Part V
Quantum Mathematics

Chapter 13
Functional Analysis

Abstract The synthesis of linear algebra with calculus opens the door to functional analysis, and a brief introduction is given to the central ideas of functional analysis based on the earlier discussions on linear algebra and integration. In functional analysis functions $f(x)$ are viewed as elements of an **infinite dimensional linear vector space**, also called function space; the space is denoted by $\mathcal{V}$. Functional analysis is studied based on Dirac's notation for linear algebra. Calculus is based on functions of a independent continuous variable. In functional analysis, and in particular in path integration, the continuous independent variable is superseded by *functions* of a continuous variable *themselves being the independent variables.*

13.1 Introduction

Linear algebra has been studied from the point of view of mapping a linear vector space E_N into another vector space E_M, and in most cases we have studied $M = N$. The case of $N \to \infty$ is now studied to obtain E_∞. The index set for E_∞ are the integers, with a vector $\mathbf{v}$ having infinitely many components v_n labeled by $n = 0, \pm 1, \pm 2, \ldots, \pm\infty$.

One would like to further generalize the index set so that it becomes a continuous variable, denoted by $x \in [-\infty, +\infty]$; a vector is then denoted by $v(x)$ and one can see that the vector can also be interpreted as a function of x. It is this convergence of linear algebra and continuous space that is termed as functional analysis. The synthesis of linear algebra with calculus leads to the generalization of the underlying linear vector space to an infinite dimensional linear vector space (function) space.

One can see that in generalizing linear algebra to functional analysis, the function $f(x)$ in endowed with the superstructure of a linear vector space. One can conversely interpret a function $f(x)$ as components of an element $|f\rangle$ of a infinite dimensional linear function space, such that

$$|f\rangle = \int_{-\infty}^{+\infty} dx f(x)|x\rangle = \lim_{a \to 0} a \sum_n f(x_n)|x_n\rangle \ ; \ x_n = na$$

B. E. Baaquie, *Mathematical Methods and Quantum Mathematics for Economics and Finance*, https://doi.org/10.1007/978-981-15-6611-0_13

The interpretation of function $f(x)$ as an infinite dimensional vector $|f\rangle$ is arrived at by studying how linear algebra can be generalized to functional analysis.

13.2 Dirac Bracket: Vector Notation

Euclidean space E_N is an N-dimensional linear vector space. Vectors are elements of E_N and matrices are linear transformations from E_N to E_M. The synthesis of linear algebra with calculus is most easily carried out by expressing vectors and matrices in Dirac's notation. Although the notation can also be applied to the general case, the discussion is confined to square matrices that are linear transformations from E_N to E_N.

The basic ingredient of the Dirac **bracket** notation is the following

$$\text{ket}: \ |..\rangle = \text{vector} \ ; \quad \text{bra}: \ \langle ..| = \text{dual vector}$$

Taken together they form the complete bracket given by

$$\text{bracket}: \ \langle ..|..\rangle = \text{scalar product} = \text{complex number}$$

Consider, for generality, a complex valued vector $\mathbf{v}$, which is represented by the ket-vector

$$\mathbf{v} = |v\rangle \ ; \quad \mathbf{v}_i = |v_i\rangle$$

There is no need to make the symbol v boldface inside the ket since the notation makes it clear. Hermitian conjugation yields the dual bra-vector that is given by

$$\mathbf{v}^\dagger = \langle v|$$

The canonical basis and its dual vectors are given by

$$\mathbf{e}_i = |i\rangle \ ; \quad \mathbf{e}_i^\dagger = \langle i|$$

Linear superposition is written as expected

$$\mathbf{u} = a\mathbf{v} + b\mathbf{w} \ \Rightarrow \ |u\rangle = a|v\rangle + b|w\rangle$$

with the dual expression given by

$$\mathbf{u}^\dagger = a^*\mathbf{v}^\dagger + b^*\mathbf{w}^\dagger \ \Rightarrow \ \langle u| = a^*\langle v| + b^*\langle w|$$

The expansion of a vector into its components is

$$\mathbf{v} = \sum_i v(i)\mathbf{e}_i \;\Rightarrow\; |v\rangle = \sum_i v(i)|i\rangle$$

In Dirac's notation, the scalar product is given by

$$\mathbf{v}^\dagger \cdot \mathbf{w} = ((\langle v|)(|w\rangle)) = \langle v||w\rangle = \langle v|w\rangle$$

Furthermore

$$\mathbf{w}^\dagger \cdot \mathbf{v} = (\mathbf{v}^\dagger \cdot \mathbf{w})^* = \langle v|w\rangle^* \Rightarrow \mathbf{w}^\dagger \cdot \mathbf{v} = \langle v|w\rangle^* = \langle w|v\rangle$$

For the orthonormal basis states

$$\mathbf{e}_i^\dagger \cdot \mathbf{e}_j = \langle i|j\rangle = \delta_{i-j} = \begin{cases} 0 & i \neq j \\ 1 & i = j \end{cases}$$

Hence, the component of a vector is given by

$$v(i) = \mathbf{e}_i^\dagger \cdot \mathbf{v} = \langle i|v\rangle$$

The outer product is the following

$$\mathbf{u} \otimes \mathbf{v}^\dagger = |u\rangle\langle v|$$

A matrix has the following expression

$$A = \sum_{ij} A_{ij}\mathbf{e}_i \otimes \mathbf{e}_j^\dagger = \sum_{ij} A_{ij}|i\rangle\langle j| \tag{13.2.1}$$

Matrix multiplication, using $\langle j|k\rangle = \delta_{j-k}$, is given by

$$\begin{aligned} AB &= \sum_{ij}\sum_{kl} A_{ij}B_{k\ell}|i\rangle\langle j|k\rangle\langle \ell| = \sum_{ij}\sum_{kl} A_{ij}B_{k\ell}|i\rangle\delta_{j-k}\langle \ell| \\ &= \sum_{ijk} A_{ij}B_{jk}|i\rangle\langle k| = \sum_{ik} C_{ik}|i\rangle\langle k| \;;\; C_{ik} = \sum_j A_{ij}B_{jk} \\ &\Rightarrow\; AB = C \end{aligned} \tag{13.2.2}$$

The completeness equation for the canonical basis is given by

$$\mathbb{I} = \sum_{I=1}^{N} \mathbf{e}_I \otimes \mathbf{e}_I^\dagger = \sum_{I=1}^{N} |I\rangle\langle I| \tag{13.2.3}$$

Any complete basis states $\mathbf{v}_I$ yield

$$\mathbb{I} = \sum_{I=1}^{N} \mathbf{v}_I \otimes \mathbf{v}_I^\dagger = \sum_{I=1}^{N} |v_I\rangle\langle v_I| \tag{13.2.4}$$

Suppose $\mathbf{v}_I$ are eigenfunctions of the matrix S; the eigenfunction equation is given by

$$S\mathbf{v}_I = \lambda_I \mathbf{v}_I \ \Rightarrow \ S|v_I\rangle = \lambda_I |v_I\rangle \ \ : I = 1, \ldots N$$

The spectral decomposition of matrix S is given by

$$S = \sum_{I=1}^{N} \lambda_I \mathbf{v}_I \otimes \mathbf{v}_I^\dagger = \sum_{I=1}^{N} \lambda_I |v_I\rangle\langle v_I| \tag{13.2.5}$$

Unitary matrix $\mathcal{U}$ yields

$$|e_I\rangle = \mathcal{U}^\dagger |v_I\rangle \ ; \ \langle e_I| = \langle v_I|\mathcal{U} \ \Rightarrow \ \mathcal{U}|e_I\rangle = |v_I\rangle \ ; \ \langle e_I|\mathcal{U}^\dagger = \langle v_I| \tag{13.2.6}$$

A Hermitian matrix S is diagonalized by a unitary matrix $\mathcal{U}$ that is given by

$$\mathcal{U}^\dagger S \mathcal{U} = \sum_{I=1}^{N} \lambda_I \mathbf{e}_I \otimes \mathbf{e}_I^\dagger = \sum_{I=1}^{N} \lambda_I |I\rangle\langle I|$$

Conversely, Eq. 13.2.6 yields the spectral decomposition

$$S = \sum_{I=1}^{N} \lambda_I \, \mathcal{U}|e_I\rangle\langle e_I|\mathcal{U}^\dagger = \sum_{I=1}^{N} \lambda_I \, |v_I\rangle\langle v_I| \tag{13.2.7}$$

Consider the expression

$$S^2 = \sum_{I,J=1}^{N} \lambda_I \lambda_J |v_I\rangle\langle v_I|v_J\rangle\langle v_J| = \sum_{I=1}^{N} \lambda_I^2 |v_I\rangle\langle v_I| \tag{13.2.8}$$

Hence, one has in Dirac's notation

$$S^n = \sum_{I=1}^{N} \lambda_I^n |v_I\rangle\langle v_I| \tag{13.2.9}$$

In general, for any function $F(S)$, one has

$$F(S) = \sum_{I=1}^{N} F(\lambda_I)|v_I\rangle\langle v_I| \tag{13.2.10}$$

In particular

$$\exp(S) = \sum_{I=1}^{N} \exp(\lambda_I)|v_I\rangle\langle v_I| \tag{13.2.11}$$

13.3 Continuous Basis States

As shown in Fig. 13.1, let the continuous variable x, $-\infty \leq x \leq +\infty$, take only *discrete* values at points $x = na$ with lattice spacing a and with $n = 0, \pm 1, \pm 2, \ldots$; in other words, the lattice is embedded in the continuous real line $\mathbb{R}$ and the lattice point n identified with the point na in $\mathbb{R}$. To obtain the continuous limit, let $a \to 0$ and the allowed values of the index for the basis vectors x can take any real value, that is, $x \in \mathbb{R}$.

Similar to the finite case given in Eq. 2.6.2, the basis states are labeled by $|n\rangle$ and the dual basis states by $\langle n|$. The linear *function space* for the continuous variable x is obtained by considering the limit of $N \to \infty$ of the linear vector space $E_{(2N+1)}$. Note one needs *first* to take $N \to \infty$ *and then* take $a \to 0$ for obtaining the continuous variable, also called the degree of freedom x.[1]

The discrete basis vectors of the linear function vector space $\mathcal{V}$ are represented by infinite column vectors with the only non-zero entry being unity in the nth position. Hence, taking $N \to \infty$ in Eq. 2.6.2 yields

$$|n\rangle \ : \ n = 0, \pm 1, \pm 2, \ldots \pm \infty$$

where the column vector is given by

$$|n\rangle = \begin{bmatrix} .. \\ 0 \\ 1 \\ 0 \\ .. \end{bmatrix} \ : \ n\text{−th position}$$

[1]For compact degrees of freedom such as the circle S^1 with radius r, the infinitesimal a is related to N by $2\pi r = Na$.

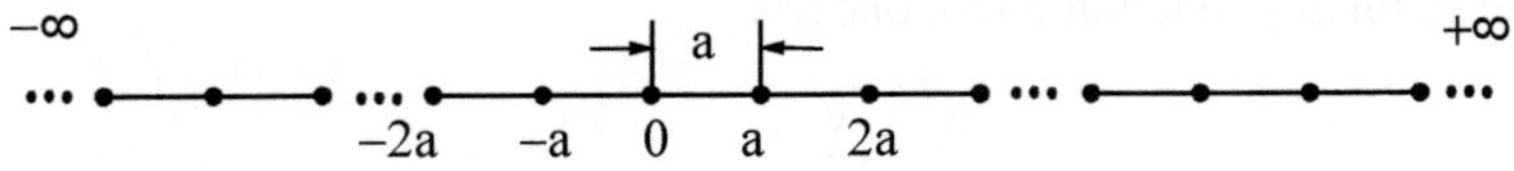

Fig. 13.1 Infinite collection of discrete basis states as a precursor for the continuum basis states. Published with permission of © Belal E. Baaquie 2020. All Rights Reserved

The basis vectors for the dual state space $\mathcal{V}_D$ are given by

$$\langle n| = \underbrace{\left[\cdots 0\ 1\ 0 \cdots\right]}_{n\text{th position}}$$
$$\Rightarrow \langle n|m\rangle = \delta_{n-m} \tag{13.3.1}$$

The completeness of the basis states, taking $N \to \infty$ of the case for finite N Eq. 3.7.1, yields the following[2]

$$\sum_{n=-\infty}^{+\infty} |n\rangle\langle n| = \text{diagonal}(.., 1, 1, \ldots) = \mathbb{I} \; : \; \text{Completeness Equation} \tag{13.3.2}$$

where $\mathbb{I}$ above is the infinite dimensional unit matrix.

The limit of $a \to 0$ needs to be taken to obtain a continuous x; in terms of the underlying lattice, the continuous point x is related to the discrete lattice point n by the following

$$-\infty \le x \le +\infty \; : \; x = \lim_{a\to 0}[na] \; ; \; n = 0, \pm 1, \pm \cdots \pm \infty$$

The basis vector for the function space $\mathcal{V}$ is given by the ket vector $|x\rangle$, with its dual vector is given by the bra vector $\langle x|$. The basis state $|n\rangle$ is dimensionless; the ket vector $|x\rangle$ has a dimension of $1/\sqrt{a}$.[3] Hence due to dimensional consistency and to have the correct continuum limit the continuous basis vector and its dual are given by

$$|x\rangle = \lim_{a\to 0} \frac{1}{\sqrt{a}}|n\rangle \; ; \; \langle x| = \lim_{a\to 0} \frac{1}{\sqrt{a}}\langle n| \tag{13.3.3}$$

---

[2]Equation 3.7.1 has to be extended to E_{2N+1} and written as

$$\sum_{i=-N}^{+N} \mathbf{e}_i \otimes \mathbf{e}_i^T = \mathbb{I}$$

[3]It will be seen later, in Eq. 13.4.2, that the Dirac δ−function has dimension of $1/a$, consistent with the dimension of $|x\rangle$.

The position projection operator is given by the outer product of the position ket vector with the bra vector and is

$$|x\rangle\langle x| = \lim_{a\to 0}\frac{1}{a}|n\rangle\langle n| \tag{13.3.4}$$

The scalar product, for $x = na$ and $x' = ma$, in the limit of $a \to 0$, is given, from Eqs. 13.3.1, 13.3.3 and 13.4.2, by the Dirac delta function

$$\langle x|x'\rangle = \lim_{a\to 0}\frac{1}{a}\delta_{m-n} \Rightarrow \langle x|x'\rangle = \delta(x - x') \tag{13.3.5}$$

where $\delta(x - x')$ is the Dirac delta-function. The properties of the Dirac delta-function are discussed in Sect. 13.4.

The completeness equation given in Eq. 13.3.2 has the following continuum limit

$$\mathbb{I} = \sum_{n=-\infty}^{+\infty}|n\rangle\langle n| = \lim_{a\to 0} a\sum_{n=-\infty}^{+\infty}|x\rangle\langle x| \tag{13.3.6}$$

$$\Rightarrow \int_{-\infty}^{\infty} dx|x\rangle\langle x| = \mathbb{I}\ :\ \text{Completeness Equation} \tag{13.3.7}$$

Equation 13.3.6 shows that the projection operators given in Eq. 13.3.4 are complete and span the entire of state space $\mathcal{V}$.

$\mathbb{I}$ is the *identity operator* on function space $\mathcal{V}$; namely for any functional vector $|f\rangle \in \mathcal{V}$, it follows from the completeness equation that

$$\mathbb{I}|f\rangle = |f\rangle$$

The completeness equation given by Eq. 13.3.7 is a key equation in the analysis of function space $\mathcal{V}$, and yields the following

$$\langle x|\mathbb{I}|x'\rangle = \int_{-\infty}^{\infty} dz\langle x|z\rangle\langle z|x'\rangle = \int_{-\infty}^{\infty} dz\delta(x - z)\delta(z - x') = \delta(x - x')$$

The above follows from the definition of the Dirac delta function $\delta(x - x')$, showing only the diagonal elements of the identity operator in function space are non-zero. The above equation shows that $\delta(x - x')$ is the matrix element of the identity operator $\mathbb{I}$ for the continuous variable $x \in \mathbb{R}$ in the x basis.

A function space $\mathcal{V}(\mathbb{R})$ of a continuous space $\mathbb{R}$ consists *of all possible functions* $f(R)$: it is for this reason that the subject of functional analysis is a generalization of linear algebra and calculus.

13.4 Dirac Delta Function

The Dirac delta function is required in the study of continuum variables, and some of its essential properties are reviewed. The Dirac delta functions are not ordinary functions, but rather are generalized functions and called distributions. The $\delta(x)$ always appears inside an integration, and heuristically speaking, $\delta(x) \simeq 1/dx$.

In essence, the Dirac delta function is the continuum generalization of the discrete Kronecker delta function given in Eq. 2.6.3. For a countable set of integers, the Kronecker delta function picks out a single value for the integer.

The Dirac $\delta-$function, denoted by $\delta(x-a)$, has an infinite value at a *single* point, namely $x = a$, and is zero at all other points.

As discussed in the Appendix, the real line is shown to have uncountably many points, with infinitely many points in the neighborhood of any given point; a single point has 'measure' zero in the language of set theory. In this dense set of points, the Dirac delta function provides an infinite weight to a single point, which has 'measure' zero, and hence, in effect, gives it a finite value.

The $\delta-$function, in arriving at the final answer, always appears inside an integration and is never a property of an observable quantity. It should be thought of as a function that picks out a single point from the continuous line.

Consider a continuous line labeled by coordinate x such that $-\infty \leq x \leq +\infty$, and let $f(x)$ be an infinitely differentiable function. The Dirac $\delta-$function is defined by the following properties

$$\begin{aligned}
&\delta(x-y) = \delta(y-x) \equiv \begin{cases} \infty & x = y \\ 0 & x \neq y \end{cases} \\
&\delta(c(x-y)) = \frac{1}{|c|}\delta(x-y) \\
&\int_{-\infty}^{+\infty} dx f(x)\delta(x-y) = f(y) \\
&\int_{-\infty}^{+\infty} dx f(x)\frac{d^n}{dx^n}\delta(x-a) = (-1)^n \frac{d^n}{dx^n} f(x)\,|_{x=a}
\end{aligned} \tag{13.4.1}$$

A representation of the $\delta-$function, based on the Gaussian distribution, is given by Eq. 13.7.5

$$\delta(x-a) = \lim_{\sigma \to 0} \frac{1}{\sqrt{2\pi\sigma^2}} \exp\left\{-\frac{1}{2\sigma^2}(x-a)^2\right\}$$

To make the connection with the discrete Kronecker delta function given in Eq. 2.6.3, consider the discretization of the continuous line into a discrete lattice with spacing a. As shown in Fig. 13.1, the continuous variable x, $-\infty \leq x \leq +\infty$, takes only discrete values at points $x = na$ with $n = 0, \pm1, \pm2, \ldots$. The discretization of Eq. 13.4.1, for $x = na$ and $y = ma$, yields the following

$$a \sum_{n=-\infty}^{\infty} f(na)\delta(x-y) = f(ma) \;\Rightarrow\; \delta(x-y) = \lim_{a\to0}\frac{1}{a}\delta_{m-n} \tag{13.4.2}$$

Note from above that the Dirac δ−function has dimension of $1/a$. The following is another representation of the delta-function is derived later, in Eq. 13.7.5

$$\delta(x-y) = \int_{-\infty}^{\infty}\frac{dp}{2\pi}e^{ip(x-y)} \tag{13.4.3}$$

13.5 Basis States for Function Space

The bra and ket vectors $\langle x|$ and $|x\rangle$ are the basis vectors of the $\mathcal{V}_D$ and $\mathcal{V}$ respectively. For $2N+1$-dimensional Euclidean space, it can readily be shown that the $2N+1$ vectors $|n\rangle, n = \pm1, \pm2, \ldots \pm N$ given in Eq. 2.6.2 are linearly independent and form a complete basis for the space: any vector in the $2N+1$-dimensional Euclidean space can be expanded as a linear sum of these basis states. The crucial property of these basis states is that the complete basis vectors satisfy the *completeness equation* given in Eq. 13.3.2. Any infinite-dimensional vector $\mathbf{v}$, denoted by $|v\rangle$, can be expanded as in Eq. 3.7.2, as follows

$$|v\rangle = \mathbb{I}|v\rangle = \sum_{n=-\infty}^{+\infty} v_n|n\rangle \;;\; v_n = \langle n|v\rangle \tag{13.5.1}$$

For the infinite dimensional function space, a complete basis set of vectors must satisfy the completeness equation; for the co-ordinate basis $|x\rangle$, completeness is given by Eqs. 13.3.7 and 13.3.5,

$$\int_{-\infty}^{\infty} dx|x\rangle\langle x| = \mathbb{I} \;;\; \langle x|x'\rangle = \delta(x-x')$$

Consider the case of $x \in \mathbb{R}$. An element of the function space $\mathcal{V}$ is a ket vector a vector $|f\rangle$ and can be thought of as an infinite dimensional vector with *components* that are given by $f(x) = \langle x|f\rangle$; hence all functions of x, namely $f(x)$ can be thought of as elements of a function space $\mathcal{V}(\mathbb{R})$. In analogy with Eq. 13.5.1, the vector $|f\rangle$ has the following representation in the $|x\rangle$ basis

$$|f\rangle = \int_{-\infty}^{\infty} dx|x\rangle\langle x|f\rangle = \int_{-\infty}^{+\infty} dx f(x)|x\rangle \;;\; f(x) = \langle x|f\rangle$$

Being an element of a function space endows the function $f(x)$ with the additional property of linearity and all the other properties of vectors that is needed to be consistent with the properties of $f(x)$.

The vector $|f\rangle$ can be mapped to a unique *dual* vector denoted by $\langle f| \in \mathcal{V}_D$; in components $f^*(x) = \langle f|x\rangle$ and

$$\langle f| = \int_{-\infty}^{\infty} dx \langle f|x\rangle\langle x| = \int_{-\infty}^{\infty} dx f^*(x)\langle x| \ ; \ f^*(x) = \langle f|x\rangle$$

Note the vector and its dual are related by complex conjugation, namely

$$\langle g|f\rangle = \langle f|g\rangle^* \ \Rightarrow \ \langle x|f\rangle = \langle f|x\rangle^* \tag{13.5.2}$$

The scalar product of two vectors is given by[4]

$$\langle g|f\rangle \equiv \int dx g^*(x) f(x)$$

The vector $|f\rangle$ and its dual $\langle f|$ have the important property that they define the norm ('length') $\langle f|f\rangle$ of the vector. The completeness equation 13.3.7 yields the following

$$\langle f|f\rangle = \int_{-\infty}^{\infty} dx f(x)^* f(x) = \int_{-\infty}^{\infty} dx |f(x)|^2 \geq 0 \tag{13.5.3}$$

Functions that appear in finance, such as the value of a security or the price of a bond, do not have a finite norm but still belong to the function space $\mathcal{V}$. Option pricing can be formulated entirely in terms of the function space $\mathcal{V}$.

13.6 Operators on Function Space

Square matrices are mapping of E_N to E_N. The concept of a matrix carries over to functional analysis in the concept of operators $\mathcal{O}$, which are the mappings of function space $\mathcal{V}$ to itself

$$\mathcal{O} : \mathcal{V} \rightarrow \mathcal{V}$$

[4] A more direct derivation of the completeness equation is the following

$$\langle g|f\rangle = \langle g| \left\{ \int_{-\infty}^{\infty} dx |x\rangle\langle x| \right\} |f\rangle \Rightarrow \mathbb{I} = \int_{-\infty}^{\infty} dx |x\rangle\langle x|$$

.

Operators are analogous to square matrices, with the elements of $\mathcal{O}$ denoted by $O(x, y)$.[5] Similar to matrices, operators have a representation given by

$$\mathcal{O} = \int dxdy\mathcal{O}(x, y)|x\rangle\langle y|$$

where $\mathcal{O}(x, y)$ is a scalar function of x, y.

A Hermitian operator $\mathcal{K}$, which is the generalization of a Hermitian matrix, is define as follow

$$\mathcal{K}^\dagger = \mathcal{K} \;\Rightarrow\; \int dxdyK(x, y)|x\rangle\langle y| = \int dxdyK^*(y, x)|x\rangle\langle y|$$

Hence, for a Hermitian operator

$$\mathcal{K}^\dagger = \mathcal{K} \;\Rightarrow\; K(x, y) = K^*(y, x)$$

The eigenfunction equation for the operator $\mathcal{K}$ is defined by

$$\mathcal{K}|\psi_n\rangle = \lambda_n|\psi_n\rangle \;\; ; \;\; n = 0, \pm 1, \pm 2, \ldots, \pm\infty$$

Using the completeness equation yields the following eigenfunction equation

$$\int dyK(x, y)\psi_n(y) = \lambda_n\psi_n(x) \tag{13.6.1}$$

Similar to Hermitian matrices, all the eigenvalues of Hermitian operators are real. The index set $n \in \mathcal{Z}$ for the eigenfunctions can be both discrete or continuous. In general, for eigenfunctions (eigenvectors) $|\psi_n\rangle$ (with components given by $\psi_n(x) = \langle x|\psi_n\rangle$) of a Hermitian operator, similar to the case of Hermitian matrices, form a complete basis. Hence

$$\sum_{n=-\infty}^{+\infty} |\psi_n\rangle\langle\psi_n| = \mathbb{I} \;\Rightarrow\; \sum_{n=-\infty}^{+\infty} \psi_n(x)\, \psi_n^*(x') = \delta(x - x')$$

The completeness equation is also termed a *resolution of the identity* since only a complete set of basis states can yield the identity operator on function space. Since the eigenfunctions $\psi_n(x)$ form a complete basis, we have

$$|f\rangle = \sum_{n=-\infty}^{\infty} |\psi_n\rangle\langle\psi_n|f\rangle = \sum_{n=-\infty}^{\infty} f_n|\psi_n\rangle \;\; ; \;\; f_n = \langle\psi_n|f\rangle$$

[5]The analogy non-square matrices can be defined for operators, but will not be pursued.

13.7 Gaussian Kernel

A Hermitian operator is equal to a symmetric operator if all its components are real. The Gaussian kernel is a symmetric operator given by

$$K(x, y) = \frac{1}{\sqrt{2\pi\sigma^2}} \exp\left\{-\frac{1}{2\sigma^2}(x-y)^2\right\}$$

The eigenfunctions are given by

$$\psi_p(x) = \frac{1}{\sqrt{2\pi}} e^{ipx} \ ; \ p \in [-\infty, +\infty]$$

To verify this consider

$$\int_{-\infty}^{\infty} dy K(x, y)\psi_p(y) = \frac{1}{\sqrt{2\pi\sigma^2}} \int_{-\infty}^{\infty} dy \exp^{-\frac{1}{2\sigma^2}(x-y)^2} \frac{e^{ipy}}{\sqrt{2\pi}}$$
$$= \frac{e^{ipx}}{2\pi\sqrt{\sigma^2}} \int_{-\infty}^{\infty} dy \exp^{-\frac{1}{2\sigma^2}y^2} e^{ipy}$$
$$\Rightarrow \int_{-\infty}^{\infty} dy K(x, y)\psi_p(y) = e^{-\frac{\sigma^2}{2}p^2} \frac{e^{ipx}}{\sqrt{2\pi}} = e^{-\frac{\sigma^2}{2}p^2}\psi_p(x) \quad (13.7.1)$$

and Eq. 13.7.1 is the eigenfunction equation.

Given the importance of the plane wave basis, the eigenvector is written in an abbreviated notation as (the factor of $1/\sqrt{2\pi}$ is included for convenience)

$$|\psi_p\rangle \equiv \frac{1}{\sqrt{2\pi}}|p\rangle$$

In this notation

$$\psi_p(x) = \langle x|\psi_p\rangle \equiv \frac{1}{\sqrt{2\pi}}\langle x|p\rangle$$

Hence, the eigenfunctions and eigenvalues of the Gaussian kernel are

$$\frac{1}{\sqrt{2\pi}}|p\rangle \ ; \ \frac{1}{\sqrt{2\pi}}\langle x|p\rangle = \frac{1}{\sqrt{2\pi}}e^{ipx} \ ; \ \lambda_p = e^{-\frac{\sigma^2}{2}p^2} \ : \ p \in [-\infty, +\infty]$$

The completeness equation is given by

$$\delta(x-y) = \langle x|\mathbb{I}|y\rangle = \int_{-\infty}^{\infty} dp \psi_p(x)\psi_p^*(y) = \int_{-\infty}^{\infty} \frac{dp}{2\pi}\langle x|p\rangle\langle p|y\rangle$$
$$\Rightarrow \delta(x-y) = \int_{-\infty}^{\infty} \frac{dp}{2\pi} e^{ip(x-y)} \quad (13.7.2)$$

In functional vector notation, the completeness equation is

$$\int_{-\infty}^{\infty} \frac{dp}{2\pi} |p\rangle\langle p| = \mathbb{I} \ : \ \langle x|p\rangle = e^{ipx} \ ; \ \langle p|x\rangle = e^{-ipx} \tag{13.7.3}$$

The resolution of the identity of $\mathbb{I}$ can be expressed in terms of the coordinate $|x\rangle$ and momentum basis $|p\rangle$ since both are complete. This is the same as the resolution of the N-dimensional unit matrix in terms of the canonical basis $\mathbf{e}_I$ and the basis in terms of the eigenfunctions $\mathbf{v}_I$ of a $N \times N$ dimensional symmetric since

$$\sum_{n=1}^{N} \mathbf{e}_I \otimes \mathbf{e}_I^T = \mathbb{I}_{N\times N} = \sum_{n=1}^{N} \mathbf{v}_I \otimes \mathbf{v}_I^T$$

These are two different representations of the same entity, namely $\mathbb{I}$, but expressed in two different complete basis states.

The spectral representation of the operator K with matrix elements $K(x, y) = \langle x|K|y\rangle$ is given by

$$K = \int_{-\infty}^{\infty} \frac{dp}{2\pi} \lambda_p |p\rangle\langle p|$$

In terms of the matrix elements, the Gaussian kernel has the following spectral representation

$$\begin{aligned} K(x, y) &= \langle x|K|y\rangle = \int_{-\infty}^{\infty} \frac{dp}{2\pi} \lambda_p \langle x|p\rangle\langle p|y\rangle \\ &= \int_{-\infty}^{\infty} \frac{dp}{2\pi} e^{-\frac{\sigma^2}{2}p^2} e^{ip(x-y)} = \frac{1}{\sqrt{2\pi\sigma^2}} \exp\left\{-\frac{1}{2\sigma^2}(x-y)^2\right\} \end{aligned} \tag{13.7.4}$$

Since $\lambda_p \to 1$ as $\sigma \to 0$, Eqs. 13.7.2 and 13.7.4 yield a Gaussian representation of the Dirac $\delta-$function given by

$$\delta(x-y) = \lim_{\sigma\to 0} \frac{1}{\sqrt{2\pi\sigma^2}} e^{-\frac{1}{2\sigma^2}(x-y)^2} = \int_{-\infty}^{\infty} \frac{dp}{2\pi} \langle x|p\rangle\langle p|y\rangle \tag{13.7.5}$$

13.8 Fourier Transform

The complete set of the *plane wave* basis states given by

$$\psi_p(x) = \frac{1}{\sqrt{2\pi}} e^{ipx} \ ; \ p \in [-\infty, +\infty]$$

is the basis of the procedure of Fourier transform. Since the plane wave basis states are complete, any function $f(x)$ that is an element of function space has an expansion in terms of the plane wave basis $\mathcal{V}$.[6] From Eq. 13.7.3

$$|f\rangle = \int_{-\infty}^{\infty} \frac{dp}{2\pi}|p\rangle\langle p|f\rangle = \int_{-\infty}^{\infty} \frac{dp}{2\pi}|p\rangle \hat{f}(p)$$

In terms of the components

$$f(x) = \langle x|f\rangle = \int_{-\infty}^{\infty} \frac{dp}{2\pi}\langle x|p\rangle\langle p|f\rangle$$

$$\Rightarrow \ f(x) = \int_{-\infty}^{\infty} \frac{dp}{2\pi} e^{ipx} \hat{f}(p) \tag{13.8.1}$$

The inverse Fourier transform is given by using the completeness equation given in Eq. 13.7.2, and yields

$$\int_{-\infty}^{\infty} dx e^{-iqx} f(x) = \int_{-\infty}^{\infty} \frac{dp}{2\pi} \int_{-\infty}^{\infty} dx e^{i(p-q)x} \hat{f}(p) = \int_{-\infty}^{\infty} \frac{dp}{2\pi} 2\pi\delta(p-q)\hat{f}(p)$$

$$\Rightarrow \ \hat{f}(q) = \int_{-\infty}^{\infty} dx e^{-iqx} f(x) \tag{13.8.2}$$

From Eq. 13.7.4, the spectral representation of the Gaussian kernel can be interpreted in the language of Fourier transform, with

$$f(x) = \frac{1}{\sqrt{2\pi\sigma^2}} \exp\left\{-\frac{1}{2\sigma^2}x^2\right\} \ ; \ \hat{f}(p) = e^{-\frac{\sigma^2}{2}p^2}$$

Example.
Consider the function

$$f(x) = e^{-m|x|} \ ; \ m > 0$$

The inverse Fourier transform given in Eq. 13.8.2 yields

$$\begin{aligned}
\hat{f}(p) &= \int_{-\infty}^{\infty} dx e^{-ipx} e^{-m|x|} = \int_{-\infty}^{0} dx e^{-ipx} e^{mx} + \int_{0}^{\infty} dx e^{-ipx} e^{-mx} \\
&= \frac{1}{m-ip}(e^{(m-ip)x})\Big|_{-\infty}^{0} - \frac{1}{m+ip}(e^{-(m+ip)x})\Big|_{0}^{+\infty} \\
&= \frac{1}{m-ip} + \frac{1}{m+ip} = \frac{2m}{p^2+m^2}
\end{aligned} \tag{13.8.3}$$

[6]There are some restrictions on $f(x)$ that do not concern us.

since $e^{i\infty} = 0$. From Eqs. 13.8.1 and 13.8.3, we obtain

$$\frac{1}{2m}e^{-m|x|} = \int_{-\infty}^{\infty}\frac{dp}{2\pi}\hat{f}(p) = \int_{-\infty}^{\infty}\frac{dp}{2\pi}\frac{e^{ipx}}{p^2+m^2} \tag{13.8.4}$$

13.8.1 Taylor Expansion

The Fourier transform of $f(x+a)$, from Eq. 13.8.1, is given by

$$f(x+a) = \int_{-\infty}^{\infty}\frac{dp}{2\pi}e^{ip(x+a)}\hat{f}(p)$$

Note that

$$e^{ip(x+a)} = e^{a\frac{d}{dx}}e^{ipx} = e^{iD}e^{ipx}$$

where the displacement Hermitian operator is given by

$$D = -ia\frac{d}{dx} \tag{13.8.5}$$

and yields

$$f(x+a) = e^{a\frac{d}{dx}}\int_{-\infty}^{\infty}\frac{dp}{2\pi}e^{ipx}\hat{f}(p) = e^{a\frac{d}{dx}}f(x)$$

Using

$$\left(a\frac{d}{dx}\right)^n = a^n\frac{d^n}{dx^n}\ ;\ \left(a\frac{d}{dx}\right)^0 = 1$$

yields the Maclaurin expansion

$$\begin{aligned} f(x+a) &= e^{a\frac{d}{dx}}f(x) = \sum_{n=0}^{\infty}\frac{a^n}{n!}\frac{d^nf(x)}{dx^n} \\ &= f(x) + a\frac{df(x)}{dx} + \frac{a^2}{2!}\frac{d^2f(x)}{dx^2} + \cdots + \frac{a^n}{n!}\frac{d^nf(x)}{dx^n} + \cdots \end{aligned} \tag{13.8.6}$$

$$\Rightarrow\ d_n(x) = \frac{1}{n!}\frac{d^nf(x)}{dx^n} \tag{13.8.7}$$

Consider the polynomial expansion for $f(x)$ given by

$$f(x) = \sum_{n=0}^{\infty}c_nx^n \tag{13.8.8}$$

Shifting x to $x + a$ yields the Maclaurin expansion

$$f(x + a) = \sum_{n=0}^{\infty} c_n (x + a)^n = \sum_{n=0}^{\infty} d_n(x) a^n \tag{13.8.9}$$

where $d_n(x)$ is a polynomial in x.

From Eq. 13.8.7 $d_n(x)$ is given

$$d_n(x) = \frac{1}{n!} \frac{d^n f(x)}{dx^n}$$

The Taylor expansion is obtained by shifting a to $a - x$; Eq. 13.8.6 yields

$$\begin{aligned} f(a) &= f(x) + (a - x)\frac{df(x)}{dx} + \frac{(a - x)^2}{2!}\frac{d^2 f(x)}{dx^2} + \cdots + \frac{(a - x)^n}{n!}\frac{d^n f(x)}{dx^n} + \cdots \\ &= \sum_{n=0}^{\infty} \frac{(a - x)^n}{n!}\frac{d^n f(x)}{dx^n} \end{aligned} \tag{13.8.10}$$

The Fourier transform for a function $f(\mathbf{x})$ of N-variables given by vector $\mathbf{x}$ is the following

$$f(\mathbf{x} + \mathbf{a}) = \int_{-\infty}^{+\infty} \frac{d^N p}{(2\pi)^N} \exp\{i\mathbf{p} \cdot (\mathbf{x} + \mathbf{a})\} \hat{f}(\mathbf{p}) = e^{\mathbf{a} \cdot \frac{\partial}{\partial \mathbf{x}}} f(\mathbf{x})$$

The generalization of the Maclaurin expansion given in Eq. 13.8.6 is the following

$$f(\mathbf{x} + \mathbf{a}) = e^{\mathbf{a} \cdot \frac{\partial}{\partial \mathbf{x}}} f(\mathbf{x}) = \exp\left\{\sum_i a_i \frac{\partial}{\partial x_i}\right\} f(\mathbf{x})$$

$$\Rightarrow f(\mathbf{x} + \mathbf{a}) = f(\mathbf{x}) + \sum_i a_i \frac{\partial f(\mathbf{x})}{\partial x_i} + \frac{1}{2!}\sum_{ij} a_i a_j \frac{\partial^2 f(\mathbf{x})}{\partial x_i \partial x_j} + \cdots \tag{13.8.11}$$

The Taylor expansion is obtained by shifting $\mathbf{a}$ to $\mathbf{a} - \mathbf{x}$; Eq. 13.8.11 yields

$$f(\mathbf{a}) = f(\mathbf{x}) + \sum_i (\mathbf{a} - \mathbf{x})_i \frac{\partial f(\mathbf{x})}{\partial x_i} + \frac{1}{2!}\sum_{ij} (\mathbf{a} - \mathbf{x})_i (\mathbf{a} - \mathbf{x})_j \frac{\partial^2 f(\mathbf{x})}{\partial x_i \partial x_j} + \cdots \tag{13.8.12}$$

13.8.2 Green's Function

It is instructive to obtain the solution for the inhomogeneous linear differential equation using the technique of Green's function. From Eq. 7.9.2, one can write the differential equations as follows

$$L[x(t)] = R(t) \ ; \ \ L = D^2 + \beta D + \omega^2 \ ; \ D = \frac{d}{dt} \ ; \ t \in [a, b] \quad (13.8.13)$$

where L is a linear differential operator. From Eq. 7.9.5

$$L = (D + r_1)(D + r_2) \quad (13.8.14)$$

Define the homogeneous solution x_H by the following

$$L(x_H(t)) = L(Ae^{-r_1 t} + Be^{-r_2 t}) = 0 \ \Rightarrow \ x_H(t) = Ae^{-r_1 t} + Be^{-r_2 t}$$

From Eq. 13.8.13

$$(D + r_1)(D + r_2)x = R \ ; \ \ r_{1,2} = \frac{\beta}{2} \pm \sqrt{\frac{\beta^2}{4} - \omega^2} \quad (13.8.15)$$

The solution of Eq. 13.8.13 can formally be solved as follows

$$x = x_H + L^{-1}R \ ; \ \ Lx_H = 0$$

The question is, how is L^{-1} defined? Define Green's function $G(t, t')$ by

$$L(t, d/dt)G(t, t') = \delta(t - t') \quad (13.8.16)$$

The solution of the differential equation is given by

$$x(t) = x_H(t) + x_{IH}(t) \quad (13.8.17)$$

where

$$x_{IH}(t) = \int_a^b d\xi G(t, \xi)R(\xi)$$

Applying L to both sides of Eq. 13.8.17 above yields the expected result

$$L(t, d/dt)x(t) = L(t, d/dt)x_H(t) + \int_a^b d\xi L(t, d/dt)G(t, \xi)R(\xi)$$

$$= \int_a^b d\xi \delta(t - \xi)R(\xi) = R(t) \quad (13.8.18)$$

Note the definition of $G(t,\xi)$ is ambiguous since

$$\widetilde{G}(t,\xi) = G(t,\xi) + x_H(t-\xi) \;\Rightarrow\; L(t, d/dt)\widetilde{G}(t,t') = \delta(t-t')$$

The ambiguity leads to the solution

$$\begin{aligned}\tilde{x}_{IH}(t) &= \int_a^b d\xi[G(t,\xi) + x_H(t-\xi)]R(\xi)\\ &= x_{IH}(t) + \int_a^b d\xi x_H(t-\xi)R(\xi)\end{aligned} \tag{13.8.19}$$

The ambiguity in $G(t,\xi)$ leads to the freedom to choose different boundary conditions for the Green's function. This becomes more important for higher dimensional Green's functions. The ambiguous term

$$\int_a^b d\xi x_H(t-\xi)R(\xi)$$

for the case of a second order linear differential equation leads to a redefinition of the parameters of the solution that need to be fixed by the boundary conditions.

To evaluate the Green's function, define the Fourier transform of $G(t,t')$ by the following

$$G(t,t') = \int_{-\infty}^{+\infty} \frac{dp}{2\pi} e^{ip(t-t')}\hat{G}(p) \tag{13.8.20}$$

Then, from Eq. 13.8.16

$$L(t, d/dt)G(t,t') = \int_{-\infty}^{+\infty} \frac{dp}{2\pi} e^{ip(t-t')}[-p^2 + i\beta p + \omega^2]\hat{G}(p) = \delta(t-t')$$

Hence

$$\hat{G}(p) = \frac{1}{-p^2 + i\beta p + \omega^2}$$

and which yields the expected result given in Eq. 13.8.16 since

$$\int_{-\infty}^{+\infty} \frac{dp}{2\pi} e^{ip(t-t')} = \delta(t-t')$$

From Eq. 13.8.20, the Green's function is given by

$$G(t,t') = \int_{-\infty}^{+\infty} \frac{dp}{2\pi} \frac{e^{ip(t-t')}}{(-p^2 + i\beta p + \omega^2)} = \int_{-\infty}^{+\infty} \frac{dp}{2\pi} \frac{e^{ip(t-t')}}{(ip + r_1)(ip + r_2)}$$

$$= \frac{1}{r_2 - r_1} \int_{-\infty}^{+\infty} \frac{dp}{2\pi} e^{ip(t-t')} \left[\frac{1}{ip + r_1} - \frac{1}{ip + r_2} \right] \tag{13.8.21}$$

where r_1, r_2 are defined in Eq. 7.9.3. Using the identity

$$\frac{1}{a} = \int_0^{\infty} dz e^{-za}$$

the integral is given by the following

$$I = \int_{-\infty}^{+\infty} \frac{dp}{2\pi} e^{ip(t-t')} \frac{1}{ip + r_1} = \int_0^{\infty} dz \int_{-\infty}^{+\infty} \frac{dp}{2\pi} e^{ip(t-t')} e^{-z(ip + r_1)}$$

$$= \int_0^{\infty} dz e^{-zr_1} \delta(z - (t - t'))$$

$$= \begin{cases} e^{-r_1(t-t')} & : (t - t') > 0 \\ 0 & : (t - t') < 0 \end{cases} = e^{-r_1(t-t')} \theta(t - t') \tag{13.8.22}$$

where the theta function, denoted by $\theta(t - t')$, is defined by

$$\theta(t - t') = \begin{cases} 1 & t > t' \\ 0 & t < t' \end{cases} \tag{13.8.23}$$

and yields the Green's function

$$G(t,t') = \frac{1}{r_2 - r_1} \left(e^{-r_1(t-t')} - e^{-r_2(t-t')} \right) \theta(t - t')$$

Hence, from Eqs. 13.8.17, 13.8.21 and 13.8.22, the solution of the differential equation is given by

$$x(t) = x_H(t) + \int_a^b d\xi G(t,\xi) R(\xi)$$

$$= x_H(t) + \frac{1}{r_2 - r_1} \int_a^b dt' \left(e^{-r_1(t-t')} - e^{-r_2(t-t')} \right) \theta(t - t') R(t')$$

$$= x_H(t) + \frac{1}{r_1 - r_2} \int_a^t dt' \left(e^{-r_2(t-t')} - e^{-r_1(t-t')} \right) R(t') \tag{13.8.24}$$

where, from Eq. 13.8.15

$$r_1 - r_2 = \sqrt{\beta^2 - 4\omega^2}$$

The result obtained in Eq. 13.8.24 is the equal to the solution obtained earlier in Eq. 7.9.5.

13.8.3 Black–Scholes Option Price

Fourier transform can be used to solve linear partial differential equations. To illustrate this, consider the Black–Scholes equation given by Eq. 11.7.5

$$\frac{\partial C}{\partial t} = -\frac{1}{2}\sigma^2\frac{\partial^2 C}{\partial x^2} + \left(\frac{1}{2}\sigma^2 - r\right)\frac{\partial C}{\partial x} + rC \tag{13.8.25}$$

Fourier transforming only the stock price variable for the option price C and payoff function $g(x)$ yields

$$C(t,x) = \int_{-\infty}^{+\infty}\frac{dp}{2\pi}e^{ipx}\hat{C}(t,p) \ ; \ g(x) = \int_{-\infty}^{+\infty}\frac{dp}{2\pi}e^{ipx}\hat{g}(p)$$

Fourier transforming Eq. 13.8.25 yields the following ordinary differential equation

$$\frac{\partial \hat{C}(t,p)}{\partial t} = \left[\frac{1}{2}\sigma^2 p^2 + i\left(\frac{1}{2}\sigma^2 - r\right)p + r\right]\hat{C}(t,p) \equiv f(p)\hat{C}(t,p)$$

Solving the equation above yields

$$\hat{C}(t,p) = \exp\{tf(p)\}\hat{C}(0,p) \ ; \ f(p) = \frac{1}{2}\sigma^2 p^2 + i\left(\frac{1}{2}\sigma^2 - r\right)p + r$$

The option is equal to the payoff function for $t = T$ and hence yields

$$\hat{C}(t,p) = \exp\{-\tau f(p)\}\hat{g}(p) \ : \ \tau = T - t$$

The Black–Scholes option price is given by

$$C(t,x) = \int_{-\infty}^{+\infty}\frac{dp}{2\pi}e^{ipx}e^{-\tau f(p)}\hat{g}(p)$$

Using the inverse Fourier transform of the payoff function yields

$$\hat{g}(p) = \int_{-\infty}^{+\infty} dx' e^{-ipx'} g(x)$$

and hence

$$C(t,x) = \int_{-\infty}^{+\infty} \frac{dp}{2\pi} dx' e^{ip(x-x')} \exp\left\{-\tau\left[\frac{1}{2}\sigma^2 p^2 + i\left(\frac{1}{2}\sigma^2 - r\right)p + r\right]\right\} g(x')$$

Performing the Gaussian integration yields the expected pricing kernel, given in Eq. 11.10.1, that

$$C(t,x) = \frac{e^{-r\tau}}{\sqrt{2\pi\sigma^2\tau}} \int_{-\infty}^{+\infty} dx' \exp\left\{-\frac{1}{2\sigma^2\tau}\left[x - x' + \tau\left(r - \frac{\sigma^2}{2}\right)\right]^2\right\} g(x')$$

13.9 Functional Differentiation

Consider variables f_n that satisfy

$$\frac{\partial f_n}{\partial f_m} = \delta_{n-m}$$

Let $t' = m\epsilon$ and $t = n\epsilon$, with $m, n = 0, \pm 1, \cdots, \pm N$. For $N \to \infty$ we have $t, t' \in [-\infty, +\infty]$, and the limit $\epsilon \to 0$ yields the *functional derivative*

$$\frac{\partial f_n}{\partial f_m} \to \frac{\delta f(t)}{\delta f(t')} \equiv \lim_{\epsilon \to 0} \frac{1}{\epsilon} \frac{\partial f_n}{\partial fm}$$
$$\Rightarrow \frac{\delta f(t)}{\delta f(t')} = \lim_{\epsilon \to 0} \frac{1}{\epsilon} \delta_{n-m} \to \delta(t - t') \qquad (13.9.1)$$

Note the extra $1/\epsilon$ is required in the definition of the functional derivative to produce the Dirac delta function given in Eq. 13.9.1.

For a continuous index σ and function $f(\sigma)$, let a *functional* of f be denoted by $G[f]$. The functional derivative is defined by

$$\frac{\delta G[f]}{\delta f(\sigma)} = \lim_{\epsilon \to 0} \frac{G[f(\sigma') + \epsilon\delta(\sigma' - \sigma)] - G[f]}{\epsilon} \qquad (13.9.2)$$

Consider a function $V(\varphi(x))$ that depends only on the function $\varphi(x)$; the functional derivative of $V(\varphi(x))$ is given by

$$\frac{\delta V(\varphi(x))}{\delta \varphi(y)} = \frac{1}{\epsilon}\Big(V\big(\varphi(x) + \epsilon\delta(x - y)\big) - V(\varphi(x))\Big) = \frac{\partial V(\varphi(x))}{\partial \varphi(x)} \delta(x - y)$$

More generally, using the chain rule yields

$$\frac{\delta V(h(x))}{\delta\varphi(y)} = \frac{\partial V(h(x))}{\partial\varphi(x)}\frac{\delta h(x)}{\delta\varphi(y)} \tag{13.9.3}$$

For $G = f^n$, the chain rule yields ($n = 1$ gives Eq. 13.9.1)

$$\frac{\delta f^n(\sigma)}{\delta f(\sigma')} = n f^{n-1}(\sigma)\delta(\sigma' - \sigma) \tag{13.9.4}$$

For the following, Eq. 13.9.3 yields

$$V[\varphi] = \int dz\varphi^n(z) \Rightarrow \frac{\delta V[\varphi]}{\delta\varphi(x)} = n\int dz\varphi^{n-1}(z)\delta(z - x) = n\varphi^{n-1}(x)$$

In general, for $f(z) = f(\varphi(z))$

$$V[\varphi] = \int dz f(z) \Rightarrow \frac{\delta V[\varphi]}{\delta\varphi(x)} = \int dz\frac{\partial f(z)}{\partial\varphi(z)}\delta(z - x) = \frac{\partial f(x)}{\partial\varphi(x)} \tag{13.9.5}$$

Consider the function

$$e^{G[f]} = \exp\left\{-\frac{m}{2}\int d\sigma f^2(\sigma)\right\}$$

The definition of the functional derivative yields

$$\frac{\delta e^{G[f]}}{\delta f(\sigma)} = \frac{1}{\epsilon}\Big[\exp\{-\frac{m}{2}\int d\sigma'[f(\sigma') + \epsilon\delta(\sigma' - \sigma)]^2 - e^{G[f]}\Big]$$
$$= \frac{1}{\epsilon}\left[\exp\left\{-\frac{m}{2}\left(\int d\sigma' f^2(\sigma') + 2\epsilon f(\sigma)\right)\right\} - e^{G[f]}\right] = -mf(\sigma)e^{G[f]}$$

The chain rule yields

$$\frac{\delta e^{G[f]}}{\delta f(\sigma)} = e^{G[f]}\frac{\delta}{\delta f(\sigma)}\left\{-\frac{m}{2}\int d\sigma' f^2(\sigma')\right\} = -mf(\sigma)e^{G[f]}$$

Maclaurin expansion can be defined using functional differentiation. Consider a functional G of two functions $f(\sigma)$, $h(\sigma)$ and let λ be a real parameter.

$$G[f + \lambda h] = \sum_{n=0}^{\infty}\frac{\lambda^n}{n!}\int d\sigma_1\cdots d\sigma_n h(\sigma_1)\cdots h(\sigma_n)\frac{\delta^n G[f]}{\delta f(\sigma_1)\cdots\delta f(\sigma_n)} \tag{13.9.6}$$

In more compact notation

$$G[f + \lambda h] = \exp\left\{\lambda\int d\sigma h(\sigma)\frac{\delta}{\delta f(\sigma)}\right\}G[f]$$

To derive the translation operator, let $\sigma \to \sigma + \epsilon$; then

$$f(\sigma + \epsilon) = f(\sigma) + \epsilon f'(\sigma) \ ; \ f'(\sigma) \equiv df(\sigma)/d\sigma$$

Hence, for $h(\sigma) = f'(\sigma)$

$$e^{i\epsilon\mathcal{P}} G[f] \equiv \exp\left\{\epsilon \int d\sigma f'(\sigma) \frac{\delta}{\delta f(\sigma)}\right\} G[f] = G(f + \epsilon f') = G[f(\sigma + \epsilon)]$$

The above yields the Hermitian translation operator that is a generalization of the displacement operator given in Eq. 13.8.5

$$\mathcal{P} = -i \int d\sigma f'(\sigma) \frac{\delta}{\delta f(\sigma)}$$

The finite translation of $\sigma \to \sigma + a$ is given by composing the infinitesimal translations and yields

$$e^{ia\mathcal{P}} G[f] \equiv \exp\left\{a \int d\sigma f'(\sigma) \frac{\delta}{\delta f(\sigma)}\right\} G[f] = G[f(\sigma + a)]$$

13.10 Functional Integration

Gaussian functional integration is one of the cornerstones of quantum mathematics, and for showing how functional integration emerges the continuum limit is taken of an ordinary multiple integral. The N-dimensional Gaussian integration discussed in Sect. 5.7 is used to introduce the idea of Gaussian functional integration. Let $t = n\epsilon$, $n = 0, \pm 1, \pm 2 \cdots \pm N$. From Eq. 5.7.2

$$\begin{aligned} S &= -\frac{1}{2} \sum_{i,j=1}^{N} x_i A_{ij} x_j + \sum_i J_i x_i \\ &= -\frac{1}{2}\epsilon^2 \sum_{i,j=1}^{N} x_i \left(\frac{1}{\epsilon^2} A_{ij}\right) x_j + \epsilon \sum_i \left(\frac{1}{\epsilon} J_i\right) x_i \end{aligned} \qquad (13.10.1)$$

Define the continuum quantities by

$$A(t, t') = \frac{1}{\epsilon^2} A_{ij} \ ; \ j(t) = \frac{1}{\epsilon} J_i \ ; \ x(t) = x_i \ ; \ t = i\epsilon \ , \ t' = j\epsilon \ , \ t'' = k\epsilon$$

Taking the limit of $N \to \infty$, $\epsilon \to 0$ yields

$$x_n \to x(t) \ ; \ t \in [-\infty, +\infty]$$

To find the inverse of $A(t,t')$ given by $A^{-1}(t,t')$, consider

$$\int_{-\infty}^{+\infty} dt' A(t,t')A^{-1}(t',t'') \simeq \epsilon \sum_{j=-\infty}^{+\infty} \frac{1}{\epsilon^2} A_{ij}(A^{-1})_{jk} = \frac{1}{\epsilon}\delta_{i-k} \to \delta(t-t'')$$

Hence

$$A(t,t') = \frac{1}{\epsilon^2}A_{ij} \; ; \; A^{-1}(t,t') = (A^{-1})_{jk}$$

and

$$\int_{-\infty}^{+\infty} dt' A(t,t')A(t',t'') = \delta(t-t'') \qquad (13.10.2)$$

The result of Gaussian functional integration is given by $N \to \infty$, $\epsilon \to 0$ limit of Eq. 5.7.4 and give by the following

$$\begin{aligned} Z[J] &= \prod_{i=1}^{N} \int_{-\infty}^{+\infty} dx_i \exp\left\{-\frac{1}{2}\sum_{i,j=1}^{N} x_i A_{ij} x_j + \sum_i J_i x_i\right\} \\ &= \exp\left\{\frac{1}{2}\sum_{jk} J_j (A^{-1})_{jk} J_k\right\} = \exp\left\{\frac{1}{2}\epsilon^2 \sum_{jk} j(t)A^{-1}(t,t')j(t')\right\} \\ &\to \exp\left\{\frac{1}{2}\int_{-\infty}^{+\infty} dt dt' j(t)A^{-1}(t,t')j(t')\right\} \end{aligned} \qquad (13.10.3)$$

Hence, in the limit for $N \to \infty$ and $\epsilon \to 0$, the continuum action S_0 and the Gaussian functional integral is given by

$$S_0 = -\frac{1}{2}\int_{-\infty}^{+\infty} dt dt' x(t)A(t,t')x(t') + \int_{-\infty}^{+\infty} dt j(t)x(t) \qquad (13.10.4)$$

$$Z = \int Dx e^{S_0} \; ; \; \int Dx = \mathcal{N} \prod_{t=-\infty}^{+\infty} \int_{-\infty}^{+\infty} dx(t)$$

where $\mathcal{N}$ is a normalization constant.

The generating functional is given by[7]

$$Z[j] = \frac{1}{Z}\int Dx \exp\left\{S_0 + \int dt j(t)(t)x(t)\right\} \qquad (13.10.5)$$

$$= \exp\left\{\frac{1}{2}\int_{-\infty}^{+\infty} dt dt' j(t)A^{-1}(t,t')j(t')\right\} \qquad (13.10.6)$$

[7] The term generating functional is used instead as generating function as in Eq. 5.7.5 to indicate that one is considering a system with infinitely many variables.

The auto-correlation function is given by

$$E[x(t)x(t')] = \frac{\delta^2}{\delta j(t)\delta j(t')}\ln(Z[0]) = A^{-1}(t,t') \tag{13.10.7}$$

13.11 Gaussian White Noise

The simplest case of Gaussian functional integration is the case of white noise since its action functional is ultra-local with all the variables being decoupled.

Recall from Eq. 11.3.1, the fundamental properties of Gaussian white noise are that

$$E[R(t)] = 0 \quad ; \quad E[R(t)R(t')] = \delta(t-t')$$

Figure 11.2 shows that there is an independent (Gaussian) random variable $R(t)$ for each instant of time t.

White noise R_n for each n has the probability distribution given Eq. 11.3.2. Since white noise is an independent random variable for each n, the probability measure for the white noise random variables is the given by

$$P[R] = \prod_{n=1}^{\infty} P(R_n) = \prod_{n=1}^{N} e^{-\frac{\epsilon}{2}R_n^2} = \exp\left\{-\frac{\epsilon}{2}\sum_{n=1}^{\infty} R_n^2\right\} \tag{13.11.1}$$

$$\int dR = \prod_{n=1}^{\infty}\sqrt{\frac{\epsilon}{2\pi}}\int_{-\infty}^{+\infty} dR_n \quad \Rightarrow \quad \int dRP[R] = 1$$

To take the continuum limit of Eq. 13.11.1, let $N\epsilon = t_2 - t_1$, and let $t = n\epsilon$. Taking the limit $\epsilon \to 0$ of Eq. 13.11.1 yields

$$\mathcal{P}[R] \to e^{S_0} \quad ; \quad S_0 = -\frac{1}{2}\int_{t_1}^{t_2} dt R^2(t) \tag{13.11.2}$$

$$Z = \int DRe^{S_0} \quad ; \quad \int dR \to \prod_{t=t_1}^{t_2}\int dR(t)$$

The action functional S_0 is ultra-local. Gaussian integration, given in Eq. 5.7.5, yields the generating functional

$$Z[j] = \frac{1}{Z}\int DRe^{\int_{t_1}^{t_2} dtj(t)R(t)}e^{S_0} \tag{13.11.3}$$

$$= \exp\left\{\frac{1}{2}\int_{t_1}^{t_2} dt\int_{t_1}^{t_2} dt' j(t)\delta(t-t')j(t')\right\} = \exp\left\{\frac{1}{2}\int_{t_1}^{t_2} dtj^2(t)\right\}$$

The correlation functions, using functional differentiation, are given by

$$E[R(t)] = 0 \ ; \ E[R(t)R(t')] = \frac{1}{Z}\int DR\ R(t)R(t')e^{S_0}$$

$$\Rightarrow \ E[R(t)R(t')] = \frac{\delta^2}{\delta j(t)\delta j(t')}\ln Z[0] = \frac{\delta}{\delta j(t)}j(t') = \delta(t-t')$$

and yields the result given in Eq. 11.3.1.

The result given in Eqs. 13.11.2 and 13.11.3 show that white noise is represented by a path integral with an ultra local action S_0 .

Noteworthy 13.1: Generating function for lognormal $x(T)$

Consider the lognormal random variable for the stock price given in Eq. 11.5.3

$$x(T) = x(t) + \left(\phi - \frac{\sigma^2}{2}\right)(T-t) + \sigma\int_t^T dt' R(t')$$

The generating function for $x(T)$ is given by the path integral for white noise $R(t)$. Equation 13.11.3 yields ($m = x(t) + (\phi - \frac{\sigma^2}{2})(T-t)$)

$$\begin{aligned} E[e^{jx(T)}] &= e^{jm}\int DR\exp\left\{j\sigma\int_t^T dt' R(t') + S_0\right\} \\ &= e^{jm}\exp\left\{\frac{1}{2}j^2\sigma^2\int_t^T dt'\int_t^T dt''\delta(t'-t'')\right\} \\ &= \exp\left\{jm + \frac{1}{2}j^2\sigma^2(T-t)\right\} \end{aligned}$$

which is the result obtained earlier in Eq. 11.5.4. Hence

$$x(T) = N(m, v) \ : \ m = x(t) + \left(\phi - \frac{\sigma^2}{2}\right)(T-t) \ ; \ v^2 = \sigma^2(T-t)$$

13.12 Simple Harmonic Oscillator

To go beyond the white noise ultra-local action given in Eq. 13.11.2, one can couple the integration variables in the action by using the first derivative dx/dt and this yields the 'harmonic oscillator' action given by

$$\begin{aligned} S &= -\frac{m}{2}\int_{-\infty}^{+\infty} dt \left[\left(\frac{dx(t)}{dt}\right)^2 + \omega^2 x^2(t) \right] + \int_{-\infty}^{+\infty} dt j(t)x(t) \\ &= -\frac{m}{2}\int_{-\infty}^{+\infty} dt x(t)\left(-\frac{d^2}{dt^2} + \omega^2\right)x(t) + \int_{-\infty}^{+\infty} dt j(t)x(t) \\ \Rightarrow A(t,t') &= m\left(-\frac{d^2}{dt^2} + \omega^2\right)\delta(t-t') \end{aligned}$$

where an integration by parts was done, discarding boundary terms at $\pm\infty$, to obtain the second expression for S above. Define the Fourier transform of $x(t), j(t)$ by $\tilde{x}(k), \tilde{j}(k)$ given by

$$x(t) = \int_{-\infty}^{+\infty} \frac{dk}{2\pi} e^{ikt}\tilde{x}(k) \ ; \ j(t) = \int_{-\infty}^{+\infty} \frac{dk}{2\pi} e^{ikt}\tilde{j}(k)$$

The Fourier transformed action given in Eq. 13.12.1 yields

$$\begin{aligned} S &= -\frac{m}{2}\int_{-\infty}^{+\infty} dt \frac{dkdk'}{(2\pi)^2} e^{it(k+k')}\tilde{x}(k')(k^2+\omega^2)\tilde{x}(k) + \int_{-\infty}^{+\infty} dt \frac{dkdk'}{(2\pi)^2}\tilde{j}(k')\tilde{x}(k) \\ &= -\frac{m}{2}\int_{-\infty}^{+\infty} \frac{dk}{2\pi}\tilde{x}(-k)(k^2+\omega^2)\tilde{x}(k) \int_{-\infty}^{+\infty} \frac{dk}{2\pi}\tilde{j}(-k)\tilde{x}(k) \end{aligned} \quad (13.12.1)$$

since

$$\int_{-\infty}^{+\infty} dt e^{it(k+k')} = 2\pi\delta(k+k')$$

The generating functional is given by

$$Z[j] = \int Dx e^S = \exp\left\{\frac{1}{2m}\int_{-\infty}^{+\infty} \frac{dk}{2\pi}\tilde{j}(-k)\frac{1}{k^2+\omega^2}j(k)\right\}$$

Doing the inverse transform on $\tilde{j}(k)$ yields generating functional

$$Z[j] = \int Dx e^S = \exp\left\{\frac{1}{2}\int_{-\infty}^{+\infty} dt j(t)A^{-1}(t,t')j(t)\right\}$$

where, from Eq. (13.8.4)

$$A^{-1}(t,t') = \frac{1}{m}\int_{-\infty}^{+\infty} \frac{dk}{2\pi}\frac{e^{ik(t-t')}}{k^2+\omega^2} = \frac{1}{2m\omega}e^{-\omega|t-t'|}$$

The auto-correlation function is given by

$$E[x(t)x(t')] = \frac{\delta^2}{\delta j(t)\delta j(t')}\ln Z[0] = A^{-1}(t,t') = \frac{1}{2m\omega}e^{-\omega|t-t'|}$$

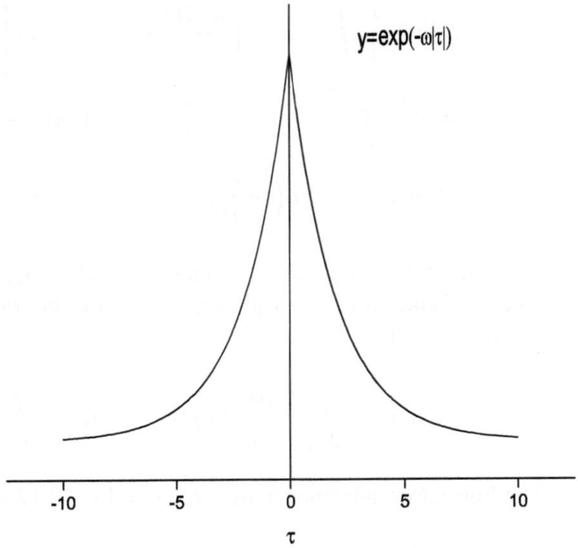

Fig. 13.2 The auto-correlation function $E[x(0)x(\tau)]$ for the simple harmonic oscillator. Published with permission of © Belal E. Baaquie 2020. All Rights Reserved

and the integral is given in Eq. 13.8.4. The auto-correlation function is derived again using the Fourier transform (dropping tilde from the Fourier transformed variables for notation's simplicity)

$$
\begin{aligned}
E[x(t)x(t')] &= \int_{-\infty}^{+\infty} \frac{dk}{2\pi}\frac{dk'}{2\pi} e^{ikt+ik't'} E[x(k)x(k')] \\
&= \int_{-\infty}^{+\infty} \frac{dk}{2\pi}\frac{dk'}{2\pi} e^{ikt+ik't'} \frac{\delta^2}{\delta j(k)\delta j(k')} \ln Z[0] \\
&= \int_{-\infty}^{+\infty} \frac{dk}{2\pi}\frac{dk'}{2\pi} e^{ikt+ik't'} \frac{1}{m}\frac{1}{k^2+\omega^2} 2\pi\delta(k+k') \\
\Rightarrow E[x(t)x(t')] &= \frac{1}{m}\int_{-\infty}^{+\infty} \frac{dk}{2\pi}\frac{e^{ik(t-t')}}{k^2+\omega^2} = \frac{1}{2m\omega} e^{-\omega|t-t'|}
\end{aligned}
\tag{13.12.2}
$$

The auto-correlation function is shown in Fig. 13.2. Note that the derivative of the auto-correlation function $E[x(0)x(\tau)]$ is singular at $\tau = 0$.

13.13 Acceleration Action

Consider the acceleration action functional given by [7]

$$
S[x] = -\frac{1}{2}\int_{-\infty}^{+\infty} dt \left[L\left(\frac{d^2x(t)}{dt^2}\right)^2 + \tilde{L}\left(\frac{dx(t)}{dt}\right)^2 + \gamma^2 x^2(t) \right] \tag{13.13.1}
$$

All the steps taken in Sect. 13.12 to derive the auto-correlation function are repeated for the more complex acceleration action. Define the Fourier transform of $x(t)$, $j(t)$ by $\tilde{x}(k)$, $\tilde{j}(k)$ and given by

$$x(t) = \int_{-\infty}^{+\infty} \frac{dk}{2\pi} e^{ikt} \tilde{x}(k) \ ; \ j(t) = \int_{-\infty}^{+\infty} \frac{dk}{2\pi} e^{ikt} \tilde{j}(k)$$

Equation 13.13.1 yields

$$S[\tilde{x}] = -\frac{1}{2} \int_{-\infty}^{+\infty} \frac{dk}{2\pi} \left[Lk^4 + \tilde{L}k^2 + \gamma^2 \right] \tilde{x}(-k) \tilde{x}(k) \tag{13.13.2}$$

In terms of the Fourier transformed variables, the generating functional given in Eq. 13.10.5 yields, for action given in Eq. 13.13.2, the following

$$\begin{aligned} Z[j] &= \int Dx \exp\left\{ S[x] + \int_{-\infty}^{+\infty} \frac{dk}{2\pi} \tilde{j}(-k) \tilde{x}(k) \right\} \\ &= \exp\left\{ \frac{1}{2} \int \frac{dk}{2\pi} \tilde{j}(-k) \frac{1}{Lk^4 + \tilde{L}k^2 + \gamma^2} \tilde{j}(k) \right\} \\ \Rightarrow D(t-t') &= E[x(t)x(t')] = \int_{-\infty}^{+\infty} \frac{dk}{2\pi} \frac{e^{ik(t-t')}}{Lk^4 + \tilde{L}k^2 + \gamma^2} \end{aligned} \tag{13.13.3}$$

The auto-correlation function $D(t-t')$ given in Eq. 13.13.3 plays a central role is the study of commodities and interest rates [8]. The auto-correlation function is analyzed for its various branches.

$$\begin{aligned} D(t-t') &= \int_{-\infty}^{\infty} \frac{dk}{2\pi} \frac{e^{ik(t-t')}}{Lk^4 + \widetilde{L}k^2 + \gamma^2} = \frac{1}{L} \int_{-\infty}^{\infty} \frac{dk}{2\pi} \frac{e^{ik(t-t')}}{(k^2 + a_+)(k^2 + a_-)} \\ &= \frac{1}{L(a_+ - a_-)} \int_{-\infty}^{\infty} \frac{dk}{2\pi} \left[\frac{1}{k^2 + a_-} - \frac{1}{k^2 + a_+} \right] e^{ik(t-t')} \end{aligned} \tag{13.13.4}$$

since

$$\frac{1}{(k^2 + a_+)(k^2 + a_-)} = \frac{1}{(a_+ - a_-)} \left[\frac{1}{k^2 + a_-} - \frac{1}{k^2 + a_+} \right]$$

with

$$a_\pm = \frac{\widetilde{L}}{2L} \pm \left| \frac{\widetilde{L}}{2L} \right| \sqrt{1 - \frac{4L\gamma^2}{\widetilde{L}^2}}$$

Equation 13.13.4 yields[8]

$$D(t-t') = \frac{1}{2L(a_+ - a_-)}\Big(\frac{1}{\sqrt{a_-}}e^{-\sqrt{a_-}|t-t'|} - \frac{1}{\sqrt{a_+}}e^{-\sqrt{a_+}|t-t'|}\Big) \quad (13.13.5)$$

The correlation function has been studied in detail by [7], including its state space interpretation.

Case I: Complex branch.

$4L\gamma^2 > \widetilde{L}^2$ and $a_\pm$ are complex; let

$$\omega = \left(\frac{\gamma^2}{L}\right)^{\frac{1}{4}}, \; a_\pm = \sqrt{\frac{\gamma^2}{L}}e^{\pm i2\phi} \; ; \; \cos(2\phi) = \frac{\widetilde{L}}{\gamma\sqrt{4L}} \quad (13.13.6)$$

Note for the complex branch, from Eq. 13.13.6 the allowed domain for $\widetilde{L}$ is given by

$$-\gamma\sqrt{4L} \le \widetilde{L} \le +\gamma\sqrt{4L}$$

In fact, it is the hall mark of the complex branch that for most cases $\widetilde{L} < 0$; the action functional yields a convergent path integral because both $L, \gamma > 0$ for all branches. Hence, for the complex branch, one has the limits

$$-1 \le \cos(2\phi) \le +1 \;\Rightarrow\; -\pi/2 \le \phi \le +\pi/2$$

From Eqs. 13.13.5 and 13.13.6

$$D_C(t-t') = \frac{1}{4iL\sin(2\phi)}\left(\frac{L}{\gamma^2}\right)^{3/4}\left(\exp\{-\omega|t-t'|e^{-i\phi} + i\phi\} - c.c.\right)$$

$$= \frac{1}{2L\sin(2\phi)}\left(\frac{L}{\gamma^2}\right)^{3/4} \exp\left\{-\omega|t-t'|\cos(\phi)\right\}\sin\left[\phi + \omega|t-t'|\sin(\phi)\right]$$

The normalization constant is

$$\mathcal{N} = \frac{1}{2L\sin(2\phi)}\left(\frac{L}{\gamma^2}\right)^{3/4} = \frac{\omega}{2\gamma\sin 2\phi} \quad (13.13.7)$$

and the complex branch propagator is given by

[8] Note that from Eq. 13.8.4

$$\int_{-\infty}^{\infty}\frac{dk}{2\pi}\frac{e^{ik(t-t')}}{k^2+\omega^2} = \frac{1}{2\omega}e^{-\omega|t-t'|} \; ; \; \omega \, : \, \text{complex}$$

.

$$D_C(t-t') = \frac{\omega}{2\gamma \sin 2\phi} e^{-\omega|t-t'|\cos(\phi)} \sin\left[\phi + \omega|t-t'|\sin(\phi)\right] \tag{13.13.8}$$

The structure of the auto-correlation is that of an exponential dampening multiplied by the sine of the time lag $|t-t'|$—and is a reflection of the fourth derivative in the Lagrangian and is essential for applying the complex branch to the empirical correlation of commodity prices [8].

Case II: Real branch

$$4L\gamma^2 < \widetilde{L}^2 \text{ and } a_\pm \ : \text{real} \tag{13.13.9}$$

Choose the following parametrization

$$\omega = \left(\frac{\gamma^2}{L}\right)^{\frac{1}{4}}, \ a^\pm = \sqrt{\frac{\gamma^2}{L}} e^{\pm 2\vartheta} \ ; \ \cosh(2\vartheta) = \frac{\widetilde{L}}{\gamma\sqrt{4L}} \tag{13.13.10}$$

Note for the real branch, in addition to the constraint on the parameters, since $\cosh(2\vartheta) > 1$ there is an additional constraint that

$$\widetilde{L} > \sqrt{4L}\gamma > 0$$

Hence $D(t-t')$ is given by

$$D_R(t-t') = \frac{\omega}{2\gamma \sinh(2\vartheta)} e^{-\omega|t-t'|\cosh(\vartheta)} \sinh\left[\vartheta + \omega|t-t'|\sinh(\vartheta)\right] \tag{13.13.11}$$

The real branch is required for describing the behavior of coupon bond and interest rate models, discussed in Chap. 16.

Case III: Critical branch

$$4L\gamma^2 = \widetilde{L}^2 \ ; \quad \text{and } a_\pm = \frac{\widetilde{L}}{2L} \equiv \omega^2 \ : \ \text{real} \tag{13.13.12}$$

From Eq. 13.13.4, for $\tau = t - t'$

$$D(\tau) = \frac{1}{L}\int_{-\infty}^{\infty} \frac{dk}{2\pi} \frac{e^{ik\tau}}{(k^2+\omega^2)^2} = -\frac{1}{L}\frac{\partial}{\partial\omega^2}\int_{-\infty}^{\infty} \frac{dk}{2\pi} \frac{e^{ik\tau}}{(k^2+\omega^2)}$$

Using the result from Eq. 13.8.4 yields

$$D(\tau) = -\frac{1}{L}\frac{\partial}{\partial\omega^2}\left[\frac{1}{2\omega}e^{-\omega|\tau|}\right] = \frac{1}{4L\omega^3}\left[1 + \omega|\tau|\right] e^{-\omega|\tau|} \tag{13.13.13}$$

The auto-correlation function for the critical branch is shown in Fig. 13.3.

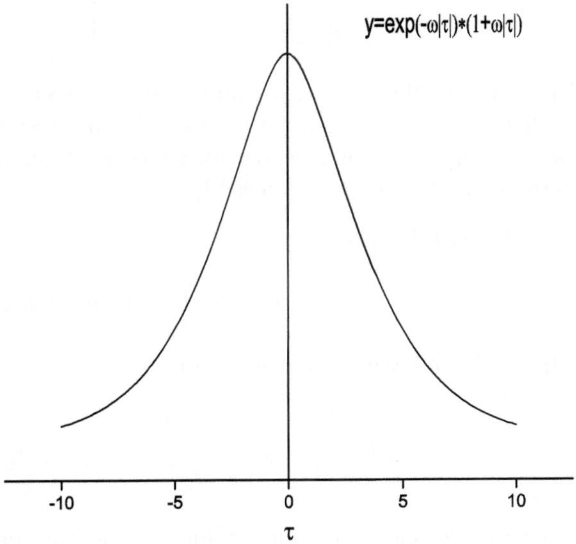

Fig. 13.3 The auto-correlation function $E[x(0)]x(\tau)]$ for the critical branch of the acceleration Lagrangian. Published with permission of © Belal E. Baaquie 2020. All Rights Reserved

For $\tau \approx 0$, the auto-correlation function is

$$D(\tau) \simeq \frac{1}{4L\omega^3}[1 - \omega^2\tau^2 + \cdots]$$

The dependence on $|\tau|$, which could give rise to a 'kink' at $\tau = 0$ for $D(\tau)$, has been removed in Eq. 13.13.13 and hence $D(\tau)$ is smooth (differentiable) for all values of τ. In particular, the derivative of $D(\tau)$ is finite at $\tau = 0$, unlike the case of the simple harmonic oscillator given in Fig. 13.2.

The higher derivative of the acceleration Lagrangian smooths out the functional integral, removing configurations that would give rise to the 'kink' at $\tau = 0$, and yields $D(\tau)$ that is differentiable at $\tau = 0$.

$D(\tau)$ being differentiable for all values of τ is a feature for all the three branches, and is the main reason why the higher derivative term was added to the simple harmonic Lagrangian [4].

Chapter 14
Hamiltonians

Abstract The formalism of Hamiltonians and path integrals forms one of the cornerstones of theoretical physics. The branch of knowledge that studies Hamiltonians and path integrals is a vast subject that a single chapter can hardly do justice to [7]. As the study of quantum mechanics shows, there are no theorems or general results that one can focus on; instead, one has important examples that illustrate the general techniques of the subject. A similar approach is taken in this Chapter: important exemplars of finance, which are the Black–Scholes and Merton equations, are analyzed to illustrate how quantum mathematics can be employed to analyze these models. The 'free particle' is seen to be the system that the Black–Scholes equation requires, and the simple harmonic oscillator model describes one version of the Merton model.

14.1 Introduction

The Black–Scholes option pricing equation and Merton equations of corporate finance have been analyzed using the mathematics of stochastic processes. Both the Black–Scholes and Merton equations yield homogeneous linear partial differential equations—which can be expressed in terms of a Hamiltonian operator driving the time evolution. The pricing kernel for both the equations is expressed by the matrix elements of the exponential of the Hamiltonian.

The Black–Scholes Hamiltonian is a linear differential operator and yields an exact solution for the pricing kernel. It is shown that the eigenfunctions of the Black–Scholes Hamiltonian with appropriate boundary conditions can be used for exactly solving a class of barrier options.

The Merton Hamiltonian is a nonlinear and non-Hermitian differential operator. It is shown that the Merton Hamiltonian is equivalent to an effective Hermitian Hamiltonian, making it more amenable for further analysis. A linear approximation is made of the Merton Hamiltonian and shown to be equivalent to the quantum harmonic oscillator. The quantum oscillator Hamiltonian is one of most important

B. E. Baaquie, *Mathematical Methods and Quantum Mathematics for Economics and Finance*, https://doi.org/10.1007/978-981-15-6611-0_14

exemplar in quantum mechanics, and the linearized Merton Hamiltonian opens the way for applying many results of quantum mechanics to corporate finance.

Starting from the definition of a martingale, it is shown that the time-evolution of a security obeys the martingale condition only if the security is annihilated by the Hamiltonian that is driving its time-evolution.

14.2 Black–Scholes and Merton Hamiltonian

The Black–Scholes Eq. 11.7.4 for option price with constant volatility is given by

$$\frac{\partial C}{\partial t} = -\frac{1}{2}\sigma^2 S^2 \frac{\partial^2 C}{\partial S^2} - rS\frac{\partial C}{\partial S} + rC \tag{14.2.1}$$

From Eq. 11.7.5 Black–Scholes equation is given by

$$\frac{\partial C}{\partial t} = -\frac{1}{2}\sigma^2 \frac{\partial^2 C}{\partial x^2} + \left(\frac{1}{2}\sigma^2 - r\right)\frac{\partial C}{\partial x} + rC$$

This yields the Black–Scholes-Schrödinger equation

$$\frac{\partial C}{\partial t} = H_{BS} C \tag{14.2.2}$$

with the Black–Scholes Hamiltonian given by

$$H_{BS} = -\frac{\sigma^2}{2}\frac{\partial^2}{\partial x^2} + \left(\frac{1}{2}\sigma^2 - r\right)\frac{\partial}{\partial x} + r \tag{14.2.3}$$

The option price C is the analog of the Schrödinger state function of quantum mechanics—and unlike the case of the quantum mechanical state function, C can be directly observed.

Merton equation given in Eq. 12.5.1 is interpreted in the formalism of quantum mechanics. The time evolution equation is analyzed to obtain the underlying Hamiltonian that is driving the contingent claim. Recall from Eq. 12.5.1, Merton equation is given by

$$\frac{\partial C}{\partial t} = -\frac{1}{2}\sigma^2 \frac{\partial^2 C}{\partial x^2} + \left(\frac{1}{2}\sigma^2 - r - v\right)\frac{\partial C}{\partial x} + (r+q)C \tag{14.2.4}$$

Equation 14.2.4 is re-written as the Merton–Schrödinger equation

$$\frac{\partial C}{\partial t} = H_M C \tag{14.2.5}$$

The Merton Hamiltonian H_M, from Eq. 14.2.4, is given by [14]

$$H_M(v, q) = -\frac{\sigma^2}{2}\frac{\partial^2}{\partial x^2} + \left(\frac{1}{2}\sigma^2 - r - v\right)\frac{\partial}{\partial x} + r + q \tag{14.2.6}$$

and is mathematically identical to a quantum mechanical particle moving in a (nonlinear) potential. Note the Merton Hamiltonian H_M is non-Hermitian since

$$H_M^\dagger \neq H_M$$

The Merton Hamiltonian $H_M(v, q)$ is equivalent to a Hermitian Hamiltonian $H_{\text{Eff}}(v, q)$, related by a similarity transformation [7]

$$H_M = e^s H_{\text{Eff}} e^{-s}$$

Hence, H_M is pseudo-Hermitian since [7]

$$H_M^\dagger = e^{-2s} H_M e^{2s}$$

The *effective Hermitian Hamiltonian* $H_{\text{Eff}}(v, q)$, from Eq. 14.2.6, is given by [14]

$$H_M(v, q) = e^s H_{\text{Eff}}(v, q) e^{-s} \tag{14.2.7}$$
$$\Rightarrow H_{\text{Eff}}(v, q) = -\frac{1}{2}\sigma^2\frac{\partial^2}{\partial x^2} + \frac{1}{2}\frac{\partial v}{\partial x} + \frac{1}{2\sigma^2}\left(\frac{1}{2}\sigma^2 - r - v(x)\right)^2 + r + q(x)$$

where s is given by [7]

$$s = \frac{1}{\sigma^2}\int_0^x dy\left(\frac{1}{2}\sigma^2 - r - v(y)\right) = \frac{1}{\sigma^2}\left(\frac{1}{2}\sigma^2 - r\right)x - \frac{1}{\sigma^2}\int_0^x dy v(y) \tag{14.2.8}$$

To obtain H_{Eff} given in Eq. 14.2.7, the following identity is required [7]

$$e^s\frac{\partial^2}{\partial x^2}e^{-s} = \left(\frac{\partial s}{\partial x}\right)^2 - \frac{\partial^2 s}{\partial x^2} - 2\left(\frac{\partial s}{\partial x}\right)\frac{\partial}{\partial x} + \frac{\partial^2}{\partial x^2}$$

Note that Merton Hamiltonian H_M is a nonlinear and non-Hermitian differential operator, whereas the effective Hamiltonian is a nonlinear Hermitian H_{Eff} operator. In Chap. 15, the path integral for the Merton Hamiltonian is based on the Hermitian H_{Eff}.

The special case of $v = 0 = q$ of the Merton Hamiltonian yields the Black–Scholes Hamiltonian given in Eq. 14.2.3

$$H_{BS} = -\frac{\sigma^2}{2}\frac{\partial^2}{\partial x^2} + \left(\frac{1}{2}\sigma^2 - r\right)\frac{\partial}{\partial x} + r \tag{14.2.9}$$

and, from Eqs. 14.2.7 and 14.2.8.

$$H_{\text{BS,Eff}}(\upsilon, q) = -\frac{1}{2}\sigma^2\frac{\partial^2}{\partial x^2} + \frac{1}{2\sigma^2}\Big(\frac{1}{2}\sigma^2 - r\Big)^2 + r$$
$$s = \frac{1}{\sigma^2}\Big(\frac{1}{2}\sigma^2 - r\Big)x \ : \ \text{Black–Scholes} \tag{14.2.10}$$

14.3 Option Pricing Kernel

For a general Hamiltonian H, the evolution equation for a contingent claim is the following

$$\frac{\partial C}{\partial t} = HC \ \Rightarrow \ C(t, x) = e^{tH}C(0, x) \tag{14.3.1}$$

where $C(0, x)$ is the initial value of the contingent claim.

In Dirac's notation, the price of the contingent claim is given by the ket vector $|C, t\rangle$. Equation 14.3.1, in Dirac's notation, is given by

$$\langle x|\frac{\partial}{\partial t}|C, t\rangle = \langle x|H|C, t\rangle$$
$$\Rightarrow \ \frac{\partial}{\partial t}|C, t\rangle = H|C, t\rangle \ \Rightarrow \ |C, t\rangle = e^{tH}|C, 0\rangle \tag{14.3.2}$$

The final value of a contingent claim maturing at time T is $|C, T\rangle$ and is equal to the payoff function (ket vector), denoted by $|g\rangle$; hence

$$\langle x|C, T\rangle = \langle x|g\rangle$$

For example, the payoff of a call option is $\langle x|g\rangle = (V - K)_+$; hence, from Eq. 14.3.2

$$|C, T\rangle = e^{TH}|C, 0\rangle = |g\rangle \ \Rightarrow \ |C, 0\rangle = e^{-TH}|g\rangle$$

and yields

$$|C, t\rangle = e^{-(T-t)H}|g\rangle \tag{14.3.3}$$

Define remaining time to maturity by $\tau = T - t$. In terms of remaining time τ, the evolution equation for the contingent claim is given by

$$-\frac{\partial C}{\partial \tau} = HC \tag{14.3.4}$$

Remaining time runs backwards: when $\tau = 0$, calendar time $t = T$ and when $\tau = T$ calendar time is given by $t = 0$. From Eq. 14.3.3

$$C(t, x) = \langle x|C, t\rangle = \langle x|e^{-\tau H}|g\rangle \tag{14.3.5}$$

Using the completeness equation given by Eq. 13.3.7

$$\mathbb{I} = \int dx|x\rangle\langle x| \tag{14.3.6}$$

yields from Eq. 14.3.5

$$C(t, x) = \int_{-\infty}^{\infty} dx'\langle x|e^{-\tau H}|x'\rangle\langle x'|g\rangle = \int_{-\infty}^{\infty} dx' p(x, x'; \tau)g(x') \tag{14.3.7}$$

where the *pricing kernel*, in terms of the Hamiltonian H, is given by [4]

$$p(x, x'; \tau) = \langle x|e^{-\tau H}|x'\rangle \tag{14.3.8}$$

For the Merton Hamiltonian given in Eq. 14.2.7, the pricing kernel is the following

$$\begin{aligned} p_{(v,q)}(x, x'; \tau) &= \langle\, x|e^{-\tau H_M}|x'\rangle = \langle\, x|e^{s}e^{-\tau H_{\rm Eff}}e^{-s}|x'\rangle \\ &= \exp\{s(x) - s(x')\}\langle\, x|e^{-\tau H_{\rm Eff}}|x'\rangle \\ &\equiv \exp\{s(x) - s(x')\}\widetilde{p}_{(v,q)}(x, x'; \tau) \end{aligned} \tag{14.3.9}$$

14.4 Black–Scholes Pricing Kernel

From Eq. 14.2.10

$$H_{BS} = e^{s}H_{BS,\rm Eff}e^{-s} \;\; ; \;\; H_{\rm BS,Eff}(v, q) = -\frac{1}{2}\sigma^2\frac{\partial^2}{\partial x^2} + c$$
$$s = \frac{1}{\sigma^2}\Big(\frac{1}{2}\sigma^2 - r\Big)x \; ; \; c = \frac{1}{2\sigma^2}\Big(\frac{1}{2}\sigma^2 - r\Big)^2 + r \;\; : \;\; \text{Black–Scholes}$$

and, from Eq. 14.3.8, the pricing kernel is given by

$$\begin{aligned} p_{BS}(x, x'; \tau) &= \langle x|e^{-\tau H_{BS}}|x'\rangle = e^{s(x)-s(x')}\langle x|e^{-\tau H_{BS,\rm Eff}}|x'\rangle \\ &= \exp\Big\{s(x) - s(x') - c\tau\Big\}\langle x|e^{\frac{1}{2}\tau\sigma^2\frac{\partial^2}{\partial x^2}}|x'\rangle \end{aligned} \tag{14.4.1}$$

The evaluation of Eq. 14.4.1 can be done efficiently by going to the 'momentum' basis in which $\partial^2/\partial x^2$ is diagonal. The Fourier transform of the $|x\rangle$ basis to momentum space, from Eq. 13.7.2, is given by

$$\langle x|x'\rangle = \delta(x-x') = \int_{-\infty}^{\infty}\frac{dp}{2\pi}\mathrm{e}^{ip(x-x')} = \int_{-\infty}^{\infty}\frac{dp}{2\pi}\langle x|p\rangle\langle p|x'\rangle \tag{14.4.2}$$

that yields, for momentum space basis $|p\rangle$ the completeness equation

$$\int_{-\infty}^{\infty}\frac{dp}{2\pi}|p\rangle\langle p| = \mathbb{I} \tag{14.4.3}$$

with the scalar product

$$\langle x|p\rangle = \mathrm{e}^{ipx}\ ;\ \langle p|x\rangle = \mathrm{e}^{-ipx}. \tag{14.4.4}$$

Note that

$$\langle x|(-\sigma^2\frac{\partial^2}{\partial x^2})|p\rangle = -\sigma^2\frac{\partial^2}{\partial x^2}\mathrm{e}^{ipx} = \sigma^2 p^2\mathrm{e}^{ipx} \tag{14.4.5}$$

Using Eq. 14.4.5 above and the completeness equation given in Eq. 14.4.3, and performing a Gaussian integration, yields

$$\begin{aligned}\langle x|e^{\frac{1}{2}\sigma^2\tau\frac{\partial^2}{\partial x^2}}|x'\rangle &= \int_{-\infty}^{\infty}\frac{dp}{2\pi}\langle x|e^{\frac{1}{2}\sigma^2\tau\frac{\partial^2}{\partial x^2}}|p\rangle\langle p|x'\rangle\\ &= \int_{-\infty}^{\infty}\frac{dp}{2\pi}e^{-\frac{1}{2}\sigma^2\tau p^2}\langle x|p\rangle\langle p|x'\rangle = \int_{-\infty}^{\infty}\frac{dp}{2\pi}e^{-\frac{1}{2}\sigma^2\tau p^2}e^{ip(x-x')}\\ \Rightarrow \langle x|e^{\frac{1}{2}\tau\sigma^2\frac{\partial^2}{\partial x^2}}|x'\rangle &= \frac{1}{\sqrt{2\pi\sigma^2\tau}}\exp\left\{-\frac{1}{2\sigma^2\tau}(x-x')^2\right\}\end{aligned} \tag{14.4.6}$$

Noteworthy 14.1: Pricing Kernel

The result obtained in Eq. 14.4.6 directly follows from the properties of the Dirac δ-function. From Eqs. 14.4.1 and 14.4.2

$$\begin{aligned}\langle x|e^{\frac{1}{2}\tau\sigma^2\frac{\partial^2}{\partial x^2}}|x'\rangle &= e^{\frac{1}{2}\tau\sigma^2\frac{\partial^2}{\partial x^2}}\delta(x-x')\\ &= e^{\frac{1}{2}\tau\sigma^2\frac{\partial^2}{\partial x^2}}\int_{-\infty}^{\infty}\frac{dp}{2\pi}e^{ip(x-x')} = \int_{-\infty}^{\infty}\frac{dp}{2\pi}e^{ip(x-y)}e^{-\frac{1}{2}\tau\sigma^2p^2}\\ &= \frac{1}{\sqrt{2\pi\tau\sigma^2}}\exp\left\{-\frac{1}{2\sigma^2\tau}(x-x')^2\right\}\end{aligned} \tag{14.4.7}$$

Hence, from Eqs. 14.4.1 and 14.4.6

$$p_{BS}(x, x'; \tau) = \frac{1}{\sqrt{2\pi\sigma^2\tau}} \exp\left\{s(x) - s(x') - c\tau - \frac{1}{2\sigma^2\tau}(x - x')^2\right\}$$
$$= \frac{1}{\sqrt{2\pi\sigma^2\tau}} \exp\left\{-r\tau - \frac{1}{2\sigma^2\tau}\left(x - x' + \tau(r - \frac{1}{2}\sigma^2)\right)^2\right\} \quad (14.4.8)$$

Equation 14.4.8 is the result for Black–Scholes, which has been obtained in Eq. 11.10.1 using the martingale condition.

Given the importance of the Black–Scholes pricing kernel, another derivation is given using the eigenfunctions of the Black–Scholes Hamiltonian H_{BS}.

From the definition of the Hamiltonian given in Eq. 14.2.3

$$\langle x|H_{BS}|p\rangle \equiv H_{BS}\langle x|p\rangle \equiv E_{BS}(p)e^{ipx} \quad (14.4.9)$$
$$\Rightarrow E_{BS}(p) = \left\{\frac{1}{2}\sigma^2 p^2 + i\left(\frac{1}{2}\sigma^2 - r\right)p + r\right\}$$

where $E_{BS}(p)$ are the (complex valued) eigenvalues of the operator H_{BS} labeled by the 'momentum' index p; from Eq. 14.4.9, e^{ipx} are the eigenfunctions. Equation 14.4.3 shows that the eigenfunctions of H_{BS} are complete. Hence, using Eq. 14.4.9

$$p_{BS}(x, \tau; x') = \langle x|\mathrm{e}^{-\tau H_{BS}}|x'\rangle = \int_{-\infty}^{\infty} \frac{dp}{2\pi}\langle x|\mathrm{e}^{-\tau H_{BS}}|p\rangle\langle p|x'\rangle$$
$$= \int_{-\infty}^{\infty} \frac{dp}{2\pi}\mathrm{e}^{-\tau E_{BS}(p)}\langle x|p\rangle\langle p|x'\rangle$$
$$= e^{-r\tau}\int_{-\infty}^{\infty} \frac{dp}{2\pi}e^{-\frac{1}{2}\tau\sigma^2 p^2 - i\tau(\sigma^2/2 - r)p}e^{ip(x-x')} \quad (14.4.10)$$

Performing the Gaussian integration in Eq. 14.4.10 above gives the pricing kernel for the Black–Scholes equation

$$p_{BS}(x, \tau; x') = \langle x|\mathrm{e}^{-\tau H_{BS}}|x'\rangle$$
$$= \frac{e^{-r\tau}}{\sqrt{2\pi\tau\sigma^2}} \exp\left[-\frac{1}{2\tau\sigma^2}\{x - x' + \tau(r - \sigma^2/2)\}^2\right] \quad (14.4.11)$$

which is result obtained earlier in Eq. 14.4.7.

14.5 Merton Oscillator Hamiltonian

Consider the oscillator potential given by

$$v(x) = \omega x \;\; ; \;\; q(x) = \lambda x \;\; ; \;\; V = e^x V_0 \tag{14.5.1}$$

For the linear potential, Eq. 14.2.3 yields the oscillator Merton Hamiltonian

$$H_M(\omega, \lambda) = -\frac{\sigma^2}{2}\frac{\partial^2}{\partial x^2} + \left(\frac{1}{2}\sigma^2 - r - \omega x\right)\frac{\partial}{\partial x} + r + \lambda x \tag{14.5.2}$$

The price of the contingent claim C, from Eq. 14.3.4, is given by

$$-\frac{\partial C}{\partial \tau} = H(\omega, \lambda)C = -\frac{\sigma^2}{2}\frac{\partial^2 C}{\partial x^2} + \left(\frac{1}{2}\sigma^2 - r - \omega x\right)\frac{\partial C}{\partial x} + (r + \lambda x)C \tag{14.5.3}$$

From Eq. 14.2.8 the *effective oscillator Hamiltonian* $H_{\text{Eff}}(\omega, \lambda)$ is given by

$$\begin{aligned} H_M(\omega, \lambda) &= e^s H_{\text{Eff}}(\omega, \lambda)e^{-s} \\ \Rightarrow H_{\text{Eff}}(\omega, \lambda) &= -\frac{1}{2}\sigma^2\frac{\partial^2}{\partial x^2} + \frac{1}{2}\omega + \frac{1}{2\sigma^2}\left(\frac{1}{2}\sigma^2 - r - \omega x\right)^2 + r + \lambda x \\ &= -\frac{1}{2}\sigma^2\frac{\partial^2}{\partial x^2} + \frac{1}{2\sigma^2}\omega^2 x^2 + hx + c \end{aligned} \tag{14.5.4}$$

where

$$h = \frac{\omega}{\sigma^2}\left(r - \frac{1}{2}\sigma^2 + \sigma^2\frac{\lambda}{\omega}\right) \;\; ; \;\; c = r + \frac{1}{2}\omega + \frac{1}{2\sigma^2}\left(r - \frac{1}{2}\sigma^2\right)^2 \tag{14.5.5}$$

Equations 14.2.8 and 14.5.1 yield

$$s = \left(\frac{1}{2} - \frac{r}{\sigma^2}\right)x - \frac{1}{2\sigma^2}\omega x^2 \tag{14.5.6}$$

14.6 Hamiltonian: Martingale Condition

Consider an option on a security $S = e^x$ that matures at time T and has a payoff function given by $g(x)$. The risk-free evolution of the security is given by the Hamiltonian H, with the value of the option at time $t < T$ being given by Eq. 14.3.5 as follows

$$C(t, x) = \int_{-\infty}^{\infty} dx' \langle x | e^{-(T-t)H} | x' \rangle \, g(x') \tag{14.6.1}$$

The martingale condition for the risk-free evolution of the security is that the price of the security at some future time, say t_*, when discounted by a martingale measure, is equal, on the average, to the price of the security at earlier time t. The equation for the martingale condition, from Eq. 11.9.2, is given by

$$S(t) = E[e^{-(t_*-t)r} S(t_*)|S(t)] \tag{14.6.2}$$

Writing the martingale condition given in Eq. 14.6.2 in terms of the pricing kernel yields

$$e^x = \int_{-\infty}^{\infty} dx' p(x, x'; \tau) e^{x'} \ ; \ \tau = t_* - t \tag{14.6.3}$$

Clearly, the martingale condition is an instantaneous condition since it is valid for any t_*. In terms of the Hamiltonian

$$S(x) = \int_{-\infty}^{\infty} dx' \langle x | e^{-(t_*-t)H} | x' \rangle \, S(x')$$

and in Dirac's notation

$$\langle x | S \rangle = \int_{-\infty}^{\infty} dx' \langle x | e^{-(t_*-t)H} | x' \rangle \, \langle x' | S \rangle \tag{14.6.4}$$

Using the completeness equation for a single security given in Eq. 13.3.7 yields

$$\mathbb{I} = \int_{-\infty}^{\infty} dx |x\rangle\langle x|$$

yields from Eq. 14.6.4, the (eigenstate) equation

$$|S\rangle = e^{-(t_*-t)H} |S\rangle$$

Since time t_* is arbitrary, the instantaneous expression of the martingale condition is given by [4]

$$H|S\rangle = 0 \ \Rightarrow \ H\left(x, \frac{\partial}{\partial x}\right) S(x) = 0 \tag{14.6.5}$$

Hence it can be seen that the security $S = e^x$ is a very special *eigenstate* of H, namely having zero energy eigenvalue; the equity is an element of the state space that is not normalizable. The equity having zero eigenenergy means that under a martingale evolution driven by H, the underlying security changes at the risk-free rate.

Note that the only property of S that was employed in deriving the Hamiltonian condition given in Eq. 14.6.5 is that S has a *martingale* evolution. For *any financial instrument* $\chi(t)$ that has its evolution driven by a Hamiltonian $\mathcal{H}$, the martingale condition is given by

$$\mathcal{H}\chi = 0 \tag{14.6.6}$$

The Black–Scholes Hamiltonian, from Eq. 14.6.12, is given by

$$H_{BS} = -\frac{\sigma^2}{2}\frac{\partial}{\partial x^2} + \left(\frac{\sigma^2}{2} - r\right)\frac{\partial}{\partial x} + r$$

and it follows that

$$H_{BS}e^x = 0$$

From the derivation of the martingale condition, it can be seen that the form of the Hamiltonian in option pricing is constrained by the requirement of annihilating the stock price $S(t)$, so as to fulfill the martingale condition given in Eq. 14.6.5.

Recall from Eq. 14.2.6, the Merton Hamiltonian H_M is given by

$$H_M(\upsilon, q) = -\frac{\sigma^2}{2}\frac{\partial^2}{\partial x^2} + \left(\frac{1}{2}\sigma^2 - r - \upsilon\right)\frac{\partial}{\partial x} + r + q \tag{14.6.7}$$

The Merton Hamiltonian H_M *does not obey* the martingale condition since

$$H_M e^x = (-\upsilon + q)e^x \neq 0$$

The fact that the Merton Hamiltonian does not obey the martingale does not imply that the derivative security C priced using the Merton Hamiltonian has arbitrage opportunities. The reason being that there is an inflow (or outflow depending on the sign) of $\upsilon(x)$ to the underlying, and similarly $q(x)$ for the derivative. In contrast, for the Black–Scholes option the martingale condition is based on the value of a dynamic portfolio—equivalent to the option and discussed in Sect. 9.2—with no inflow or outflow of capital.

Consider the special case of the Merton Hamiltonian, which is given by

$$\upsilon = q \text{ and define } U(x) \equiv r + \upsilon(x) = r + q(x) \tag{14.6.8}$$

The special case of the Merton Hamiltonian, from Eq. 14.6.7, is given by

$$H_U = -\frac{\sigma^2}{2}\frac{\partial^2}{\partial x^2} + \left(\frac{1}{2}\sigma^2 - U(x)\right)\frac{\partial}{\partial x} + U(x) \tag{14.6.9}$$

where the potential $U(x)$ is an arbitrary function of x. The Hamiltonian fulfills the martingale condition since

$$H_U e^x = 0$$

The Hamiltonian given in Eq. 14.6.9 was written in [4] based solely on satisfying the martingale condition. The interpretation from finance for the Hamiltonian H_U is the following: the derivative security C obeys the martingale condition if the rate of inflow of payments $v(x)$ equals the rate of increase in the value $q(x)$ of the derivative security C .

The Hamiltonian H_U can in principle be used for pricing derivative securities of V. Similar to the analysis for the Black–Scholes case, the pricing kernel, from Eq. 14.3.8, is given by

$$p_U(x,\tau|x') = \langle x|\mathrm{e}^{-\tau H_U}|x'\rangle = \int_{BS} DX\mathrm{e}^{S_U}$$

where

$$S_U = -\int_0^\tau dt\left[\frac{1}{2\sigma^2}\left(\frac{dx}{dt} - U(x) + \frac{1}{2}\sigma^2\right)^2 + U(x)\right] \tag{14.6.10}$$

H_U is equivalent via a similarity transformation to a Hermitian Hamiltonian H_{Eff} given by [4]

$$H_U = e^s H_{\text{Eff}} e^{-s} \tag{14.6.11}$$

where, from Eqs. 14.2.7 and 14.2.8

$$H_{\text{Eff}} = -\frac{\sigma^2}{2}\frac{\partial^2}{\partial x^2} + \frac{1}{2}\frac{\partial U}{\partial x} + \frac{1}{2\sigma^2}\left(U + \frac{1}{2}\sigma^2\right)^2 \quad ; \quad s = \frac{1}{2}x - \frac{1}{\sigma^2}\int_0^x dy U(y)$$

H_{Eff} is Hermitian and hence its eigenfunctions form a complete basis; from this it follows that the Hamiltonian H_U can also be diagonalized using the eigenfunctions of H_{Eff}. In particular

$$H_{\text{Eff}}|\phi_n\rangle = E_n|\phi_n\rangle \quad \Rightarrow \quad H_U|\psi_n\rangle = E_n|\psi_n\rangle$$

where

$$|\psi_n\rangle = e^s|\phi_n\rangle \quad ; \quad \langle\,\tilde{\psi}_n| = e^{-s}\langle\,\phi_n| \neq \langle\,\psi_n|$$

The Black–Scholes Hamiltonian H_{BS} has $U(x) = r$ and hence

$$H_{BS} = e^s H_{\text{Eff}} e^{-s} = e^{\alpha x}\left[-\frac{\sigma^2}{2}\frac{\partial^2}{\partial x^2} + \gamma\right]e^{-\alpha x} \tag{14.6.12}$$

where

$$H_{\text{Eff}} = -\frac{\sigma^2}{2}\frac{\partial^2}{\partial x^2} + \gamma\ ;\ \alpha = \frac{1}{\sigma^2}\left(\frac{1}{2}\sigma^2 - r\right)\ ;\ \gamma = \frac{1}{2\sigma^2}\left(r + \frac{1}{2}\sigma^2\right)^2 \tag{14.6.13}$$

14.7 Hamiltonian and Potentials

Can one, in principle, include a potential term in the option pricing Hamiltonian? The answer is yes: one can in fact include a potential term in the Black–Scholes formalism as given in Eq. 14.6.9, and the potential can also be used to represent a certain class of path-dependent options, as has been considered in [8]. In particular, some of the barrier options, as seen in Sect. 14.8, can be expressed as a problem of a Hamiltonian with a potential.

Path dependent options, such as the barrier options, are all evolved by the Black–Scholes Hamiltonian H_{BS}; the entire effect of barrier options (discussed in Sect. 14.7), and of some path dependent options can be effectively realized by a potential $V(x)$ that is **added** to the Black–Scholes Hamiltonian H_{BS}, and yields an effective Hamiltonian given by

$$H_{\text{effective}} = H_{BS} + U$$

The Lagrangian given in Eq. 15.3.1 for a class of path dependent options yields

$$H_{\text{Path Dependent Option}} = H_{BS} + igf(x)$$

where the function $f(x)$ encodes the path dependent option. In particular, from Eq. 15.3.2 for the (path dependent) Asian option, its non-Hermitian Hamiltonian is given by

$$H_{\text{Asian Option}} = H_{BS} + ige^x$$

and has been discussed in [4, 8].

14.8 Double Knock Out Barrier Option

A double barrier option confines the value of the stock to lie between two barriers, denoted by e^a and e^b as in shown in Figure 14.1, with the value of the option being zero if the stock price takes a value outside the barrier.

The Black–Scholes Hamiltonian is used to solve the barrier option problem. Consider the Hamiltonian $H_{\text{BS}} + U$ for the double knock out barrier option. The fundamental idea is that the option becomes worthless the moment the stock price 'hits'

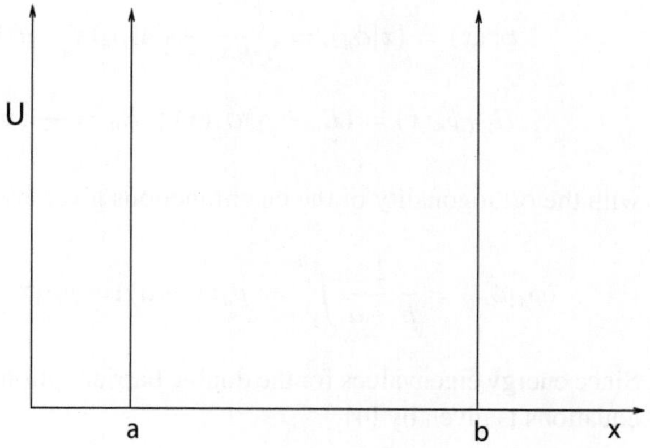

Fig. 14.1 Potential Barrier U for Up and Out Barrier Option. Published with permission of © Belal E. Baaquie 2020. All Rights Reserved

the barrier—that is, equals the barrier value. *Boundary condition* are imposed on the stock price by choosing its *eigenfunctions* to be zero on the boundary. This is achieved by choosing a potential $U(x)$ so that the eigenfunctions vanish outside the barrier.

The double knock out barrier's Hamiltonian is given by

$$H_{DB} = H_{BS} + U(x) = e^{s}[H_{\mathrm{Eff}} + U(x)]e^{-s} \tag{14.8.1}$$

The Black–Scholes Hamiltonian, from Eq. 14.6.12, is given by

$$\begin{aligned} H_{BS} &= -\frac{\sigma^2}{2}\frac{\partial}{\partial x^2} + \left(\frac{\sigma^2}{2} - r\right)\frac{\partial}{\partial x} + r \\ &= e^{s} H_{\mathrm{Eff}} e^{-s} \\ &= e^{\alpha x}\left[-\frac{\sigma^2}{2}\frac{\partial^2}{\partial x^2} + \gamma\right]e^{-\alpha x} \end{aligned} \tag{14.8.2}$$

The potential barrier for double knock-out barrier option $U(x)$ is given by

$$U(x) = \begin{cases} \infty & x < a \\ 0 & a < x < b \\ \infty & x > b \end{cases} \tag{14.8.3}$$

The potential results in eigenfunctions that vanish on both of the boundaries. This is the well known problem of a particle in an infinitely deep quantum well. The Hamiltonian H_{Eff} has (normalized) eigenfunctions $|\phi_n\rangle$, *vanishing* at both boundaries, with eigenenergies $E_n + \gamma$ and given by

$$\phi_n(x) = \langle x|\phi_n\rangle = \sqrt{\frac{2}{b-a}}\sin[p_n(x-a)] \ ; \ p_n = \frac{n\pi}{b-a} \tag{14.8.4}$$
$$H_{\text{Eff}}\phi_n(x) = (E_n+\gamma)\phi_n(x) \ ; \ E_n = \frac{\sigma^2}{2}(p_n)^2 \ ; \ n = 1, 2, \ldots \infty$$

with the orthogonality of the eigenfunctions given by

$$\langle\phi_n|\phi_{n'}\rangle = \frac{2}{b-a}\int_a^b \sin p_n(x-a)\sin p_{n'}(x-a)dx = \delta_{n-n'} \tag{14.8.5}$$

Since energy eigenvalues for the double barrier option are discrete, the completeness equations is given by [4]

$$\sum_{n=1}^{+\infty}|\phi_n\rangle\langle\phi_n| = \mathbb{I} \ \Rightarrow \ \sum_{n=1}^{+\infty}\langle x|\phi_n\rangle\langle\phi_n|x'\rangle = \delta(x-x')$$

Hence the pricing kernel is given by [4]

$$\begin{aligned} p_{DB}(x,x') &= \langle x|e^{-\tau H_{DB}}|x'\rangle = \langle x|e^{s}e^{-\tau H_{\text{Eff}}}e^{-s}|x'\rangle \\ &= e^{-\gamma\tau}e^{\alpha(x-x')}\sum_{n=1}^{\infty}e^{-\tau E_n}\phi_n(x)\phi_n(x') \\ &= \frac{e^{-\gamma\tau+\alpha(x-x')}}{2(b-a)}\sum_{n=-\infty}^{\infty}e^{-\frac{\tau\sigma^2p_n^2}{2}}(e^{ip_n(x-x')} - e^{ip_n(x+x'-2a)}) \end{aligned} \tag{14.8.6}$$

The price, at time t, of a double knock out barrier call option expiring at time T and with strike price K—provided it has not already been knocked out—is given by $(\tau = T-t)$

$$C_{DB}(x;\tau;K) = e^{-r\tau}E[(e^{x(T)}-K)_+\mathbf{1}_{a<x(t')<b,\ t<t'<T}] \tag{14.8.7}$$

where **1** stands for the indicator function that is non-zero only if the path is within the barrier. The price of the double knock out barrier call option, using Eq. 14.8.6, is given by

$$C_{DB}(x;\tau;K) = \int_{-\infty}^{\infty}dx' p_{DB}(x,x')(e^{x'}-K)_+$$

The pricing of other barrier options such as the down-and-in, up-and-out, the soft barrier options, exotic options and so on can be defined by appropriate choices of $U(x)$.

Chapter 15
Path Integrals

Abstract The Hamiltonian formalism for pricing options as well as contingent claims on a firm's valuation is complemented by the path integral. Only a handful of problems in finance are directly tractable using the Hamiltonian since in general nonlinear systems, such as the Merton Hamiltonian, are too difficult to solve analytically. One can recast the problem in the language of path integration; this allows for a different approach and point of view to attack otherwise intractable problems, and in some cases leads to simplifications that make the problem easier to solve.

15.1 Introduction

The *Feynman Path Integral* [7, 12, 13, 21, 37] is a functional integral. However, it has the distinguishing feature that unlike functional integration in general, the path integral has an underlying infinite dimensional linear vector space on which the Hamiltonian acts; the path integral is built out of the linear vector space. It is this combination of functional integration and linear vector space that makes the path integral exceptionally powerful—in both modeling random phenomena as well as in providing mathematical and theoretical tools for analytical and numerical studies of random phenomena. The path integral has proven to be a powerful tool for studying random systems.

Path integration is a vast subject [7], and instead of trying to address the subject in general, the focus of this Chapter is on analyzing the path integrals that are of direct relevance to finance. The two leading models of finance, which are the Black-Scholes equation for option pricing and the Merton equation of corporate finance, are re-expressed in the formalism of path integration. This Chapter concentrates on developing the path integral for pricing options and modeling a firm's contingent claims.

The path integral is defined using the Hamiltonian. The Hamiltonian H is a differential operator and recasting the matrix elements of $\exp\{-\tau H\}$ in terms of the path integral provides another distinct analytical tool—as well as another point of view—for studying the problem on hand. Since the matrix elements of $\exp\{-\tau H\}$

B. E. Baaquie, *Mathematical Methods and Quantum Mathematics for Economics and Finance*, https://doi.org/10.1007/978-981-15-6611-0_15

directly yields the pricing kernel required for option pricing, the path integral is particularly suitable for studying option theory—and other subjects like risky corporate bonds that use the formalism of option theory. The Black-Scholes option pricing path integral is studied in great detail.

The general Merton model is nonlinear and yields an intractable path integral. A linear approximation is made of the Merton model, and leads to the harmonic oscillator path integral, which is one of the cornerstones of path integration. The linearized Merton model is solved using the harmonic oscillator path integral.

The path integral formalism entails a shift away the Hamiltonian operator to the properties of the *Lagrangian* and the *action*. Similar to the role played by the Hamiltonian in the Chap. 14, the Lagrangian L is the central mathematical structure in the study of the path integrals [7]. Unlike the Hamiltonian, which is a differential operator, the action depends on the *entire path* of a security, from its initial value to its final value at the payoff function.

15.2 Feynman Path Integral

Recall the pricing kernel is a central quantity for solving the problem of mathematical finance. For a single security with remaining time $\tau = T - t$, Eq. 14.3.8 gives

$$p(x, \tau; x') = \langle x|e^{-\tau H}|x'\rangle$$

For a Hamiltonian of the form $H = T + U$ where T is a differential operator, and U is a potential term (function of the underlying security), in general $TU - UT \equiv [T, U] \neq 0$. Since $e^{-\tau H} \neq e^{-\tau T}e^{-\tau U}$, to compute $e^{-\tau H}$ is usually quite intractable.

Feynman's approach to the path integral rests on the fact that even though it may not be possible to evaluate the matrix elements of $e^{-\tau H}$, what one can evaluate exactly are the matrix elements of $e^{-\epsilon H}$, where $\epsilon \to 0$. This is because $\tau = \epsilon$ allows for the simplification that to leading order $e^{-\epsilon H} \simeq e^{-\epsilon T}e^{-\epsilon U}$, and this is the fundamental reason for going over to the path integral formulation. To determine the *pricing kernel*, discretize time so that there are N time steps to maturity, with each time step equal to $\epsilon = \tau/N$. The continuous time label of the variables $x(t)$ is discretized to x_i, where $0 \leq i \leq N$. The matrix element $\langle x|e^{-\tau H}|x'\rangle$ can then be written as an N-fold product of $e^{-\epsilon H}$in the following manner

$$\begin{aligned} p(x, \tau; x') &= \lim_{N\to\infty} \langle x| \left[\mathrm{e}^{-\epsilon H}\right]^N |x'\rangle \\ &= \lim_{N\to\infty} \langle\, x|\underbrace{\mathrm{e}^{-\epsilon H}\cdots \mathrm{e}^{-\epsilon H}}_{N-fold\ product}|x'\rangle \end{aligned} \tag{15.2.1}$$

The limit of $N \to \infty$ is required to obtain results for continuous time and is eventually taken in most calculations.

Recall the completeness equation for $|x\rangle$ is given by Eq. 13.3.7

$$\mathbb{I} = \int_{-\infty}^{\infty} dx|x\rangle\langle x| \tag{15.2.2}$$

Inserting the completeness equation $(N-1)$ times in Eq. 15.2.1 yields

$$p(x; \tau|x') = \left(\prod_{i=1}^{N-1}\int dx_i\right)\prod_{i=1}^{N}\langle x_i|e^{-\epsilon H}|x_{i-1}\rangle \tag{15.2.3}$$

with boundary conditions

$$x_N = x, \ x_0 = x' \tag{15.2.4}$$

Consider the following generic Hamiltonian, which includes the Merton Hamiltonian

$$H = T + U \ ; \ T = \frac{1}{2}\sigma^2\frac{\partial^2}{\partial x^2} \ ; \ U = U(x) \tag{15.2.5}$$

The Campbell-Baker-Hausdorff formula yields, for non-commuting operators A and B such that $[A, B] \neq 0$, the following infinite expansion [13]

$$e^A e^B = e^{A+B+\frac{1}{2}[A,B]+\cdots}$$

Hence

$$e^{-\epsilon H} = e^{-\epsilon(T+U)} \simeq e^{-\epsilon T}e^{-\epsilon U} + O(\epsilon^2)$$

Let $x = x_i, x' = x_{i-1}$ and consider a single time step; we hence have

$$\langle x|e^{-\epsilon H}|x'\rangle = \langle x|e^{-\epsilon(T+U)}|x'\rangle \simeq \langle x|e^{-\epsilon T}e^{-\epsilon U}|x'\rangle + O(\epsilon^2) \tag{15.2.6}$$

Hence, to leading order in ϵ

$$\langle x|e^{-\epsilon H}|x'\rangle \simeq \langle x|e^{-\epsilon T}e^{-\epsilon U}|x'\rangle = \langle x|e^{-\epsilon T}|x'\rangle e^{-\epsilon U(x')}$$

Equation 14.4.6 yields

$$\langle x|e^{-\epsilon T}|x'\rangle = \langle x|e^{\frac{1}{2}\sigma^2\epsilon\frac{\partial^2}{\partial x^2}}|x'\rangle = \frac{1}{\sqrt{2\pi\sigma^2\epsilon}}\exp\{-\frac{1}{2\sigma^2\epsilon}(x-x')^2\} \tag{15.2.7}$$

and hence

$$\langle x|e^{-\epsilon H}|x'\rangle = \frac{1}{\sqrt{2\pi\sigma^2\epsilon}}\exp\left\{-\frac{\epsilon}{2\sigma^2}(\frac{x-x'}{\epsilon})^2 - \epsilon U(x')\right\} + O(\epsilon^2) \tag{15.2.8}$$

The Dirac-Feynman formula relates the matrix elements of the Hamiltonian's exponential to the Lagrangian and is given by [6]

$$\langle x_i|e^{-\epsilon H}|x_{i-1}\rangle = \langle x|e^{-\epsilon H}|x'\rangle \equiv \mathcal{N}(\epsilon)e^{\epsilon L(x;x';\epsilon)} = \mathcal{N}(\epsilon)e^{\epsilon L(x_i;x_{i-1};\epsilon)} \quad (15.2.9)$$

yields the Lagrangian for the discrete time lattice given by

$$L(x;x';\epsilon) = -\frac{1}{2\sigma^2}\left(\frac{x-x'}{\epsilon}\right)^2 - U(x') = L(x_i;x_{i-1};\epsilon) \quad (15.2.10)$$

A more symmetric form for the Lagrangian, correct to leading order in ϵ, is

$$L(x_i;x_{i-1};\epsilon) = -\frac{1}{2\sigma^2}\left(\frac{x_i-x_{i-1}}{\epsilon}\right)^2 - \frac{1}{2}(U(x_i)+U(x_{i-1}) \quad (15.2.11)$$

The discrete time *action* follows from Eq. 15.2.3

$$S = \epsilon\sum_{n=1}^{N} L(x_n;x_{n-1};\epsilon) \quad (15.2.12)$$

From Eqs. 15.2.3 and 15.2.12, the discrete time *Feynman Path Integral* representation for the pricing kernel is given by

$$p(x;\tau,x') \equiv \int DXe^S = \left[\frac{1}{\sqrt{2\pi\sigma^2\epsilon}}\right]^N \prod_{n=1}^{N-1}\int dx_n e^S \quad (15.2.13)$$

with, from Eq. 15.2.4, the following boundary conditions

$$x_N = x,\ x_0 = x'$$

Note the important fact that there are N-matrix elements in the path integral, leading to a summation in action from $n = 1$ to N and to a normalization. In contrast, there are only $N-1$-integration variables.

$$\int DX = \left[\frac{1}{\sqrt{2\pi\sigma^2\epsilon}}\right]^N \prod_{n=1}^{N-1}\int dx_n = \frac{1}{\sqrt{2\pi\sigma^2\epsilon}}\prod_{n=1}^{N-1}\int\frac{dx_n}{\sqrt{2\pi\sigma^2\epsilon}} \quad (15.2.14)$$

This mis-match of the pre-factor with the number of integrations will play a key role in providing a finite path integral when the limit of $N\to\infty$ is taken.

Equation 15.2.13 is a finite N-dimensional multiple integral and is suitable for numerical studies. One can see from Eq. 15.2.13 that at a given instant of time $t = n\epsilon$, the system has only one stochastic variable given by x_n.

In many cases, the normalization $\mathcal{N}$ is independent of the integration variable x_i; in such cases the normalization constant $\mathcal{N}$ can be ignored, and the limit of $N \to \infty$ then exists. Taking the continuum limit yields

$$\lim_{\epsilon \to 0} \frac{x - x'}{\epsilon} = \lim_{\epsilon \to 0} \frac{x_i - x_{i-1}}{\epsilon} \to \frac{dx(t)}{dt}$$

and hence the Lagrangian is given by [6]

$$L(t) = -\frac{1}{2\sigma^2} \left(\frac{dx}{dt} \right)^2 - U(x) \tag{15.2.15}$$

The pricing kernel is given by the continuous time Feynman Path Integral

$$p(x; \tau, x') = \langle x | e^{-\tau H} | x' \rangle = \int DX e^S = \mathcal{N} \prod_{t=0}^{\tau} \int_{-\infty}^{+\infty} dx(t) e^S$$

$$\text{Boundary Conditions } x(\tau) = x \ ; \ x(0) = x' \tag{15.2.16}$$

with action given by

$$S = \int_0^{\tau} dt L \left(x, \frac{dx}{dt} \right)$$

In the continuum limit, Eq. 15.2.16 shows that the path integration entails performing infinitely many integrations, one integration over $x(t)$ for each instant of time $t \in [0, \tau]$. The path integral for white noise is known as the Wiener integral [27, 40] in probability theory.

Note the important fact that the stochastic variables $x(t)$ are the integration variables and the path integral is given by integrating over all the stochastic variables, and as shown in Fig. 15.1, is a summation over all the random paths taken by the security. More precisely, the path integral is a summation over all possible continuous functions (paths) that connect the initial value of the stock price x_i at initial time t_i to the final value of the stock price x_f at initial time t_f, as shown in Fig. 15.1.

The probability distribution function for the occurrence of a specific random path is given by

$$\exp\{S\} \Big/ \int DX \exp\{S\}$$

The path integral is a realization of the *second continuum limit*, mentioned earlier, that emerges in functional analysis: a one-dimensional integral of calculus is the sum over all values of an independent continuous variable; in contrast, a functional integral is the sum over all possible continuous functions.

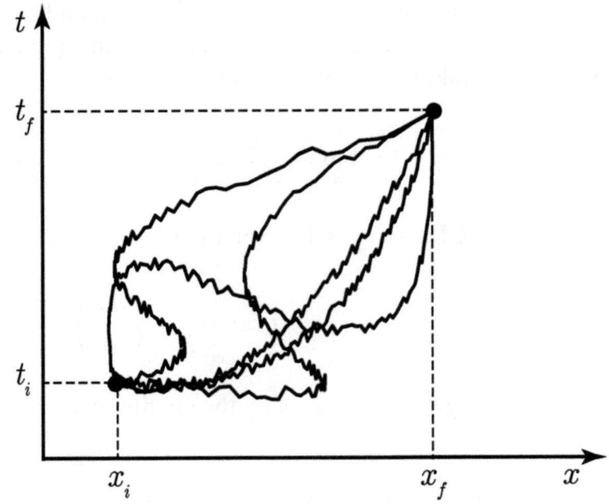

Fig. 15.1 Stock price having initial value x_i at time t_i propagating to final value x_f at time t_f taking all possible paths. Published with permission of © Belal E. Baaquie 2020. All Rights Reserved

For the case where there are many independent variables, for example a stock price having stochastic volatility, the path integral can be extended in a straightforward manner [4, 7].

15.3 Path Dependent Options

Path dependent options come in many varieties. The two main categories are options that are typically like barrier options that restrict the values of the stock price to take values within a pre-specified range for the option to have a non-zero value. The other category is options with payoff functions that depend on the all the values of the stock price from its initial value till the maturity of the option [8].

Consider for concreteness the pricing kernel of the Double Knock Out Barrier Option considered in Sect. 14.8; the value of the stock price leads to the restriction that $a \le x(t') \le b$; $t \le t' \le T$, where t is current time and T is the time for the maturity of the option. The path integration measure, similar to Eq. 15.5.4 is given by

$$\int_{DB} DX = \left[\frac{1}{2\pi\epsilon\sigma^2}\right]^{N/2} \prod_{i=1}^{N-1} \int_a^b dx_i$$

and, similar to the Black-Scholes case, the pricing kernel for the double knock out option is given by

$$p_{DB}(x, \tau; x') = \int_{DB} DX e^{S_{BS}}$$

Note the important fact that the action for the double knock out option is given by the Black-Scholes action S_{BS}, and only the domain of integration for the path integral changes to account for the barrier option. The path integral for the barrier option is evaluated more efficiently using the Hamiltonian, and the pricing kernel $p_{DB}(x, \tau; x')$ is given in Eq. 14.8.6, where

$$p_{DB}(x, \tau; x') = \int_{DB} DXe^{S_{BS}} = \langle x|e^{-\tau H_{DB}}|x'\rangle$$

where H_{DB} is the double knock out barrier option Hamiltonian given in Eq. 14.8.1.

The payoff function for many path dependent options is given by

$$g[x, K] = \max\left(K, \frac{1}{\tau}\int_0^\tau dsf(x(s))\right)_+$$

For the Asian option $f(x(s)) = e^{x(s)}$. A number of such path dependent options, using a path integral representation, have been analyzed in [8].

The price of the option at present time t is given by the Black-Scholes path integral, namely that

$$C(x_0; \tau; K) = \int_{BS} DXe^{S_{BS}} g(x, K)$$
$$= \int_{-\infty}^{+\infty} d\xi g(\xi, K)\left[\int_{BS} DX\delta[\xi - \frac{1}{\tau}\int_0^\tau dsf(x(s))]e^{S_{BS}}\right]$$
$$\text{Mixed Boundary Conditions}: x(\tau) = x_0\ ;\quad \frac{dx(0)}{dt} = 0$$

Using the Fourier representation of the delta-function given in Eq. 13.4.3 yields

$$C(x_0, \tau; K) = \int_{-\infty}^{+\infty} \frac{d\xi dp}{2\pi} e^{ip\xi} Z(p) g(\xi, K)$$
$$\Rightarrow Z(p) = \int_{BS} DXe^{S_{\text{Eff}}(p)}$$
$$\text{where}\quad S_{\text{Eff}}(p) = S_{BS} - i\frac{p}{\tau}\int_0^\tau dsf(x(s)) \tag{15.3.1}$$

In particular for the Asian option

$$L_{\text{Asian Option}} = L_{BS} - i\frac{p}{\tau}e^x \tag{15.3.2}$$

15.4 Merton Lagrangian

The Lagrangian for the Merton Hamiltonian H_M is evaluated; the Black-Scholes Hamiltonian H_{BS} is a special case of the Merton Hamiltonian. Although one can in principle find the Lagrangian directly for H_M, it is more convenient to use the effective Hamiltonian, given below from Eq. 14.2.7

$$H_M(v, q) = e^s H_{\text{Eff}}(v, q)e^{-s}$$
$$H_{\text{Eff}}(v, q) = -\frac{1}{2}\sigma^2\frac{\partial^2}{\partial x^2} + \frac{1}{2}\frac{\partial v}{\partial x} + \frac{1}{2\sigma^2}\left(\frac{1}{2}\sigma^2 - r - v(x)\right)^2 + r + q(x)$$

Note that the Merton Hamiltonian, for nonlinear v, q is an intractable problem; casting it in the path integral formulation for linear v, q can, in fact, be solved analytically—and shown in Sect. 15.12. For the general nonlinear case, the path integral can give insight on how to analyze it and also provide a platform for numerical studies.

The first step in studying the Merton path integral is to obtain the Merton Lagrangian.

Recall from Eq. 14.3.9

$$\begin{aligned} p_{(v,q)}(x, x'; \tau) &= \langle\, x|e^{-\tau H}|x'\rangle = \exp\{s(x) - s(x')\}\langle x|e^{-\tau H_{\text{Eff}}}|x'\rangle \\ &= \exp\left\{s(x) - s(x')\right\}\widetilde{p}_{(v,q)}(x, x'; \tau) \\ \widetilde{p}_{(v,q)}(x, x'; \tau) &\equiv \langle x|e^{-\tau H_{\text{Eff}}}|x'\rangle \end{aligned} \tag{15.4.1}$$

The Feynman path integral can be used to calculate

$$\widetilde{p}_{(v,q)}(x, x'; \tau) = \langle x|e^{-\tau H_{\text{Eff}}}|x'\rangle$$

Writing the Merton Hamiltonian in the notation of Eq. 14.2.6 yields

$$H_{\text{Eff}} \equiv T + U \;\; ; \;\; T = -\frac{1}{2}\sigma^2\frac{\partial^2}{\partial x^2}$$
$$U = \frac{1}{2}\frac{\partial v}{\partial x} + \frac{1}{2\sigma^2}\left(\frac{1}{2}\sigma^2 - r - v(x)\right)^2 + r + q(x) \tag{15.4.2}$$

From Eqs. 15.2.15 and 15.4.2, the effective Merton Lagrangian is [14]

$$\mathcal{L}_{\text{Eff}}(t) = -\frac{1}{2\sigma^2}\left(\frac{dx}{dt}\right)^2 - U(x) \tag{15.4.3}$$
$$= -\frac{1}{2\sigma^2}(\frac{dx}{dt})^2 - \left\{\frac{1}{2}\frac{\partial v}{\partial x} + \frac{1}{2\sigma^2}\left(\frac{1}{2}\sigma^2 - r - v(x)\right)^2 + r + q(x)\right\}$$

The Feynman path integral representation of the Merton pricing kernel, from Eq. 15.2.16, is given by the following [14]

$$\widetilde{p}_{(v,q)}(x, x'; \tau) = \langle\, x|e^{-\tau H_{\text{Eff}}}|x'\rangle = \int DX \exp\left\{\int_0^{\tau} dt \mathcal{L}_{\text{Eff}}(t)\right\}$$

$$\int DX = \mathcal{N} \prod_{t=0}^{\tau} \int dx(t)\ ;\ \text{B.C.}\ :\ x(0) = x'\ ;\ x(\tau) = x \tag{15.4.4}$$

As mentioned earlier, the Merton path integral, similar to the Merton Hamiltonian, is nonlinear and is an intractable problem. Instead, a *linear approximation* is made for the Merton Hamiltonian that leads to a tractable path integral. Recall from Eq. 14.5.1 the linear oscillator potential for the Merton Hamiltonian is given by

$$v(x) = \omega x\ ;\ q(x) = \lambda x\ ;\ V = e^x \tag{15.4.5}$$

The effective Merton Lagrangian, from Eq. 15.4.3, is the oscillator potential given by

$$\mathcal{L}_{\text{Eff}}(t) = -\frac{1}{2\sigma^2}(\frac{dx}{dt})^2 - \frac{1}{2\sigma^2}\omega^2 x^2 - hx - c \tag{15.4.6}$$

where h, c are given in Eq. 14.5.5.

The Merton oscillator Lagrangian given in Eq. 15.4.6 shows that one needs to evaluate the pricing kernel for the harmonic oscillator, which is discussed later in Sect. 15.11.

15.5 Black-Scholes Discrete Path Integral

The path integral is derived for the case of the Black-Scholes Hamiltonian from first principles. The reason that this is possible is because the Hamiltonian H_{BS} essentially consists of only the differential operator T, which is defined in Eq. 15.4.2. The derivation of the path integral from first principles does not need any approximations and hence provides independent verification to the steps taken to derive path integral for the nonlinear Hamiltonian.

The discrete Black-Scholes path integral, consisting of a finite number of time steps N, is given by a N-dimensional multiple integral. The integration is performed exactly for the multiple integral and the limit of $N \to \infty$ is then taken to recover the continuous time path integral, showing some of the inner workings of the path integral.

Consider the case of risk neutral evolution of a stock price driven by the Black-Scholes Hamiltonian. For stock price $S = e^x$, Eq. 14.2.3 yields, from the Dirac-Feynman equation given in Eq. 15.2.9, the following

$$\langle x|e^{-\epsilon H_{BS}}|x_{i-1}\rangle = N_{BS}(\epsilon)e^{\epsilon L_{BS}}$$

where $N_{BS}(\epsilon)$ is a normalization constant. The function L_{BS} is the Black-Scholes Lagrangian. The pricing kernel for the Black-Scholes process given in Eq. 14.4.11 is exact, and setting $\tau = \epsilon$ yields

$$\begin{aligned}
\langle x|e^{-\epsilon H_{BS}}|x_{i-1}\rangle &= \frac{e^{-r\epsilon}}{\sqrt{2\pi\epsilon\sigma^2}} \exp\left\{-\frac{1}{2\epsilon\sigma^2}\left\{x_i - x_{i-1} + \epsilon(r - \sigma^2/2)\right\}^2\right\} \\
&= \frac{e^{-r\epsilon}}{\sqrt{2\pi\epsilon\sigma^2}} \exp\left\{-\frac{\epsilon}{2\sigma^2}\left\{\frac{\delta x_i}{\epsilon} + (r - \sigma^2/2)\right\}^2\right\} \\
&= \mathcal{N}_{BS}(\epsilon)e^{\epsilon L_{BS}}
\end{aligned}$$

where $\delta x_i = x_i - x_{i-1}$. Hence, the Black-Scholes Lagrangian is give by

$$L_{BS}(i) = -\frac{1}{2\sigma^2}\left(\frac{\delta x_i}{\epsilon} + r - \frac{\sigma^2}{2}\right)^2 - r \tag{15.5.1}$$

$$S_{BS} = \epsilon\sum_{i=1}^{N} L_{BS}(i)\ ;\ \mathcal{N}_{BS}(\epsilon) = \frac{1}{\sqrt{2\pi\epsilon\sigma^2}} \tag{15.5.2}$$

Keeping in mind that the index i in x_i labels remaining time with $N\epsilon = \tau = T - t$, where t is real time, the boundary conditions are given by

$$x_N = x,\ x_0 = x' \tag{15.5.3}$$

The completeness equation yields for the integration measure

$$\int DX = \left[\frac{1}{2\pi\epsilon\sigma^2}\right]^{N/2} \prod_{i=1}^{N-1} \int_{-\infty}^{+\infty} dx_i \tag{15.5.4}$$

The Black-Scholes pricing kernel is given by the path integral

$$p_{BS}(x, \tau; x') = \langle x|e^{-\tau H_{BS}}|x'\rangle = \int DX\ e^{S_{BS}} \tag{15.5.5}$$

The path integration for the Black-Scholes pricing kernel is explicitly performed. From Eq. 15.5.2 the action is given by

$$S_{BS} = -\frac{1}{2\epsilon\sigma^2}\sum_{i=1}^{N}\left(x_i - x_{i-1} + \epsilon\left(r - \frac{\sigma^2}{2}\right)\right)^2 - r\epsilon N \tag{15.5.6}$$

Define the following change of variables

$$\zeta_i = x_i - x_{i-1} + \epsilon(r - \frac{\sigma^2}{2}) \; ; \; d\zeta_i = dx_i \; ; \; i = 1, \ldots, N \qquad (15.5.7)$$

There are only $N - 1$ original x_i integration variables, whereas N independent integration variables have been defined for the ζ_i's. Hence, one needs to introduce one constraint on the ζ_i variables, and this will be done by implementing the boundary conditions given in Eq. 15.5.3.

The boundary condition at x_0 is automatically included in Eq. 15.5.7. To implement the boundary condition at x_N impose the following constraint on the ζ_i variables

$$\sum_{i=1}^{N} \zeta_i = x_N - x_0 + N\epsilon\left(r - \frac{\sigma^2}{2}\right) \Rightarrow x_N - x_0 + N\epsilon\left(r - \frac{\sigma^2}{2}\right) - \sum_{i=1}^{N} \zeta_i = 0$$

$$\Rightarrow m - \sum_{i=1}^{N} \zeta_i = 0 \; ; \; m \equiv x - x' + \tau\left(r - \frac{\sigma^2}{2}\right) \; : \; x_N = x \; ; \; x_0 = x' \qquad (15.5.8)$$

Implementing the boundary condition above as a delta-function inside the $\int D\zeta$ path integral, yields from Eq. 15.5.5

$$p_{BS}(x, \tau; x') = \langle x|e^{-\tau H_{BS}}|x'\rangle = \int_{BS} DXe^{S_{BS}}$$
$$= e^{-r\tau}\left[\frac{1}{2\pi\epsilon\sigma^2}\right]^{N/2} \prod_{i=1}^{N} \int_{-\infty}^{+\infty} d\zeta_i \exp\left\{-\frac{1}{2\epsilon\sigma^2}\sum_{i=1}^{N} \zeta_i^2\right\} \delta\left(m - \sum_{i=1}^{N} \zeta_i\right)$$

Using the integral representation of the delta function given in Eq. 13.4.3 yields

$$\delta\left(m - \sum_{i=1}^{N} \zeta_i\right) = \int_{-\infty}^{\infty} \frac{dp}{2\pi} \exp\left\{ip\left(m - \sum_{i=1}^{N} \zeta_i\right)\right\}$$

and leads to the factorization of the ζ_i integration into decoupled Gaussian integration. Hence

$$p_{BS}(x, \tau; x') = e^{-r\tau}\left[\frac{1}{2\pi\epsilon\sigma^2}\right]^{N/2} \int_{-\infty}^{+\infty} \frac{dp}{2\pi}$$
$$\times \prod_{i=1}^{N} \int_{-\infty}^{+\infty} d\zeta_i \exp\left\{-\frac{1}{2\epsilon\sigma^2}\sum_{i=1}^{N} \zeta_i^2 + ip\left[m - \sum_{i=1}^{N} \zeta_i\right]\right\}$$
$$= e^{-r\tau} \int_{-\infty}^{+\infty} \frac{dp}{2\pi} e^{ipm} \left[\int_{-\infty}^{+\infty} \frac{d\zeta}{\sqrt{2\pi\epsilon\sigma^2}} \exp\left\{-\frac{1}{2\epsilon\sigma^2}\zeta^2 + ip\zeta\right\}\right]^N$$
$$= e^{-r\tau} \int_{-\infty}^{+\infty} \frac{dp}{2\pi} \exp\left\{-\frac{N\epsilon\sigma^2}{2} p^2 + ipm\right\}$$

Using $m \equiv x - x' + \tau(r - \sigma^2/2)$, $N\epsilon = \tau$ and performing the Gaussian integration yields

$$p_{BS}(x, \tau; x') = \frac{e^{-r\tau}}{\sqrt{2\pi\tau\sigma^2}} \exp\left[-\frac{1}{2\tau\sigma^2}\left\{x - x' + \tau(r - \sigma^2/2)\right\}^2\right] \quad (15.5.9)$$

and we have recovered the expected Black-Scholes pricing kernel given in Eq. 14.4.11.

15.6 Path Integral: Time-Dependent Volatility

Given the importance of the Black-Scholes equation, the Black-Scholes action is generalized to the case where the volatility is time dependent. Hence

$$\sigma^2 = \sigma^2(t) \ \rightarrow \ \sigma_i^2 \equiv e^{y_i} \ : \ i = 1, \ldots, N$$

The notation of $\exp\{y_i\}$ is useful in the derivations of this section. Generalizing Eq. 15.5.6 to the case of a time-dependent volatility yields the following action

$$S_{BS} = -\frac{1}{2\epsilon}\sum_{i=1}^{N} e^{-y_i}\left(x_i - x_{i-1} + \epsilon\left(r - \frac{e^{y_i}}{2}\right)\right)^2 - r\epsilon N \quad (15.6.1)$$

The term $r\epsilon N$ is removed for notational convenience. Hence, define

$$S = -\frac{1}{2\epsilon}\sum_{i=1}^{N} e^{-y_i}\left(x_i - x_{i-1} + \epsilon\left(r - \frac{e^{y_i}}{2}\right)\right)^2 \quad (15.6.2)$$

Let

$$c_i = r - \frac{1}{2}e^{y_i}$$

Then

$$S = -\frac{1}{2\epsilon(1 - \rho^2)}\sum_{i=1}^{N} e^{-y_i}(x_i - x_{i-1} + \epsilon c_i)^2 \quad (15.6.3)$$

with boundary values given by

$$x_0 = x', \ x_N = x \quad (15.6.4)$$

Make a change of variables

$$x_i = z_i - \epsilon \sum_{j=1}^{i} c_j, \quad dx_i = dz_i, \quad i = 1, 2, \ldots, N-1$$

with boundary values

$$z_0 = x' \ ; \ z_N = x + \epsilon \sum_{j=1}^{N} c_j$$

Hence, from Eq. (15.6.3)

$$S = -\frac{1}{2\epsilon} \sum_{i=1}^{N} \mathrm{e}^{-y_i} (z_i - z_{i-1})^2 \tag{15.6.5}$$

The path integral with a time dependent volatility is given by, from Eqs. 15.5.5 and 15.6.5

$$\langle x| e^{-\tau H_{BS}} |x'\rangle = \frac{\mathrm{e}^{-y_N/2}}{\sqrt{2\pi\epsilon}} \prod_{i=1}^{N-1} \int_{-\infty}^{+\infty} dz_i \frac{\mathrm{e}^{-y_i/2}}{\sqrt{2\pi\epsilon}} \mathrm{e}^S \tag{15.6.6}$$

Note in Eq. 15.5.4 there is one extra normalization factor besides a normalization with each integration, as given in Eq. 15.2.14 for the case of constant volatility; hence, the following extra term is outside the path integral

$$\frac{\mathrm{e}^{-y_N/2}}{\sqrt{2\pi\epsilon}}$$

All the z_i integrations can be performed exactly; one starts from the boundary by first integrating over z_1, and then over z_2, ... and finally over z_{N-1}. The z_1 -integration yields

$$\begin{aligned} &\int_{-\infty}^{+\infty} \frac{dz_1 \mathrm{e}^{-y_1/2}}{\sqrt{2\pi\epsilon}} \exp\left\{-\frac{1}{2\epsilon}\left[\mathrm{e}^{-y_2}(z_2 - z_1)^2 + \mathrm{e}^{-y_1}(z_1 - z_0)^2\right]\right\} \\ &= \frac{\mathrm{e}^{y_2/2}}{\sqrt{\mathrm{e}^{y_1} + \mathrm{e}^{y_2}}} \exp\left\{-\frac{1}{2\epsilon}\frac{1}{\mathrm{e}^{y_1} + \mathrm{e}^{y_2}}(z_2 - z_0)^2\right\} \end{aligned} \tag{15.6.7}$$

Repeating this procedure $(N-1)$ times, and restoring the factor of $r\epsilon N$, yields

$$\langle x| e^{-\tau H_{BS}} |x'\rangle = \frac{1}{\sqrt{2\pi\epsilon \sum_{i=1}^{N} \mathrm{e}^{y_i}}} \exp\{S_1 - r\epsilon N\} \tag{15.6.8}$$

where, for $\tau = \epsilon N$

$$S_1 = -\frac{1}{2\epsilon \sum_{i=1}^{N} \mathrm{e}^{y_i}}(z_N - z_0)^2 = -\frac{1}{2\epsilon \sum_{i=1}^{N} \mathrm{e}^{y_i}}\left\{x - x' + \epsilon \sum_{i=1}^{N} c_i\right\}^2$$

$$= -\frac{1}{2\epsilon \sum_{i=1}^{N} \mathrm{e}^{y_i}}\left\{x - x' + r\tau - \frac{1}{2}\epsilon \sum_{i=1}^{N} e^{y_i}\right\}^2 \tag{15.6.9}$$

Taking the continuum limit of $\epsilon \to 0$ yields

$$\langle x|e^{-\tau H_{BS}}|x'\rangle = \frac{1}{\sqrt{2\pi \int_0^\tau dt \sigma^2(t)}} \exp\{S_1 - r\tau\}$$

$$S_1 = -\frac{1}{2\int_0^\tau dt\sigma^2(t)}\left\{x - x' + r\tau - \frac{1}{2}\int_0^\tau dt\sigma^2(t)\right\}^2 \tag{15.6.10}$$

For the case of constant volatility $\mathrm{e}^{y_i} = \sigma^2$ = constant; Eq. 15.6.10 then reduces to the Black-Scholes pricing kernel given in Eq. 15.5.9.

15.7 Black-Scholes Continuous Path Integral

The $\int DX$ path integral for the Black-Scholes case, from Eq. 15.5.4, has an integration measure that is essentially the measure for the flat space $\mathbb{R}^N$; the limit of $N \to \infty$ for $\int DX$ can consequently be taken and a well defined continuous-time path integral is obtained. From Eqs. 15.5.2 and 15.5.4, taking the continuum limit of $\epsilon \to 0$, yields the following

$$S_{BS} = \int_0^\tau dt L_{BS} = -\frac{1}{2\sigma^2}\int_0^\tau dt \left(\frac{dx}{dt} + r - \frac{1}{2}\sigma^2\right)^2 - r\tau \tag{15.7.1}$$

with boundary conditions $x(0) = x'$ and $x(\tau) = x$.

Up to an irrelevant constant, the continuous time path integral for the Black-Scholes pricing kernel is the following

$$p_{BS}(x, \tau|x') = \langle x|\mathrm{e}^{-\tau H_{BS}}|x'\rangle = \int DX \mathrm{e}^{S_{BS}} \tag{15.7.2}$$

$$\int DX = \mathcal{N} \prod_{t=0}^{\tau} \int_{-\infty}^{+\infty} dx(t)$$

The Black-Scholes pricing kernel is evaluated by a continuous functional integration by a summation over all possible values of the stock price's stochastic variables $x(t)$ in its evolution from x to x'.

Equation 15.5.9 is the result for $p_{BS}(x, \tau|x')$, and hence

$$\int DXe^{S_{BS}} = \frac{e^{-r\tau}}{\sqrt{2\pi\tau\sigma^2}} \exp\left[-\frac{1}{2\tau\sigma^2}\{x - x' + \tau(r - \sigma^2/2)\}^2\right] \quad (15.7.3)$$

From Eq. 15.7.1, the Black-Scholes action yields the following

$$S_{BS} = -\frac{1}{2\sigma^2}\int_0^\tau dt\left[\left(\frac{dx}{dt}\right)^2 + 2\left(r - \frac{1}{2}\sigma^2\right)\frac{dx}{dt} + \tau\left(r - \frac{1}{2}\sigma^2\right)^2\right] - r\tau$$

$$= -\frac{1}{2\sigma^2}\left[\int_0^\tau dt\left(\frac{dx}{dt}\right)^2 + 2(r - \frac{1}{2}\sigma^2)(x - x') + \tau\left(r - \frac{1}{2}\sigma^2\right)^2\right] - r\tau (15.7.4)$$

Define

$$S_0 = -\frac{1}{2\sigma^2}\int_0^\tau dt(\frac{dx}{dt})^2$$

Comparing Eqs. 15.7.3 and 15.7.4 yields

$$\int DXe^{S_0} = \frac{1}{\sqrt{2\pi\tau\sigma^2}} \exp\left[-\frac{1}{2\tau\sigma^2}(x - x')^2\right] \quad (15.7.5)$$

15.8 Stationary Action: Euler-Lagrange Equation

The path integral provides a representation of the pricing kernel given by

$$p(x, x'; \tau) = \langle x|e^{-\tau H}|x'\rangle = \int DXe^S$$

A complete trajectory $x(t)$, starting at x and ending at x'—as shown in Fig. 15.2—is required to yield one value for $S[x]$. If one keeps track of all the values of the action S that are generated as one runs through all the possible paths, then there is one path, denoted by $x_c(t)$, for which S has the minimum value. Finding the path $x_c(t)$ is an infinite dimensional extension of finding the minimum value a function, discussed in Sect. 6.6.

One approach for evaluating the path integral is to expand the action S about its minimum, and then perform the path integral about the minimum [7]. The minimum value of S is determined by the variation of the path $x(t)$ shown in Fig. 15.2, and is functional analogue of the minimization condition discussed in Sect. 6.6. The variation of the paths is realized by functional differentiation with respect to the path $x(t)$ and is given by

$$\frac{\delta S[x_c]}{\delta x(t)} = 0 \ ; \ \ 0 < t < \tau$$

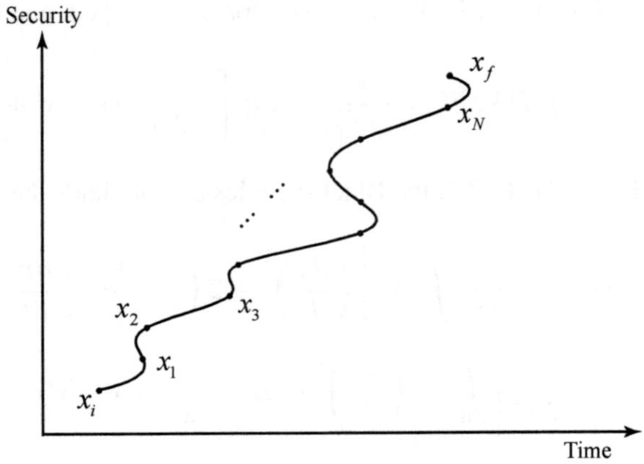

Fig. 15.2 One continuous trajectory from initial to final value of the security. Published with permission of © Belal E. Baaquie 2020. All Rights Reserved

Note the boundary conditions are not varied and hence the strict inequality $0 < t < \tau$ for the variation.

Writing the action in terms of the Lagrangian yields

$$S[x] = \int_0^\tau dt L(x, \dot{x}) \ ; \ \dot{x} = \frac{dx}{dt}$$

From Eq. 13.9.5, and doing an integration by parts, yields

$$\begin{aligned}
\frac{\delta S[x]}{\delta x(t)} &= \int_0^\tau dt' \left[\frac{\partial L(x, \dot{x})}{\partial x(t')} \frac{\delta x(t')}{\delta x(t)} + \frac{\partial L(x, \dot{x})}{\partial \dot{x}(t')} \frac{\delta \dot{x}(t')}{\delta x(t)} \right] \\
&= \int_0^\tau dt' \left[\frac{\partial L(x, \dot{x})}{\partial x(t)} \delta(t' - t) + \frac{\partial L(x, \dot{x})}{\partial \dot{x}(t)} \frac{d}{dt'} \delta(t' - t) \right] \\
&= \int_0^\tau dt' \left[\frac{\partial L(x, \dot{x})}{\partial x(t)} - \frac{d}{dt'} \frac{\partial L(x, \dot{x})}{\partial \dot{x}(t)} \right] \delta(t' - t)
\end{aligned}$$

Hence

$$\frac{\delta S[x]}{\delta x(t)} = \frac{\partial L(x, \dot{x})}{\partial x(t)} - \frac{d}{dt} \frac{\partial L(x, \dot{x})}{\partial \dot{x}(t)} \ ; \ 0 < t < \tau \tag{15.8.1}$$

The integration by parts does not generate any (surface) terms since there is no variation at the endpoints of the path: the variation is restricted to only $0 < t < \tau$. The value of the path at the endpoints is fixed by the boundary conditions.

To determine the stationary path $x_c(t)$ that minimizes the action, one has the following

$$\frac{\delta S[x_c]}{\delta x(t)} = 0 \Rightarrow \frac{\partial L(x_c, \dot{x}_c)}{\partial x(t)} - \frac{d}{dt} \frac{\partial L(x_c, \dot{x}_c)}{\partial \dot{x}(t)} = 0 \ : \ 0 < t < \tau \tag{15.8.2}$$

which is the famous Euler-Lagrange equation.[1]

The path integral is expanded around the stationary solution in the following manner. Armed with the stationary solution, define the following change of variables, where $\xi(t)$ are now the new stochastic integration variables

$$x(t) = x_c(t) + \xi(t) \Rightarrow \prod_{t=0}^{\tau} \int dx(t) = \prod_{t=0}^{\tau} \int d\xi(t) \tag{15.8.3}$$

The stationary solution has the following boundary conditions

$$x_c(0) = x' = x(0) \ ; \ \ x_c(\tau) = x = x(\tau)$$

Boundary conditions imposed on $x_c(t)$ make it agree with the stochastic variables $x(t)$ at $t = 0, \ \tau$ and yields the following boundary conditions for ξ

$$\xi(0) = 0 = \xi(\tau) \tag{15.8.4}$$

The path integral is then given by

$$p(x, x'; \tau) = \langle x|\mathrm{e}^{-\tau H}|x'\rangle = \int D\xi \mathrm{e}^{S[x_c+\xi]} \tag{15.8.5}$$

The reason for expanding the action about the stationary solution is because in many cases, as will be seen in the next few Sections, the action $S[x_c + \xi]$ has a simple form.

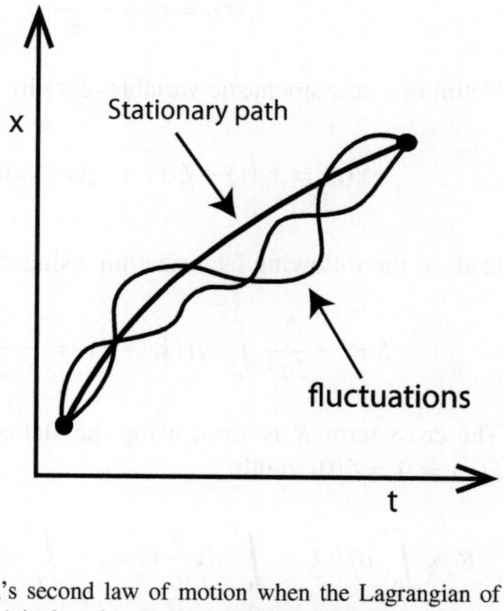

Fig. 15.3 Stationary path, with nearby fluctuations, in the path integral. The starting and ending points are fixed, and determined by the boundary conditions. Published with permission of © Belal E. Baaquie 2020. All Rights Reserved

[1]The Euler-Lagrange equation yields Newton's second law of motion when the Lagrangian of mechanics is used. The stationary solution $x_c(t)$ is then the classical solution of Newton's second law of motion.

Figure 15.3 shows the stationary path $x_c(t)$ and the fluctuations ξ in the neighborhood of the stationary path. Note that due to the boundary conditions given in Eq. 15.8.4, the fluctuations all go to zero at the boundaries of the stationary path.

15.9 Black-Scholes Euler-Lagrange Equation

To understand the workings of the Euler-Lagrange equation, the simplified case of the Black-Scholes path integral is analyzed. From Eq. 15.7.5

$$\int DX \exp\left\{-\frac{1}{2\sigma^2}\int_0^\tau dt\left(\frac{dx}{dt}\right)^2\right\} = \frac{1}{\sqrt{2\pi\tau\sigma^2}}\exp\left[-\frac{1}{2\tau\sigma^2}(x-x')^2\right] \quad (15.9.1)$$

with action and Lagrangian given by

$$S[x] = -\frac{1}{2\sigma^2}\int_0^\tau dt\left(\frac{dx}{dt}\right)^2 \ ; \ L = -\frac{1}{2\sigma^2}\left(\frac{dx}{dt}\right)^2 \quad (15.9.2)$$

The Euler-Lagrange equation yields the stationary solution $x_c(t)$

$$\frac{d^2x_c}{dt^2} = 0 \ \Rightarrow \ x_c(t) = a + bt$$

Boundary conditions $x_c(\tau) = x,\ x_c(0) = x'$ yields

$$x_c(t) = x' + \frac{x-x'}{\tau}t \ \Rightarrow \ \frac{dx_c}{dt} = \frac{x-x'}{\tau}$$

Defining a new stochastic variables $\xi(t)$ by

$$x(t) = x_c(t) + \xi(t) \ ; \ \xi(\tau) = 0 = \xi(0) \ ; \ \int Dx = \int D\xi$$

leads to the following factorization, using $dx/dt = \dot{x},\ d\xi/dt = \dot{\xi}$

$$S = -\frac{1}{2\sigma^2}\int_0^\tau dt(\dot{x}_c + \dot{\xi})^2 = -\frac{1}{2\sigma^2}\int_0^\tau dt[(\dot{x}_c)^2 + (\dot{\xi})^2] + R \quad (15.9.3)$$

The cross-term R is zero; using the stationary solution and boundary conditions $\xi(\tau) = 0 = \xi(0)$ yields

$$R \propto \int_0^\tau dt\dot{x}_c\dot{\xi} = \int_0^\tau dt\frac{d}{dt}(\dot{x}_c\xi) - \int_0^\tau dt\xi\frac{d^2x_c}{dt^2} = \dot{x}_c(\tau)\xi(\tau) - \dot{x}_c(0)\xi(0) = 0$$

and the action is given by

$$S[x] = S[x_c] + S[\xi] \ ; \ \ S[\xi] = -\frac{1}{2\sigma^2}\int_0^\tau dt\left(\frac{d\xi}{dt}\right)^2 \tag{15.9.4}$$

The value of the stationary action is given by

$$S[x_c] = -\frac{1}{2\sigma^2}\int_0^\tau (\frac{dx_c}{dt})^2 = -\frac{1}{2\sigma^2}(\frac{x-x'}{\tau})^2\int_0^\tau = -\frac{1}{2\tau\sigma^2}(x-x')^2$$

The path integral is hence given by

$$\int DXe^S = e^{S[x_c]}\int D\xi \exp\left\{-\frac{1}{2\sigma^2}\int_0^\tau dt\left(\frac{d\xi}{dt}\right)^2\right\} \tag{15.9.5}$$

Comparing Eq. 15.9.5 with Eq. 15.9.1 shows that we have recovered the expected result with the new result that

$$\int D\xi e^{S[\xi]} = \int D\xi \exp\{-\frac{1}{2\sigma^2}\int_0^\tau dt(\frac{d\xi}{dt})^2\} = \frac{1}{\sqrt{2\pi\tau\sigma^2}} \tag{15.9.6}$$

To see how the path integral $\int D\xi$ develops the result given in Eq. 15.9.6, consider the discretized case of $\xi(t) \to \xi_n \ ; \ t = n\epsilon$ with the boundary conditions for the discretized path integral is given by $\xi_1 = 0 = \xi_N$. The case of time-dependent volatility is used since one can then keep track of each integration in the path integral; using result Eq. 15.6.6 yields

$$\int D\xi e^{S[\xi]} = \langle 0|e^{-\tau H_{BS}}|0\rangle = \frac{e^{-y_N/2}}{\sqrt{2\pi\epsilon}}\prod_{i=1}^{N-1}\int_{-\infty}^{+\infty} d\xi_i\frac{e^{-y_i/2}}{\sqrt{2\pi\epsilon}}e^S \tag{15.9.7}$$

Each $\int d\xi_i$ integration cancels the singular term $e^{-y_i/2}/\sqrt{2\pi\epsilon}$ from terms in the action; each integration also contributes to the denominator which is *outside* the path integrations. Hence, for constant $e^{y_i/2} = \sigma$ and $\tau = \epsilon N$, Eq. 15.6.8 yields

$$\int D\xi e^{S[\xi]} = \frac{1}{\sqrt{2\pi\epsilon\sum_{i=1}^N e^{y_i}}} \to \frac{1}{\sqrt{2\pi\epsilon N\sigma^2}} = \frac{1}{\sqrt{2\pi\tau\sigma^2}} \tag{15.9.8}$$

15.10 Stationary Action: Time Dependent Volatility

One can extend the concept of the stationary action to evaluate the path integral for the case of a time dependent volatility. Recall from Eq. 15.6.1, the action is given by

$$S_{BS} = -\frac{1}{2\epsilon}\sum_{i=1}^{N} e^{-y_i}\left(x_i - x_{i-1} + \epsilon(r - \frac{e^{y_i}}{2})\right)^2 - r\epsilon N$$

To simplify the notation, recall $e^{y_i} = \sigma_i^2$; the $\lim \epsilon \to 0$ continuum limit of the action is taken and yields

$$S[x] = -\frac{1}{2}\int_0^{\tau} dt \frac{1}{\sigma^2(t)}\left(\frac{dx}{dt} + r - \frac{1}{2}\sigma^2(t)\right)^2 - r\tau \tag{15.10.1}$$

The following path integral is evaluated

$$\int DXe^{S[x]}$$

Ignoring the term $r\tau$ that plays no role in the derivations, the Lagrangian is given by

$$L(t) = -\frac{1}{2}\frac{1}{\sigma^2(t)}\left(\frac{dx}{dt} + \alpha(t)\right)^2 ;\ \ \alpha(t) = r - \frac{1}{2}\sigma^2(t)$$

The Euler-Lagrange equation yields the following stationary solution

$$\frac{d}{dt}\left[\frac{1}{\sigma^2(t)}\left(\frac{dx_c(t)}{dt} + \alpha(t)\right)\right] = 0$$

and yields

$$\frac{1}{\sigma^2(t)}\left(\frac{dx_c(t)}{dt} + \alpha(t)\right) = \frac{1}{\sigma^2(0)}\left(\frac{dx_c(0)}{dt} + \alpha(0)\right) \tag{15.10.2}$$

Integrating the equation above, and using the boundary condition given in Eq. 15.6.4, $x(0) = x',\ x(\tau) = x$, yields

$$x_c(t) - x_c(0) + \int_0^t dt'\alpha(t') = \frac{1}{\sigma^2(0)}\left(\frac{dx_c(0)}{dt} + \alpha(0)\right)\int_0^t dt'\sigma^2(t')$$
$$\Rightarrow \frac{1}{\sigma^2(0)}\left(\frac{dx_c(0)}{dt} + \alpha(0)\right) = \frac{x - x' + \int_0^{\tau} dt\alpha(t)}{\int_0^{\tau} dt\sigma^2(t)} \tag{15.10.3}$$

Similar to the case given in Eq. 15.9.4, the action for time dependent volatility factorizes and

$$S[x] = S[x_c] + S[\xi]\ ;\ \ S[\xi] = -\frac{1}{2}\int_0^{\tau} dt \frac{1}{\sigma^2(t)}\left(\frac{d\xi}{dt} + \alpha(t)\right)^2 \tag{15.10.4}$$

The stationary action, from Eqs. 15.10.1 and 15.10.2, is given by

$$S[x_c] = -\frac{1}{2}\left[\frac{1}{\sigma^2(0)}\left(\frac{dx_c(0)}{dt} + \alpha(0)\right)\right]^2 \int_0^\tau dt\sigma^2(t)$$

Hence, from the equation above and Eq. 15.10.3 yields the final answer

$$S[x_c] = -\frac{1}{2\int_0^\tau dt\sigma^2(t)}\left\{x - x' + r\tau - \frac{1}{2}\int_0^\tau dt\sigma^2(t)\right\}^2 \qquad (15.10.5)$$

which is the result obtained earlier in Eq. 15.6.10.

The path integral for the Black-Scholes action with time dependent volatility is given by

$$\int DXe^{S[x]} = e^{S[x_c]}\int D\xi e^{S[\xi]} = \mathcal{N}e^{S[x_c]}$$

where, from Eq. 15.6.10

$$\mathcal{N} = \int D\xi e^{S[\xi]} = \frac{1}{\sqrt{2\pi\int_0^\tau dt\sigma^2(t)}}$$

15.11 Harmonic Oscillator Pricing Kernel

The simple harmonic oscillator is one of the most important exemplars in physics, and its appearance in the linearized Merton equation opens the way to the application of many techniques from physics.

The oscillator pricing kernel is evaluated using the Feynman path integral. There are many different methods of greater generality that can be used for evaluating path integrals [7]. Instead of exploring path integrals in general, the simplest and most direct method is used for the oscillator pricing kernel [7].

Consider the harmonic oscillator action, in presence of an external function $j(t)$, given by

$$S = S_0 + \int_0^\tau j(t)x(t) \;\; ; \;\; S_0 = -\frac{1}{2}m\int_0^\tau (\dot{x}^2 + \omega^2 x^2) \qquad (15.11.1)$$

with boundary conditions

$$x(0) = x \;\; ; \;\; x(\tau) = x' \qquad (15.11.2)$$

The path integral for the harmonic oscillator is given by

$$p(x, x'; \tau) = \langle x'|\mathrm{e}^{-\tau H}|x\rangle = \int DX\mathrm{e}^{S[x]} \qquad (15.11.3)$$

Note that, unlike the rest of the sections, the notation for the path integral given in Eq. 15.11.3 has switched the role of x, x'.

From Eq. 15.11.1, the Lagrangian is given by

$$L = -\frac{1}{2}m(\dot{x}^2 + \omega^2 x^2) \quad (15.11.4)$$

The Euler-Lagrange equation is given in Eq. 15.8.2

$$\frac{\delta S_0[x_c]}{\delta x(t)} = 0 \Rightarrow \frac{\partial L(x_c, \dot{x}_c)}{\partial x(t)} - \frac{d}{dt}\frac{\partial L(x_c, \dot{x}_c)}{\partial \dot{x}(t)} = 0$$

and yields the following stationary solution for the oscillator Lagrangian

$$\ddot{x}_c - \omega^2 x_c = \frac{d^2 x_c}{dt^2} - \omega^2 x_c = 0 \ ; \ \ddot{x} \equiv \frac{d^2 x_c}{dt^2} \quad (15.11.5)$$

$$\Rightarrow \ x_c(t) = Ae^{-\omega t} + Be^{\omega t} \ ; \ \omega = |\omega| > 0 \quad (15.11.6)$$

Note

$$\dot{x}_c(t) = \omega(-Ae^{-\omega t} + Be^{\omega t}) \quad (15.11.7)$$

To determine the constants A, B the boundary conditions given in Eq. 15.11.2 are imposed on $x_c(t)$

$$x_c(0) = x \ ; \ x_c(\tau) = x' \quad (15.11.8)$$

The boundary conditions, from Eq. 15.11.12, yield

$$x_c(0) = A + B = x \ ; \ x_c(\tau) = Ae^{-\omega\tau} + Be^{\omega\tau} = x'$$

and hence

$$A = \frac{1}{2\sinh\omega\tau}(xe^{\omega\tau} - x') \ ; \ B = \frac{1}{2\sinh\omega\tau}(x' - xe^{-\omega\tau}) \quad (15.11.9)$$

The stationary solution $x_c(t)$, from Eqs. 15.11.5 and 15.11.9, is given by

$$x_c(t) = \frac{1}{2\sinh\omega\tau}\left[(xe^{\omega\tau} - x')e^{-\omega t} + (x' - xe^{-\omega\tau})e^{\omega t}\right] \quad (15.11.10)$$

The stationary solution yields the following change of variables, where $\xi(t)$ are now the new stochastic integration variables

$$x(t) = x_c(t) + \xi(t) \ \Rightarrow \ \prod_{t=0}^{\tau}\int dx(t) = \prod_{t=0}^{\tau}\int d\xi(t) \quad (15.11.11)$$

Boundary conditions imposed on $x_c(t)$ make it agree with the stochastic variables $x(t)$ at $t = 0,\ \tau$ and yields the following boundary conditions for ξ

$$\xi(0) = 0 = \xi(\tau) \tag{15.11.12}$$

The key to solving the harmonic oscillator path integral is that stationary solution leads to the following factorization

$$S_0[x_c + \xi] = S_0[x_c] + S_0[\xi] \tag{15.11.13}$$

This factorization is possible only because the oscillator action is quadratic in the stochastic integration variables. For the nonlinear case, the expansion about the stationary solution leads to the semi-classical expansion of the path integral [7].

A proof of the factorization of $S_0[x_c + \xi] = S_0[x_c] + S_0[\xi]$, as in Eq. 15.11.13, is given below. The action, for $\dot{\xi} = d\xi/dt$, is given by

$$\begin{aligned}
S_0[\xi + x_c] &= -\frac{m}{2}\int_0^\tau dt\{(\dot{\xi} + \dot{x}_c)^2 + \omega^2(\xi + x_c)^2\} \\
&= -\frac{m}{2}\int_0^\tau dt\{\dot{\xi}^2 + 2\dot{\xi}\dot{x}_c + \dot{x}_c^2 + \omega^2(\xi^2 + 2\xi x_c + x_c^2)\} \\
&= -\frac{m}{2}\int dt(\dot{x}_c^2 + \omega^2 x_c^2) - m\int(\dot{\xi}\dot{x}_c + \omega^2\xi x_c) - \frac{m}{2}\int dt(\dot{\xi}^2 + \omega^2\xi^2)
\end{aligned}$$

Note, since $\xi(0) = \xi(\tau) = 0$

$$\begin{aligned}
\int_0^\tau \dot{\xi}\dot{x}_c &= \int_0^\tau \left\{\frac{d}{dt}(\xi\dot{x}_c) - \xi\ddot{x}_c\right\} \\
&= \xi(0)\dot{x}_c(0) - \xi(\tau)\dot{x}_c(\tau) - \int \xi\ddot{x}_c dt = -\int \xi\ddot{x}_c dt
\end{aligned}$$

The cross-terms of x_c and ξ in the action are

$$m\int(\dot{\xi}\dot{x}_c + \omega^2\xi x_c) = m\int dt(-\ddot{x}_c + \omega^2 x_c)\xi = 0$$

The cancellation is due to Eq. 15.11.5. Hence

$$S_0[x_c + \xi] = S_0[x_c] + S_0[\xi]\ ;\ \ S_0[\xi] = -\frac{m}{2}\int dt(\dot{\xi}^2 + \omega^2\xi^2) \tag{15.11.14}$$

Noteworthy 15.1: Stationary Action $S_0[x_c]$

The stationary action $S_0[x_c]$ is given by

$$S_0[x_c] = -\frac{m}{2}\int_0^\tau dt(\dot{x}_c^2 + \omega^2 x_c^2) = -\frac{m}{2}(x_c\dot{x}_c)\Big|_0^\tau - \frac{m}{2}\int_0^\tau dt x_c(-\ddot{x}_c + \omega^2 x_c)$$
$$= -\frac{m}{2}[x_c(\tau)\dot{x}_c(\tau) - x_c(0)\dot{x}_c(0)]$$

where Eq. 15.11.5 makes the integral zero leaving only the boundary term. The stationary action, using Eqs. 15.11.6 and 15.11.7, is hence

$$S_0[x_c] = -\frac{m\omega}{2}\left[B^2(e^{2\omega\tau} - 1) + A^2(1 - e^{-2\omega\tau})\right] \tag{5.11.15}$$

From Eqs. (5.11.15) and 15.11.9, the stationary action is

$$S_0[x_c] = -\frac{m\omega}{2\sinh\omega\tau}\left[(x^2 + x'^2)\cosh\omega\tau - 2xx'\right] \tag{5.11.16}$$

The factorization S_0 given in Eq. 15.11.14 yields the following

$$S[x_c + \xi] = S_0[x_c + \xi] + \int dt j(t)(x_c + \xi)(t)$$
$$= S_0[x_c] + \int_0^\tau j(t)x_c(t) + S_0[\xi] + \int_0^\tau j(t)\xi(t)$$

The change of variables $x = x_c + \xi$ yields a path integral for which the stochastic variables are coupled to the source $j(t)$—and one needs to perform the path integral over the stochastic variables $\xi(t)$.

The path integral for the pricing kernel is hence given by

$$\int Dx e^S = e^{F_0[x_c;j]}\int D\xi e^{F[\xi;j]} \tag{15.11.17}$$
$$F_0[x_c; j] = S_0[x_c] + \int_0^\tau j(t)x_c(t)$$
$$F[\xi; j] = S_0[\xi] + \int_0^\tau dt j(t)\xi(t)$$

The following path integral over the stochastic variables $\xi(t)$, with boundary conditions $\xi(0) = 0 = \xi(\tau)$ given in Eq. 15.11.12, needs to performed

$$\int D\xi \exp\{F[\xi; j]\} \ ; \ \xi(0) = 0 = \xi(\tau) \tag{15.11.18}$$

Due to the boundary conditions $\xi(0) = 0 = \xi(\tau)$ the path integration over the stochastic variables $\xi(t)$ given in Eq. 15.11.18 yields a result independent of the initial conditions x, x'. The boundary conditions on $\xi(t)$ are satisfied by the Fourier sine expansion for $\xi(t)$

$$\xi(t) = \sum_{n=1}^{\infty} \sin\left[\frac{n\pi t}{\tau}\right] \xi_n \ ;$$

where, for each n, ξ_n is an independent and real integration variable. Hence, using the orthogonality of the sine functions,

$$\int_0^{\tau} dt \sin\left[\frac{n\pi t}{\tau}\right] \sin\left[\frac{m\pi t}{\tau}\right] = \frac{\tau}{2}\delta_{n-m}$$

yields, from Eq. 15.11.14 for the action $S_0[\xi]$, the following

$$F[\xi; j] = -\frac{1}{2}\sum_{n=1}^{\infty} \lambda_n \xi_n^2 + \sum_{n=1}^{\infty} j_n \xi_n$$

$$\lambda_n = \frac{1}{2} m\tau\{(\frac{n\pi}{\tau})^2 + \omega^2\} \ ; \ j_n = \int_0^{\tau} dt j(t) \sin\left[\frac{n\pi t}{\tau}\right]$$

All the ξ_n variables have decoupled and yield

$$\int D\xi = C \prod_{n=1}^{+\infty} \int_{-\infty}^{+\infty} d\xi_n$$

where C is a normalization constant. Hence

$$\begin{aligned}
\int D\xi \exp\{F[\xi; j]\} &= C \prod_{n=1}^{+\infty} \int_{-\infty}^{+\infty} d\xi_n \exp\left\{-\frac{1}{2}\sum_{n=1}^{\infty} \lambda_n \xi_n^2 + \sum_{n=1}^{\infty} j_n \xi_n\right\} \\
&= C \prod_{n=1}^{\infty} \left(\sqrt{\frac{2\pi}{\lambda_n}}\right) \exp\{\frac{1}{2}\sum_{n=1}^{\infty} j_n \frac{1}{\lambda_n} j_n\} \\
&= \mathcal{N}(\tau) \exp\{\frac{1}{2}\sum_{n=1}^{\infty} j_n \frac{1}{\lambda_n} j_n\} \qquad (15.11.19)
\end{aligned}$$

Using the identity yields

$$\sinh\theta = \theta \prod_{n=1}^{\infty}\left[1 + \frac{\theta^2}{\pi^2 n^2}\right] \ \Rightarrow \ \prod_{n=1}^{\infty} \lambda_n = C' \prod_{n=1}^{\infty}\left[1 + \frac{(\omega\tau)^2}{\pi^2 n^2}\right] = C' \frac{\sinh\omega\tau}{\omega\tau}$$

Hence, the normalization is given by

$$\mathcal{N}(\tau) = \tilde{C}\sqrt{\frac{\omega\tau}{\sinh\omega\tau}}$$

Recall from Eq. 15.11.18 that the normalization is independent of x, x'. In the limit of $\omega \to 0$, the oscillator path integral reduces to the result given in Eq. 15.9.6; in particular, the normalization of the oscillator action yields the result given in $(m = 1/\sigma^2)$

$$\lim_{\omega\to 0}\mathcal{N}(\tau) = \sqrt{\frac{m}{2\pi\tau}}$$

The constant $\tilde{C}$ is fixed by above limit and yields the result

$$\mathcal{N}(\tau) = \sqrt{\frac{m\omega}{2\pi\sinh\omega\tau}}$$

Equation 15.11.19 yields the following

$$\sum_{n=1}^{\infty} j_n\frac{1}{\lambda_n}j_n = \int_0^{\tau}dt\int_0^{\tau}dt' j(t)D(t,t')j(t')$$

$$D(t,t') = \sum_{n=1}^{\infty}\frac{1}{\lambda_n}\sin[\frac{n\pi t}{\tau}]\sin[\frac{n\pi t'}{\tau}]$$

Using the identity

$$\sum_{n=1}^{+\infty}\frac{\cos(\theta)}{a^2+n^2} = \frac{\pi}{2a}\frac{\cosh[a(\pi-|\theta|)]}{\sinh\pi a} - \frac{1}{2a^2} \tag{15.11.20}$$

yields [7]

$$D(t,t') = \frac{1}{mw\sinh w\tau}\sinh w(\tau - t)\sinh(wt')\ ;\ \ t > t' \tag{15.11.21}$$

and hence

$$\frac{1}{2}\int_0^{\tau}dt\int_0^{\tau}dt' j(t)D(t,t')j(t')$$
$$= \frac{1}{mw\sinh w\tau}\int_0^{\tau}dt\int_0^{t}dt' J(t)\sinh w(\tau-t)\sinh(wt')J(t') \tag{15.11.22}$$

Note, from Eq. 15.11.10, that

$$\int_0^\tau j(t)x_c(t) = \frac{x}{\sinh w\tau}\int_0^\tau dt J(t)\sinh w(\tau - t) + \frac{x'}{\sinh w\tau}\int_0^\tau dt J(t)\sinh(wt) \quad (15.11.23)$$

Hence, collecting the results for Eqs. 15.11.17 and 15.11.18, from Eqs. 15.11.22 and 15.11.23

$$\int Dxe^S = \mathcal{N}(\tau)\exp\{S[x_c; j]\}\ ;\ \mathcal{N}(\tau) = \sqrt{\frac{m\omega}{2\pi\sinh\omega\tau}}\ ;\ \omega > 0 \quad (15.11.24)$$

$$S[x_c; j] = S_0[x_c] + \int_0^\tau j(t)x_c(t) + \frac{1}{2}\int_0^\tau dt\int_0^\tau dt' j(t)D(t,t')j(t')$$

and, from Eq. (5.11.16)

$$S_0[x_c] = -\frac{m\omega}{2\sinh\omega\tau}\left[(x^2 + x'^2)\cosh\omega\tau - 2xx'\right]$$

Furthermore, the generating functional is given by

$$\begin{aligned} Z[j] &= \frac{1}{Z}\int Dxe^S \\ &= \exp\left\{\frac{1}{2}\int_0^\tau dt\int_0^\tau dt' j(t)D(t,t')j(t') + \int_0^\tau j(t)x_c(t)\right\} \end{aligned}$$

with the stationary solution $S_0[x_c]$ canceling out due to the normalization by Z fixed by requiring $Z[0] = 1$.

15.12 Merton Oscillator Pricing Kernel

From Eq. 14.3.9 the Merton pricing kernel is given by

$$p_{(\omega,\lambda)}(x, x'; \tau) = \exp\{s(x) - s(x')\}\widetilde{p}_{(\omega,\lambda)}(x, \tau; x') \quad (15.12.1)$$

where the oscillator pricing kernel requires the evaluation of

$$\widetilde{p}_{(\omega,\lambda)}(x, x'; \tau) \equiv \langle x|e^{-\tau H_{\text{Eff}}(\omega,\lambda)}|x'\rangle$$

The Merton pricing kernel has important applications in option pricing [11] and in the analysis of risky coupon bonds [14].

It is shown that the pricing kernel $p_{(\omega,\lambda)}(x, x'; \tau)$ depends on sign *and* magnitude of ω, unlike $\widetilde{p}_{(\omega,\lambda)}(x, \tau; x')$ that depends only on $|\omega|$. This is due to the similarity transformation that yields the pre-factor of $\exp\{s(x) - s(x')\}$ in Eq. 15.12.1, and it is this term that yields the dependence of $p_{(\omega,\lambda)}(x, x'; \tau)$ on ω.

The path integral for the pricing kernel of Merton's oscillator, from Eqs. 15.4.4 and 15.4.6, is given by

$$\widetilde{p}_{(\omega,\lambda)}(x, x'; \tau) = \int D\zeta e^{S_0[\zeta]} = \mathcal{N}(\tau) e^{\widetilde{F}_{(\omega,\lambda)}} \tag{15.12.2}$$

$$S[\zeta] = \int_0^\tau dt \mathcal{L}_{\text{Eff}}(t)$$

$$\text{Boundary conditions } : \zeta(0) = x \; ; \quad \zeta(\tau) = x'$$

Consider the general case of the oscillator Lagrangian that has a linear term, and is given by

$$\mathcal{L}(t) = -\frac{1}{2\sigma^2}\left(\frac{d\zeta}{dt}\right)^2 - \frac{1}{2\sigma^2}\omega^2\zeta^2 + j(t)\zeta(t) \tag{15.12.3}$$

The Feynman path integral for the oscillator kernel is given by

$$\int D\zeta \exp\left\{\int_0^\tau \mathcal{L}(t)\right\} = \mathcal{N}(\tau) e^{\widetilde{F}(j)}$$

The Lagrangian given in Eq. 15.12.3 is the harmonic oscillator. Performing the path integration above yields the following exact results given in Eq. 15.11.24

$$\mathcal{N}(\tau) = \sqrt{\frac{|\omega|}{2\pi\sigma^2 \sinh(|\omega|\tau)}} \tag{15.12.4}$$

and

$$\begin{aligned}
\tilde{F}(j) = &-\frac{|\omega|}{2\sigma^2 \sinh(|\omega|\tau)}\left((x^2 + (x')^2)\cosh(|\omega|\tau) - 2xx'\right) \\
&+ \frac{x}{\sinh(|\omega|\tau)}\int_0^\tau dt j(t)\sinh(|\omega|(\tau - t)) + \frac{x'}{\sinh(|\omega|\tau)}\int_0^\tau dt j(t)\sinh(|\omega|t) \\
&+ \frac{\sigma^2}{|\omega|\sinh(|\omega|\tau)}\int_0^\tau dt \int_0^t dt' j(t)\sinh(|\omega|(\tau - t))\sinh(|\omega|t') j(t')
\end{aligned} \tag{15.12.5}$$

Consider the linear oscillator potential given by

$$\upsilon(x) = \omega x \; ; \quad q(x) = \lambda x \tag{15.12.6}$$

From Eq. 15.4.6, the effective Merton oscillator Lagrangian is given by

$$\mathcal{L}_{\text{Eff}}(t) = -\frac{1}{2\sigma^2}\left(\frac{d\zeta}{dt}\right)^2 - \frac{1}{2\sigma^2}\omega^2\zeta^2 - h\zeta - c$$

and, from Eq. 14.5.5

$$h = \frac{\omega}{\sigma^2}\left(r - \frac{1}{2}\sigma^2 + \sigma^2\frac{\lambda}{\omega}\right) \;;\; c = r + \frac{1}{2}\omega + \frac{1}{2\sigma^2}\left(r - \frac{1}{2}\sigma^2\right)^2$$

For the Merton Lagrangian, $j(t) = -h$ is a constant; hence the integrals yield

$$\int_0^\tau dt j(t) \sin|\omega|(\tau - t)) = -h\frac{1}{|\omega|}(\cosh(|\omega|\tau) - 1) = \int_0^\tau dt j(t) \sinh(|\omega|\tau)$$

and

$$I = \int_0^\tau dt \int_0^t dt' j(t) \sinh(|\omega|(\tau - t)) \sinh(|\omega|t') j(t')$$
$$\Rightarrow I = \frac{h^2}{|\omega|}\left[\frac{\tau}{2}\sinh(|\omega|\tau) - \frac{1}{|\omega|}(\cosh(|\omega|\tau) - 1)\right]$$

Hence, from Eq. 15.12.5 and above $\widetilde{p}_{(\omega,\lambda)}(x, x'; \tau)$ is given by

$$\begin{aligned}\widetilde{p}_{(\omega,\lambda)}(x, x'; \tau) &= \mathcal{N}(\tau)e^{\widetilde{F}_{(\omega,\lambda)}} \\ \widetilde{F}_{(\omega,\lambda)} &= \tilde{F}(-h) - c\tau \\ &= A_{(\omega,\lambda)} - \frac{|\omega|}{2\sigma^2\sinh(|\omega|\tau)}\left((x^2 + (x')^2)\cosh(|\omega|\tau) - 2xx'\right) \\ &\quad - \frac{h}{|\omega|}\frac{\cosh(|\omega|\tau) - 1}{\sinh(|\omega|\tau)}(x + x')\end{aligned} \tag{15.12.7}$$

Hence

$$\begin{aligned}\widetilde{F}_{(\omega,\lambda)}(x, x') = A_{(\omega,\lambda)} &- \frac{|\omega|}{2\sigma^2\sinh(|\omega|\tau)}\left((x^2 + (x')^2)\cosh(|\omega|\tau) - 2xx'\right) \\ &- \frac{\omega}{|\omega|\sigma^2}(r - \frac{1}{2}\sigma^2 + \sigma^2\frac{\lambda}{\omega})\frac{\cosh(|\omega|\tau) - 1}{\sinh(|\omega|\tau)}(x + x')\end{aligned} \tag{15.12.8}$$

and

$$\begin{aligned}A_{(\omega,\lambda)} &= -c\tau + \frac{h^2\sigma^2}{\omega^2\sinh(|\omega|\tau)}\left[\frac{\tau}{2}\sinh(|\omega|\tau) - \frac{1}{|\omega|}(\cosh(|\omega|\tau) - 1)\right] \\ \Rightarrow A_{(\omega,\lambda)} &= -\frac{1}{2}\omega\tau - \frac{1}{|\omega|\sigma^2\sinh(|\omega|\tau)}\left(r - \frac{1}{2}\sigma^2 + \sigma^2\frac{\lambda}{\omega}\right)^2(\cosh(|\omega|\tau) - 1) \\ &\quad + \tau\left(\frac{\lambda}{\omega} - 1\right)\left\{(r + \frac{1}{2}\sigma^2) + \frac{1}{2}\sigma^2(\frac{\lambda}{\omega} - 1)\right\}\end{aligned} \tag{15.12.9}$$

Define

$$\epsilon(\omega) = \begin{cases} 1 & \omega > 0 \\ -1 & \omega < 0 \end{cases} \quad \Rightarrow \quad \omega = \epsilon(w)|\omega| \tag{15.12.10}$$

From Eq. 15.12.7, the oscillator component of the Merton pricing kernel is

$$\widetilde{p}_{(\omega,\lambda)}(x, x'; \tau) = \mathcal{N}(\tau) \exp\{\widetilde{F}_{(\omega,\lambda)}\}$$

Recall from Eq. 14.5.6

$$s = (\frac{1}{2} - \frac{r}{\sigma^2})x - \frac{1}{2\sigma^2}\omega x^2$$

From Eq. 15.12.1 the Merton pricing kernel is given by

$$\begin{aligned} p_{(\omega,\lambda)}(x, x'; \tau) &= e^{s(x)-s(x')}\widetilde{p}_{(\omega,\lambda)}(x, \tau; x') \\ &= \mathcal{N}(\tau) \exp\{s(x) - s(x') + \widetilde{F}_{(\omega,\lambda)}\} \equiv \mathcal{N}(\tau) \exp\{F_{(\omega,\lambda)}\} \end{aligned}$$

where, from Eqs. 15.12.8 and 15.12.10

$$\begin{aligned} F_{(\omega,\lambda)}(x, x') &= s(x) - s(x') + \widetilde{F}_{(\omega,\lambda)}(x, x') \\ &= A_{(\omega,\lambda)} - \frac{|\omega|}{2\sigma^2 \sinh(|\omega|\tau)}\Big[x^2\big\{\cosh(|\omega|\tau) + \epsilon(\omega)\sinh(|\omega|\tau)\big\} \\ &\quad +(x')^2\big\{\cosh(|\omega|\tau) - \epsilon(\omega)\sinh((|\omega|\tau)\big\} - 2xx'\Big] \\ &\quad - \frac{x}{\sigma^2 \sinh(|\omega|\tau)}\Big[\epsilon(\omega)(r - \frac{1}{2}\sigma^2 + \sigma^2\frac{\lambda}{\omega})\{\cosh(|\omega|\tau) - 1\} - \sigma^2\sinh(|\omega|\tau)(\frac{1}{2} - \frac{r}{\sigma^2})\Big] \\ &\quad - \frac{x'}{\sigma^2 \sinh(|\omega|\tau)}\Big[\epsilon(\omega)(r - \frac{1}{2}\sigma^2 + \sigma^2\frac{\lambda}{\omega})\{\cosh(|\omega|\tau) - 1\} + \sigma^2\sinh(|\omega|\tau)(\frac{1}{2} - \frac{r}{\sigma^2})\Big] \end{aligned} \tag{15.12.11}$$

Define

$$\theta = \omega\tau = \epsilon(\omega)|\omega|\tau \;\Rightarrow\; \epsilon^2(\omega) = 1 \;\; ; \;\; \sinh(|\omega|\tau) = \epsilon(\omega)\sinh(\omega\tau)$$

Hence, from 15.12.11 and after simplifications, the final result is given by

$$\begin{aligned} F_{(\omega,\lambda)}(x, x') &= A_{(\omega,\lambda)} - \frac{\omega}{2\sigma^2\sinh(\theta)}\left(x^2 e^{\theta} + (x')^2 e^{-\theta} - 2xx'\right) \\ &\quad - \frac{x}{\sigma^2\sinh(\theta)}\Big[(r - \frac{1}{2}\sigma^2 + \sigma^2\frac{\lambda}{\omega})\{\cosh(\theta) - 1\} - \sigma^2\sinh(\theta)(\frac{1}{2} - \frac{r}{\sigma^2})\Big] \\ &\quad - \frac{x'}{\sigma^2\sinh(\theta)}\Big[(r - \frac{1}{2}\sigma^2 + \sigma^2\frac{\lambda}{\omega})\{\cosh(\theta) - 1\} + \sigma^2\sinh(\theta)(\frac{1}{2} - \frac{r}{\sigma^2})\Big] \end{aligned} \tag{15.12.12}$$

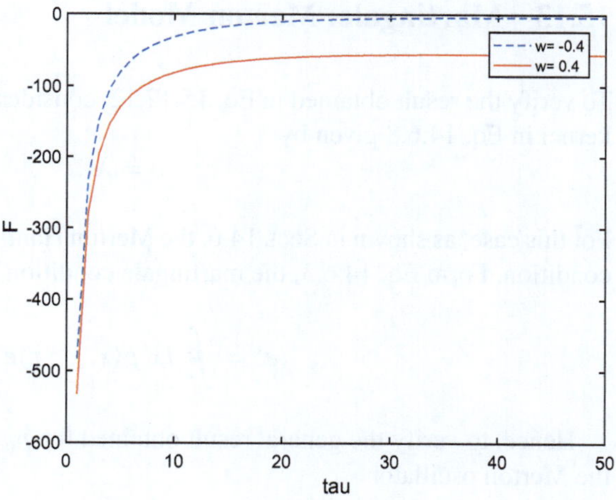

Fig. 15.4 For the two cases of $\pm\omega$, the value of $F_{(\omega,\lambda)}(x, x')$ for fixed $x \neq x'$ as a function of τ. Published with permission of © Belal E. Baaquie 2020. All Rights Reserved

and from Eq. 15.12.9

$$A_{(\omega,\lambda)} = -\frac{1}{2}\omega\tau - \frac{1}{\omega\sigma^2 \sinh(\theta)}\left(r - \frac{1}{2}\sigma^2 + \sigma^2\frac{\lambda}{\omega}\right)^2 (\cosh(\theta) - 1)$$
$$+\tau\left(\frac{\lambda}{\omega} - 1\right)\left\{(r + \frac{1}{2}\sigma^2) + \frac{1}{2}\sigma^2(\frac{\lambda}{\omega} - 1)\right\} \qquad (15.12.13)$$

The normalization, from Eq. 15.12.4, is given by

$$\mathcal{N}(\tau) = \sqrt{\frac{\omega}{2\pi\sigma^2 \sinh(\omega\tau)}}$$

Figure 15.4 shows the dependence of $F_{(\omega,\lambda)}(x, x')$ on ω. Note the important fact that, as shown in Eq. 15.12.12, that unlike the oscillator kernel $\widetilde{p}_{(\omega,\lambda)}(x, x'; \tau)$ that depends only on $|\omega|$ as shown in Eq. 15.12.7, the full pricing kernel $p_{(\omega,\lambda)}(x, x'; \tau)$ depends on the sign and magnitude of ω; this fact is crucial for the study of contingent claims since $-|\omega|$ yields the payments of coupons *made* by the firm; in contrast, a term with $|\omega|$ represents the firm *receiving* payments.

It follows from Eq. 15.12.12 that

$$F_{(\omega,\lambda)}(x, x') = F_{(-\omega,-\lambda)}(x', x)$$

Reversing the sign of ω reverses the direction of remaining time τ. Figure 15.4 shows the dependence of $F_{(\omega,\lambda)}(x, x')$ on the sign of ω. $F_{(\omega,\lambda)}(x, x')$ is negative for both signs of ω, and hence yields a convergent Gaussian integral that is required in its application to the pricing of options [11] as well as to the pricing of a coupon bond [14].

15.13 Martingale: Merton Model

To verify the result obtained in Eq. 15.12.12, consider the special case of the pricing kernel in Eq. 14.6.8 given by

$$\omega = \lambda$$

For this case, as shown in Sect. 14.6, the Merton Hamiltonian satisfies the martingale condition. Form Eq. 14.6.3, the martingale condition is given by

$$e^x = \int dx' p(x, x'; \tau) e^{x'}$$

Hence, to verify the general result obtained for $p_{(\omega,\lambda)}$, one needs to show that for the Merton oscillator

$$e^x = \int dx' p_{(\omega,\omega)}(x, x'; \tau) e^{x'} \tag{15.13.1}$$

It is explicitly verified below that for the Merton oscillator pricing kernel, the martingale condition given in Eq. 15.13.1 is, in fact, satisfied.

From Eq. 15.12.12, the following integral needs to be evaluated

$$\int dx' p_{(\omega,\omega)}(x, x'; \tau) e^{x'} = \mathcal{N}(\tau) \int dx' e^{F_{(\omega,\omega)}(x,x')+x'}$$

To perform the Gaussian integration over x', the following shift is performed $x' \to x' + e^\theta x$ and yields

$$\begin{aligned}\int dx' p_{(\omega,\omega)}(x, x'; \tau) e^{x'} &= \mathcal{N}(\tau) \int dx' \exp\{F_{(\omega,\omega)}(x, x' + e^\theta x) + x' + e^\theta x\} \\ &= \mathcal{N}(\tau) \int dx' \exp\{F'\}\end{aligned}$$

From Eq. 15.12.12

$$F' = F_{(\omega,\omega)}(x, x' + e^\theta x) + x' + e^\theta x \tag{15.13.2}$$

$$\begin{aligned}&= -\frac{\omega}{2\sigma^2 \sinh(\theta)} e^{-\theta}(x')^2 - \frac{x}{\sinh(\theta)}\left[\frac{r}{\sigma^2}(e^\theta - 1) + \frac{1}{2}(e^{-\theta} - 1)\right] \\ &\quad - \frac{x' + xe^\theta}{\sinh(\theta)}\left[\frac{r}{\sigma^2}(e^{-\theta} - 1) + \frac{1}{2}(e^\theta - 1)\right] + x' + xe^\theta + A_{(\omega,\omega)}\end{aligned} \tag{15.13.3}$$

The coefficient of the x-term in Eq. 15.13.2 above is given by

$$e^{\theta} - \frac{1}{\sinh(\theta)}\left[\frac{r}{\sigma^2}(e^{\theta}-1) + \frac{1}{2}(e^{-\theta}-1)\right] - \frac{e^{\theta}}{\sinh(\theta)}\left[\frac{r}{\sigma^2}(e^{-\theta}-1) + \frac{1}{2}(e^{\theta}-1)\right]$$
$$= e^{\theta} - \frac{1}{\sinh(\theta)}\left[\frac{1}{2}(e^{-\theta}-1) + \frac{1}{2}(e^{2\theta}-e^{\theta})\right] = 1 \tag{15.13.4}$$

From Eqs. 15.13.2 and 15.13.4

$$F' = x + x' - \frac{\omega}{2\sigma^2\sinh(\theta)}e^{-\theta}(x')^2$$
$$-\frac{x'}{\sinh(\theta)}\left[\frac{r}{\sigma^2}(e^{-\theta}-1) + \frac{1}{2}(e^{\theta}-1)\right] + A_{(\omega,\omega)} \tag{15.13.5}$$
$$= x - \frac{\omega}{2\sigma^2\sinh(\theta)}e^{-\theta}(x')^2 - \frac{x'}{\sigma^2\sinh(\theta)}(r + \frac{1}{2}\sigma^2)(e^{-\theta}-1) + A_{(\omega,\omega)}$$

From Eqs. 15.13.2 and 15.13.5

$$\mathcal{N}(\tau)\int dx' \exp\{F_{(\omega,\omega)}(x, x' + e^{\theta}x) + x' + e^{\theta}x\}$$
$$= e^{x}\mathcal{N}(\tau)\int dx' \exp\Big\{ -\frac{\omega}{2\sigma^2\sinh(\theta)}e^{-\theta}(x')^2$$
$$-\frac{x'}{\sigma^2\sinh(\theta)}(r + \frac{1}{2}\sigma^2)(e^{-\theta}-1) + A_{(\omega,\omega)}\Big\}$$
$$= e^{x}\mathcal{N}(\tau)\sqrt{\frac{2\pi e^{\theta}\sigma^2\sinh(\theta)}{\omega}}e^{-\frac{1}{2}\theta}$$
$$\times \exp\left\{\frac{1}{2\sigma^2\omega\sinh(\theta)}(r + \frac{1}{2}\sigma^2)^2\left[e^{\theta}(e^{-\theta}-1)^2 - 2(\cosh(\theta)-1)\right]\right\}$$
$$= e^{x} \tag{15.13.6}$$

and yields the expected result given in Eq. 15.13.1

$$\int dx' p_{(\omega,\omega)}(x, x'; \tau)e^{x'} = \mathcal{N}(\tau)\int dx' e^{F_{(\omega,\omega)}(x,x')+x'} = e^{x} \tag{15.13.7}$$

The martingale condition is given in Eq. 14.6.3. It is explicitly verified in Eq. 15.13.7 that the martingale condition is valid for the Merton pricing kernel $p_{(\omega,\omega)}(x, x'; \tau)$.

Hence the special case of $\omega = \lambda$ for the Merton pricing kernel defines a martingale time evolution for the security $V = e^{x}$ and could be used for pricing options.

Chapter 16
Quantum Fields for Bonds and Interest Rates

Abstract The theory of coupon bonds and of interest rates on cash deposits are based on two different types of quantum fields: coupon bonds are described by forward interest rates, modeled using a *linear* quantum field, and Libor interest rates that are modeled using a *nonlinear* quantum field. Key concepts of forward interest rates and Libor, including the Lagrangian and Hamiltonian are discussed. The martingale process for these two cases are expressed in the framework of quantum fields. Both the Lagrangian and Hamiltonian formulation of quantum fields are developed to demonstrate the utility of these two different, but complimentary, approaches.

16.1 Introduction

Coupon bonds, both sovereign and corporate, are the largest component of the capital markets. Merton's formulation of risky coupon bonds, discussed in Chap. 12, depends on the valuation of a firm, and a typical firm may issue a few bonds with varying maturity and face value. In contrast, sovereign bonds have a vast range of maturities, face values and coupons—being described by the *forward interest rates*.

Quantum field theory has been applied to a wide range of problems in economics and finance [8], and which cover topics that are outside the scope of this book. Instead of addressing all the myriad forms of applications, this Chapter concentrates on using quantum field theory to study of finance, and in particular bonds and interest rates.

The first part of this Chapter analyzes forward interest rates, which are modeled using a linear quantum field [4]. The Lagrangian and action for the forward interest rates is defined, and the formalism is used to obtain a time-dependent state space and a time dependent Hamiltonian. The path integral is used to impose the martingale condition on the evolution of the forward interest rates that, in turn, fixes its drift: the drift depends only on the volatility of the forward interest rates. The concept of the numeraire and numeraire invariance is discussed, and is utilized for a path integral derivation of the Call option price of a zero coupon bond.

Libor (London Interbank Offered Rate) is a widely used financial instrument for simple interest rates on time deposits of cash. Libor requires a nonlinear quantum field

B. E. Baaquie, *Mathematical Methods and Quantum Mathematics for Economics and Finance*, https://doi.org/10.1007/978-981-15-6611-0_16

for its description [5]. The Libor Market Model is defined and the Libor Lagrangian and time-dependent Hamiltonian are derived. The martingale condition is imposed using the Hamiltonian formalism, and the nonlinear drift of Libor is derived. Libor is a nonlinear quantum field because Libor drift, unlike the case of the forward interest rates, depends on Libor as well. Black's caplet result is derived exactly using the Libor path integral.

16.2 Coupon Bonds

Coupon bonds have been discussed briefly in Sect. 1.13 and the main results are summarized. A *zero coupon bond* is a financial instrument that gives a single predetermined payoff, of say \$1, called the principal amount or face value, when it matures at some fixed future time T; it's price at earlier time $t < T$ is given by $B(t, T)$. Note that for a zero coupon bond there are no coupon payments and hence the name.

Consider the simplest case of an economy that has a constant interest rate r. A continuously compounded fixed cash deposit of \$1 made at time t will yield, at time T in the future, a cash of amount $\exp\{(T-t)r\}$. Hence a zero coupon bond yielding \$1 at time T has a present value of

$$B(t, T) = e^{-(T-t)r}$$

In general, a real economy has an interest rate that is not constant over future time. Instead, for each future time T there is a separate effective interest rate, denoted by $r(t, T)$, called the *term structure of interest rates*. The zero coupon bond is given by

$$B(t, T) = e^{-(T-t)r(t,T)} \quad \Rightarrow r(t, T) = -\frac{1}{T-t} \ln B(t, T) \tag{16.2.1}$$

Consider a *coupon bond*, denoted by $\mathcal{B}(t)$, that pays a principal of L when it matures at time T, and pays *fixed* dividends (coupons) a_i at times T_i, $i = 0, 1, 2, ...N$, shown in Fig. 16.1. The tenor ℓ of the bond is the intervals at which the coupons are paid and yields

$$T_i = i\ell \ ; \ i = 0, 1, 2, ...N$$

The value of the coupon bond at time $t < T_i$ can be shown to be equivalent to a portfolio of zero coupon bonds with maturities coinciding with the payment dates of the coupons.

Fig. 16.1 Future time for payments of coupons. Published with permission of ©Belal E. Baaquie 2020. All Rights Reserved

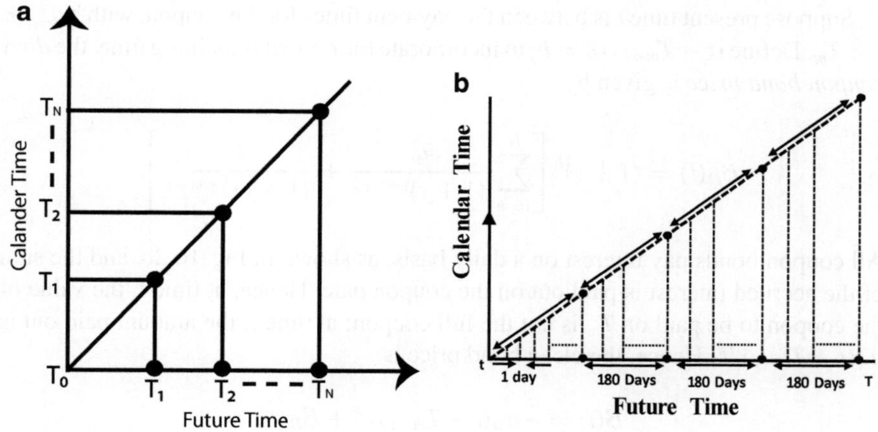

Fig. 16.2 Calendar and future times for the payments of coupons: **a** yearly and **b** daily. Published with permission of © Belal E. Baaquie 2020. All Rights Reserved

The coupon bond with pre-fixed coupons a_i is given by

$$\mathcal{B}(t) = \sum_{i=1}^{N} a_i B(t, T_i) + LB(t, T) = \sum_{i=1}^{N} c_i B(t, T_i) \qquad (16.2.2)$$

$\mathcal{B}(t)$ is called the *clean bond price*, and is the actual market price. The time of maturity of the coupon bond is taken to be the date of the last coupon payment, that is $T = T_N$. The final payment is included in the sum by setting $c_i = a_i$; $c_N = a_N + L$.

There *four types of time* for a coupon bond: present time, present future time, future calendar time and future time. This is the key for understanding fixed income securities. Figure 16.2 shows the payments of the coupon bond for future time and calendar time.

Given the wide variety of coupon bonds, with different face values L, different amounts and number of coupon payments a_i and T_N respectively, to compare the *rates of return* of two different coupon bonds a yield-to-maturity y is defined for coupon bonds: it is the annual yield such that, at time T_0, the present values of the future cash flows, discretely discounted yearly by y, equals the face value of the coupon bond.

$$\mathcal{B}(T_0) = \sum_{i=1}^{N} \frac{a_i}{(1+y)^i} + \frac{L}{(1+y)^N} \; ; \; T_i = i\ell \; : \; i = 1, \cdots, N$$

Note that the first coupon a_1 is paid on T_1 *after a full tenor time* ℓ has elapsed. No coupon is paid on the date of issue T_0. The coupon payments and bond price are shown in Fig. 16.2a and b.

Suppose present time t is between the payment times for the coupon, with $T_{m-1} < t < T_m$. Define $(t - T_{m-1})/\ell = b$; to incorporate the correct remaining time, the *dirty coupon bond price* is given by

$$\mathcal{B}_D(t) = (1+y)^b \left[\sum_{i=m}^{N} \frac{a_i}{(1+y)^{i-m+1}} + \frac{L}{(1+y)^{N-m+1}} \right]$$

All coupon bonds pay interest on a daily basis, as shown in Fig. 16.2b, and the sum of the accrued interest is paid out on the coupon date. Hence, at time t, the value of the coupon to be paid on T_m is not the full coupon; at time t, the amount paid out is $a_m(t - T_{m-1})/\ell$. Hence, the clean bond price is

$$\mathcal{B}(t) = -a_m(t - T_{m-1})/\ell + \mathcal{B}_D(t)$$

Daily data is provided for both clean and dirty bond prices. When $t \to T_m$, then

$$b = (t - T_{m-1})/\ell \to m - (m-1) = 1\ ;\ \ a_m(t - T_{m-1})/\ell \to a_m$$

In practice, at time T_m, the price of the bond is ex-dividend the coupon due on T_m; if the bond is sold at T_m, the coupon a_m is paid to the seller of the bond and the buyer get the next coupon (Fig. 16.3).

Hence, as $t \to T_m$, the dirty bond converges to the clean bond

$$\mathcal{B}_D(t) \to -a_m + (1+y)\left[\frac{a_m}{(1+y)} + \frac{a_{m+1}}{(1+y)^2} + \cdots \frac{L}{(1+y)^{N-m+1}} \right]$$

$$\Rightarrow \mathcal{B}_D(T_m) = \mathcal{B}(T_m) = \sum_{i=m+1}^{N} \frac{a_i}{(1+y)^{i-m}} + \frac{L}{(1+y)^{N-m}}$$

Note the sum for the coupon payments starts at $m+1$ and the first coupon to be paid is at T_{m+1}; no coupon is paid at time T_m. If the coupon bond is sold at time T_m, the first coupon payment for the buyer is at T_{m+1}.

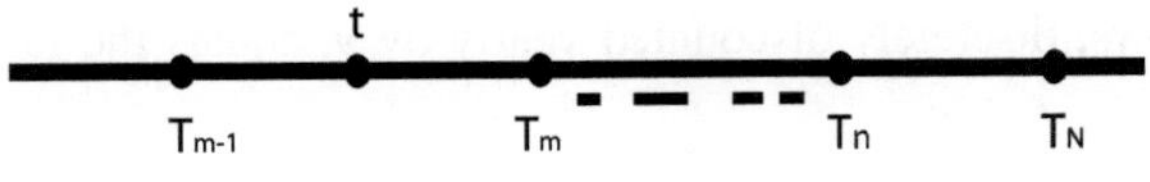

Fig. 16.3 Calendar and future times for the price of a bond at time t in between the time for the payments of the coupons. Published with permission of © Belal E. Baaquie 2020. All Rights Reserved

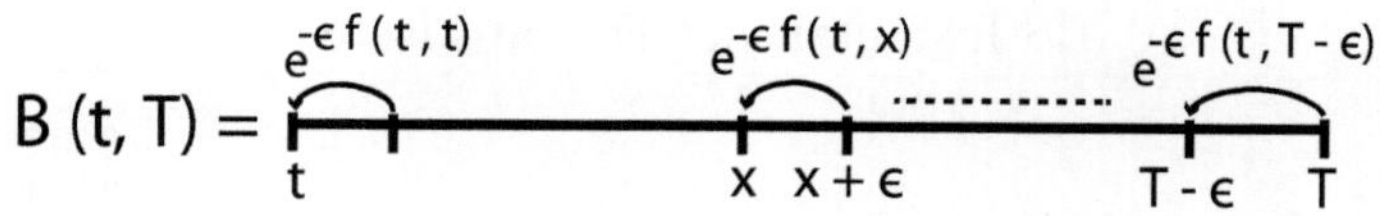

Fig. 16.4 The domain of the state space of the interest rates. The state space $\mathcal{V}_t$ is indicated for two distinct calendar times t_1 and t_2. Published with permission of © Belal E. Baaquie 2020. All Rights Reserved

16.3 Forward Interest Rates

The *forward interest rate* $f(t, x)$ is the interest rate—fixed at time $t < x$—for an instantaneous (overnight) loan taken at future time x. At every instant t, $f(t, x)$—as a function of x—is an entire curve.

The bond $B(t, T)$ is obtained by discounting the \$1 payoff taking infinitesimal backward time steps ϵ from maturity T to present time t, as shown in Fig. 16.4, and yields[1]

$$B(t, T) = e^{-\epsilon f(t,t+\epsilon)} e^{-\epsilon f(t,t+2\epsilon)} .. e^{-\epsilon f(t,x)} ... e^{-\epsilon f(t,T)} \$1$$

$$\Rightarrow B(t, T) = \exp\{-\int_t^T dx f(t, x)\}$$

Consider a zero coupon bond $B(T_1, T_2)$ that is going to be issued at some future time $T_1 > t$, with expiry at time T_2. The *forward price* locks-in its price at time t for delivery when it is issued at future time T_1; as expected, it is given by

$$F(t, T_1, T_2) = \exp\{-\int_{T_1}^{T_2} dx f(t, x)\} = \frac{B(t, T_2)}{B(t, T_1)}$$

At any instant t, there exist in the market Treasury bonds with maturity of up to future time T_{FR} and hence forward interest rates $f(t, x)$, at any instant of calendar time t, exists for all future times given by $t < x < t + T_{FR}$.[2]

Figure 16.5 shows the market data on the forward interest rates for the US\$ obtained from the US forward interest rates for the 2000–2019. Figure 16.5 plots the daily traded values of forward interest rates for only yearly maturities from one to ten. The forward interest rates' time evolution for ten maturities has the appearance of ten points randomly evolving in time, but in a very correlated manner; for example all the lines move up and down together and they never cross.

The HJM-model for the forward interest rates is the industry benchmark for pricing interest rate and bond derivatives [23]. The HJM-model, for Gaussian white noise $R(t)$, is given by [23]

[1] The fixed payoff \$1 is assumed and is not written out explicitly.

[2] T_{FR} is greater than 30 years.

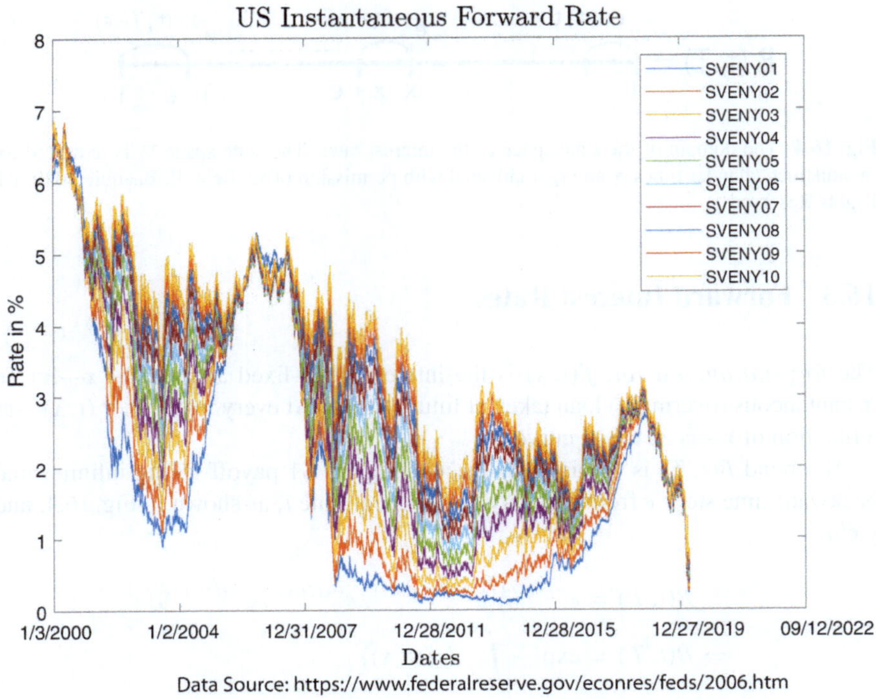

Fig. 16.5 US Forward Interest Rates f(t,10 years), f(t, 9 years), ... f(t, 2 year), and f(t,1 year). Published with permission of © Belal E. Baaquie 2020. All Rights Reserved

$$\frac{\partial f(t,x)}{\partial t} = \alpha(t,x) + \sigma(t,x)R(t) \qquad (16.3.1)$$

where $\alpha(t, x)$ is the drift and $\sigma(t, x)$ the volatility of the forward interest rates. Both are deterministic functions.

The full forward rate curve consists of infinitely many maturities, all of which evolve randomly in a *highly correlated* manner. The forward interest rates are modeled using a *quantum field*. Quantum field theory has been discussed in [8, 9], together with its application to economics and finance; quantum field theory provides the mathematical tools for the description of coupon bonds and interest rates.

$\mathcal{A}(t, x)$ is a quantum field that drives the evolution of the forward interest rates and replaces white noise $R(t)$ in the HJM-model. $\mathcal{A}(t, x)$ is an *independent* random variable for each x and each t. Hence, the generalization of HJM-model yields

$$\frac{\partial f(t,x)}{\partial t} = \alpha(t,x) + \sigma(t,x)\mathcal{A}(t,x) \qquad (16.3.2)$$

The difference between white noise and a quantum field can be seen by comparing $R(t)$ and $\mathcal{A}(t, x)$. For fixed time t_0, white noise has only one independent

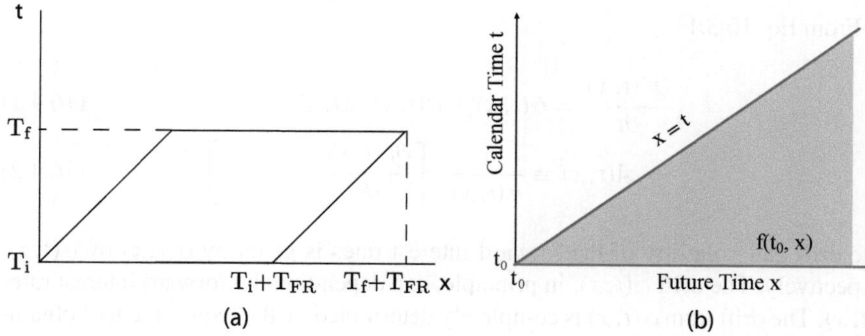

Fig. 16.6 **a** The trapezoidal domain $\mathcal{T}$ of the forward rates defined for finite calendar and finite future time. **b** The domain of forward interest rates in the limit of $T_f, T_{FR} \to \infty$. Published with permission of © Belal E. Baaquie 2020. All Rights Reserved

(Gaussian) random variable $R(t_0)$ whereas the quantum field $\mathcal{A}(t_0, x)$ has infinitely many independent random variables, one for each x, with $x \in [t_0, t_0 + T_{FR}]$. The theory of quantum fields is a mathematical framework for studying problems involving infinitely many (independent) degrees of freedom, and so one is naturally led to it's techniques.

For the sake of concreteness, consider the evolution of the forward rates starting from some initial time T_i to a future time T_f. Since all the forward rates $f(t, x)$ are always for the future, it is always true that $x > t$; hence the quantum field $\mathcal{A}(t, x)$ is defined on a domain consisting of a *trapezoid* $\mathcal{T}$ that is bounded in the maturity direction by parallel lines $x = t$ and $x = T_{FR} + t$, and in the time direction by the horizontal lines $t = T_i$ and $t = T_f$ as shown in Fig. 16.6a. Every point inside the domain $\mathcal{T}$ represents an independent integration variable $\mathcal{A}(t, x)$. Forward interest rates are modeled by a two-dimensional quantum field on a finite domain.

The forward rates starting from some time T_i can in principle be defined into the infinite future, that is with $T_f = \infty$; for simplicity the limit of T_{FR} is also taken, and yields a semi-infinite domain shown in Fig. 16.6b. Since the forward rates $f(t, x)$ are only defined for the future, one always has $x > t$. The finite domain $\mathcal{T}$ of the forward rates can be extended, as shown in Fig. 16.6b, to a semi-infinite wedge that is bounded by line $x = t$ and by the line $t = T_i$ in the time direction.

16.4 Action and Lagrangian

The quantum field driving the coupon bonds is discussed in some detail as it exemplifies the appearance of quantum fields in finance. Empirical evidence to support the modeling of forward interest rates with a two dimensional quantum field is discussed extensively in [4, 5]. The empirical study of the forward interest rates leads to the Lagrangian discussed in this section.

From Eq. 16.3.1

$$\frac{\partial f(t,x)}{\partial t} = \alpha(t,x) + \sigma(t,x)\mathcal{A}(t,x) \tag{16.4.1}$$

$$\Rightarrow \mathcal{A}(t,x) \equiv \frac{1}{\sigma(t,x)} \cdot \left[\frac{\partial f(t,x)}{\partial t} - \alpha(t,x)\right] \tag{16.4.2}$$

The drift and volatility of the forward interest rates is given by $\alpha(t,x)$ by $\sigma(t,x)$, respectively. The drift $\alpha(t,x)$, in principle, can depend on the forward interest rates $f(t,x)$. The drift term $\alpha(t,x)$ is completely determined by the requirement of obtaining a martingale evolution of the forward interest rates. In fact, the main difference between the modeling of Libor interest rates and of $f(t,x)$ is that the Lagrangian for Libor has a nonlinear drift term.

The volatility $\sigma(t,x)$ for both the forward interest rates and for Libor does not depend on $f(t,x)$ or on the Libor rates, as it can be shown that this leads to an unstable system.

An extensive empirical study of the forward interest rates leads to the following Lagrangian density $\mathcal{L}[f]$ [4]

$$\mathcal{L}[f] = \mathcal{L}_{\text{kinetic}}[f] + \mathcal{L}_{\text{rigid}}[f] + \mathcal{L}_{\text{stiff}}[f] \tag{16.4.3}$$

$$= -\frac{1}{2}\left[\mathcal{A}^2 + \frac{1}{\mu^2}\left(\frac{\partial \mathcal{A}}{\partial x}\right)^2 + \frac{1}{\lambda^4}\left(\frac{\partial^2 \mathcal{A}}{\partial x^2}\right)^2\right] \tag{16.4.4}$$

$$-\infty \le f(t,x) \le +\infty \ : \ \mathcal{A}(t,x) \equiv \frac{1}{\sigma(t,x)} \cdot [\frac{\partial f(t,x)}{\partial t} - \alpha(t,x)] \tag{16.4.5}$$

The action for the forward rates is given by

$$S[f] = \int_{T_i}^{T_f} dt \int_t^{t+T_{FR}} dx \mathcal{L}[f] \tag{16.4.6}$$

$$\equiv \int_{\mathcal{T}} \mathcal{L}[f] \tag{16.4.7}$$

Since the field theory is defined on a finite domain $\mathcal{T}$ as shown in Fig. 16.6a, to complete the definition of the model boundary conditions need to be specified on all the four boundaries of the finite trapezoid $\mathcal{T}$.

- *Fixed (Dirichlet) Initial and Final Conditions*

The initial and final (Dirichlet) conditions in the time direction are given by

$$t = T_i \ ; \ T_i < x < T_i + T_{FR} \ ; \ f(T_i, x) : \text{specified initial forward rate curve}$$
$$t = T_f \ ; \ T_f < x < T_f + T_{FR} \ ; \ f(T_f, x) : \text{specified final forward rate curve}$$

• *Free (Neumann) Boundary Conditions*

The Lagrangian in the action $\int_{\mathcal{T}} \mathcal{L}[f]$ given in Eq. 16.4.6 can be integrated by parts with respect to x; the requirement that there are no boundary terms yields, for $x = t$ or $x = t + T_{FR}$, the following Neumann boundary condition

$$T_i < t < T_f \ : \ \frac{\partial}{\partial x}\left(\frac{\frac{\partial f(t,x)}{\partial t} - \alpha(t,x)}{\sigma(t,x)}\right) = \frac{\partial \mathcal{A}(t,x)}{\partial x}\Big|_{x=t;x=t+T_{FR}} = 0 \quad (16.4.8)$$

The Neumann boundary conditions are necessary for obtaining a Hamiltonian for the forward interest rates. Doing an integration by parts in the maturity direction using the Neumann boundary conditions yields, from Eqs. 16.4.7 and 16.4.3, the action

$$S = -\frac{1}{2}\int_{\mathcal{T}} \mathcal{A}\left(1 - \frac{1}{\mu^2}\frac{\partial^2}{\partial x^2} + \frac{1}{\lambda^4}\frac{\partial^4}{\partial x^4}\right)\mathcal{A} \quad (16.4.9)$$

The quantum field theory of the forward interest rates is defined by the partition function Z, which is obtained by integrating over all configurations of $f(t,x)$, and yields the Feynman path integral

$$Z = \int Df e^{S[f]} \ ; \ \int Df \equiv \prod_{(t,x)\epsilon\mathcal{T}} \int_{-\infty}^{+\infty} df(t,x) \quad (16.4.10)$$

$e^{S[f]}/Z$ is the probability for different field configurations to occur when the functional integral over $f(t,x)$ is performed.

The change of variables given in Eq. 16.4.2, even for nonlinear drift $\alpha(t,x)$, yields [5][3]

$$\int Df = \int D\mathcal{A} \ ; \ \int D\mathcal{A} = \prod_{(t,x)\epsilon\mathcal{T}} \int_{-\infty}^{+\infty} d\mathcal{A}(t,x)$$

From Eq. 16.4.4, the equivalent path integral is given by

$$Z = \int D\mathcal{A} e^{S[\mathcal{A}]} \quad (16.4.11)$$

$$S[\mathcal{A}] = -\frac{1}{2}\int_{\mathcal{P}}\left[\mathcal{A}^2 + \frac{1}{\mu^2}\left(\frac{\partial \mathcal{A}}{\partial x}\right)^2 + \frac{1}{\lambda^4}\left(\frac{\partial^2 \mathcal{A}}{\partial x^2}\right)^2\right] \quad (16.4.12)$$

In terms of the forward interest rates, the action and Lagrangian is given by

$$S = \int_{t_1}^{t_2} dt \int_{x,x'} \mathcal{L}_f(t,x,x') \ ; \ \int_x \equiv \int_t^{t+T_{FR}} dx \ ; \ t_1 < t_2 \quad (16.4.13)$$

[3]Upto an irrelevant constant.

where a general Lagrangian density for the bond forward interest rates, from Eqs. 16.4.2 and 16.4.4, is given by

$$\mathcal{L}_f(t,x,x') \tag{16.4.14}$$
$$= -\frac{1}{2}\left[\frac{\partial f(t,x)/\partial t - \alpha(t,x)}{\sigma(t,x)}\right]\mathcal{D}^{-1}(x,x';t;T_{FR})\left[\frac{\partial f(t,x')/\partial t - \alpha(t,x')}{\sigma(t,x')}\right]$$
$$-\infty \leq f(t,x) \leq +\infty \tag{16.4.15}$$

Neumann boundary conditions, given in Eq. 16.4.8, have been incorporated into the expression for the Lagrangian by requiring that the fields obey these conditions.

The representations of the forward interest rates' path integral given in Eqs. 16.4.10 and 16.4.11 both have their own utility.

16.5 Correlation Function

Consider the Gaussian forward interest rates with a *deterministic* drift $\alpha(t,x)$. The propagator of the forward interest rates is given by the connected correlation function[4]

$$E[f(\tau,x)f(\tau',x')]_c = \int_{t_0}^{\tau} dt \int_{t_0}^{\tau'} dt' \sigma(t,x)\sigma(t',x')E[\mathcal{A}(t,x))\mathcal{A}(t',x')]$$

For simplicity, consider a semi-infinite trapezoid with $T_{FR} \to \infty$, as shown in Fig. 16.6b. From Eq. 16.4.12

$$E[\mathcal{A}(t,x)\mathcal{A}(t',x')] = \delta(t-t')\mathcal{G}(x-t;x'-t) \tag{16.5.1}$$

Similar to the singularity of white noise given in Eq. 11.3.3, the value of the equal time correlation function, due to the Dirac delta function, has the following Wilson expansion [5, 8]

$$E[\mathcal{A}(t,x)\mathcal{A}(t,x')] = \frac{1}{\epsilon}\mathcal{G}(x-t;x'-t) + \cdots \tag{16.5.2}$$

The explicit expression $\mathcal{G}(x-t;x'-t)$ for the finite trapezoid is given by [4], and is derived below.

The forward rate auto-correlation function, from Eqs. 16.8.4 and 16.5.1, is given by

$$E[f(\tau,x)f(\tau',x')]_c = \theta(\tau-\tau')\int_{t_0}^{\tau} dt\sigma(t,x)\sigma(t,x')\mathcal{G}(x-t;x'-t)\ ; \quad \tau > \tau'$$

[4] $E[AB]_c = E[AB] - E[A]E[B]$.

From Eq. 16.4.8 the propagator $\mathcal{D}(x-t; x'-t)$ satisfies the Neumann boundary conditions

$$\frac{\partial}{\partial x} E[\mathcal{A}(t,x))\mathcal{A}(t,x')\Big|_{x=t} = 0 = \frac{\partial}{\partial x'} E[\mathcal{A}(t,x))\mathcal{A}(t,x')\Big|_{x'=t}$$

The Neumann boundary condition leads to the following Fourier expansion

$$\mathcal{A}(t,x) = \int_t^{+\infty} \frac{dk}{\pi/2} e^{i\omega t} \cos(k(x-t))\mathcal{A}(t,k) \tag{16.5.3}$$

Note the propagator $\mathcal{D}(x-t; x'-t)$ depends only on $\theta = x - t > 0$ due to the trapezoidal domain and the Neumann boundary conditions. The Neumann boundary conditions at $x = t$ and $x' = t$, from Eqs. 16.4.12 and 16.5.3, yield

$$\begin{aligned} \mathcal{G}(x-t; x'-t) &= \lambda^4 \int_0^{+\infty} \frac{dk}{\pi/2} \frac{\cos(k(x-t))\cos(k(x'-t))}{\lambda^4 + \frac{\lambda^4}{\mu^2}k^2 + k^4} \\ &= \lambda^4 \int_{-\infty}^{+\infty} \frac{dk}{2\pi} \cdot \frac{e^{ik(x+x'-2t)} + e^{ik(x-x')}}{\lambda^4 + \frac{\lambda^4}{\mu^2}k^2 + k^4} \end{aligned} \tag{16.5.4}$$

The propagator given in Eq. 16.5.4 has *three branches*, with the explicit expression for $\mathcal{D}(x-t; x'-t)$ for the three branches is given in Sect. 13.13. In the notation of Eq. 13.13.3

$$L = 1\,;\ \widetilde{L} = \frac{\lambda^4}{\mu^2}\,;\ \gamma^2 = \lambda^4$$

The real branch, from Eq. 13.13 is given by

$$4L\gamma^2 < \widetilde{L}^2 \Rightarrow \sqrt{2}\mu < \lambda$$

Empirical studies show that the forward interest rates correlation functions are described by the real branch [4]. Hence

$$\mathcal{G}(x-t; x'-t) = \lambda^4 \left[D_R(x+x'-2t) + D_R(x-x')\right]\,;\ \ x, x' > t$$

where, from Eq. 13.13.11

$$D_R(\tau) = \frac{\omega}{2\gamma \sinh(2\vartheta)} e^{-\omega|\tau|\cosh(\vartheta)} \sinh\left[\vartheta + \omega|\tau|\sinh(\vartheta)\right] \tag{16.5.5}$$

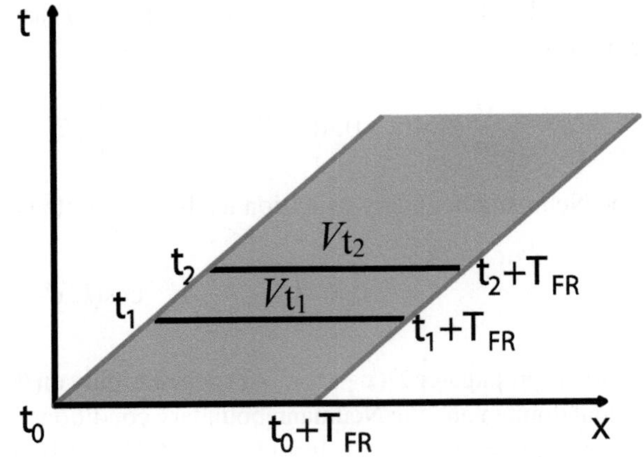

Fig. 16.7 The domain of the state space of the interest rates. The state space $\mathcal{V}_t$ is indicated for two distinct calendar times t_1 and t_2. Published with permission of © Belal E. Baaquie 2020. All Rights Reserved

16.6 Time Dependent State Space $\mathcal{V}_t$

The Hamiltonian and the state space of a system are two independent properties of a quantum theory; the Lagrangian and path integral is a result of these two ingredients. The essential features of the interest rates' Hamiltonian and state space are reviewed; a detailed discussion is given in [4].

Since $x \in [t, t + T_{FR}]$ the quantum field $f(t, x)$ exists only for future time, that is, for $x > t$. In particular, the interest rates' quantum field has a *distinct* state space $\mathcal{V}_t$ for every instant t.

The state space at time t is labeled by $\mathcal{V}_t$, and the state vectors in $\mathcal{V}_t$ are denoted by $|f_t\rangle$. For fixed time t, the state space $\mathcal{V}_t$, as shown in Fig. 16.7, consists of all possible functions of the interest rates, with future time $x \in [t, t + T_{FR}]$.

In continuum notation, the basis states of $\mathcal{V}_t$ are tensor products over the future time x and satisfy the following completeness equation

$$|f_t\rangle = \prod_{t \le x \le t+T_{FR}} |f(t, x)\rangle \tag{16.6.1}$$

$$\mathbb{I}_t = \prod_{t \le x \le t+T_{FR}} \int_{-\infty}^{+\infty} df(t, x) |f_t\rangle\langle f_t| \equiv \int Df_t |f_t\rangle\langle f_t|$$

The path integral for a time interval $[t_1, t_2]$ as shown in Fig. 16.7 can be constructed from the Hamiltonian and state space by applying the time slicing method. The time-dependent interest rate Hamiltonian $\mathcal{H}(t)$ propagates the interest rates *backwards* in time. The final state $|f_{\text{final}}\rangle$ at future calendar time t_2 is propagated backwards to an initial state $\langle f_{\text{initial}}|$ at the earlier time t_1. In terms of the boundary conditions given in Eq. 16.4.8, the final and initial states, as shown in Fig. 16.7, are

$$|f_{\text{final}}\rangle = \prod_{t_1 \le s \le T_{FR}+t_1} |f(t_1, s)\rangle \ ; \ \langle f_{\text{initial}}| = \prod_{t_2 \le s \le T_{FR}+t_2} \langle f(t_2, s)|$$

Since the state space and Hamiltonian are both time-dependent one has to use the time-ordering operator $\mathcal{T}$ to keep track of the time-dependence: $\mathcal{H}(t)$ for earlier time t is placed to the left of $\mathcal{H}(t)$ that refers to later time. The *matrix element* from a given a final (coordinate basis) state $|f_{\text{final}}\rangle$ at time t_2 to an arbitrary initial (coordinate basis) state $\langle f_{\text{initial}}|$ at time t_1 is given by the following [4]

$$\mathcal{K} = \langle f_{\text{initial}}|\mathcal{T}\left\{\exp - \int_{t_1}^{t_2} \mathcal{H}(t)\, dt\right\}|f_{\text{final}}\rangle \tag{16.6.2}$$

Due to the time dependence of the state spaces $\mathcal{V}_t$ the forward interest rates that determine $\mathcal{K}$ form a trapezoidal domain shown in Fig. 16.7.

16.7 Time Dependent Hamiltonian

The action for the forward interest rates, from Eq. 16.4.13, is given by

$$S = \int_{t_1}^{t_2} dt \int_{x,x'} \mathcal{L}(t, x, x')\ ; \quad \int_x \equiv \int_t^{t+T_{FR}} dx\ ; \ \ t_1 < t_2 \tag{16.7.1}$$

where a general Lagrangian density for the bond forward interest rates, from Eq. 16.4.14, is given by

$$\mathcal{L}(t, x, x') \tag{16.7.2}$$
$$= -\frac{1}{2}\left[\frac{\partial f(t,x)/\partial t - \alpha(t,x)}{\sigma(t,x)}\right] \mathcal{D}^{-1}(x, x'; t; T_{FR}) \left[\frac{\partial f(t,x')/\partial t - \alpha(t,x')}{\sigma(t,x')}\right]$$
$$-\infty \leq f(t,x) \leq +\infty \tag{16.7.3}$$

Neumann boundary conditions, given in Eq. 16.4.8, have been incorporated into the expression for the Lagrangian by requiring that the fields obey these conditions. The derivation for the Hamiltonian is done for an arbitrary propagator $\mathcal{D}^{-1}(t, x, x')$, although for most applications a specific choice is made.

Discretizing time into a lattice of spacing ϵ yields $t \to t_n = n\epsilon$ and the action yields

$$S = \epsilon \sum_{n=n_1}^{n_2} \int_{x,x'} \mathcal{L}(t_n, x, x') \equiv \epsilon \sum_{n=n_1}^{n_2} L(t_n) \tag{16.7.4}$$

where the Lagrangian is given by

$$\epsilon L(t_n) = \epsilon \int_{x,x'} \mathcal{L}(t_n, x, x') \tag{16.7.5}$$

Since only two slices of time are required for the Lagrangian density, the notation of t and $t + \epsilon$ is used of t_n and t_{n+1}, respectively.

$$\left[\frac{\partial f(t,x)/\partial t - \alpha(t,x)}{\sigma(t,x)}\right] \to \frac{1}{\epsilon\sigma(t,x)} \cdot [f_{t+\epsilon}(x) - f_t(x) - \epsilon\alpha(t,x)]$$
$$\equiv \frac{1}{\epsilon} H_t(x) \tag{16.7.6}$$

where $f(t,x) \equiv f_t(x)$ has been written to emphasize that time t is a parameter for the interest rates Hamiltonian. Hence, from Eq. 16.7.5, the Lagrangian $L(t)$ is given by

$$\epsilon L(t) = \epsilon \int_t^{t+T_{FR}} dx dx' \mathcal{L}(t,x,x')$$
$$= -\frac{1}{2\epsilon} \int_{x,x'} \mathcal{H}_t(x)\mathcal{D}^{-1}(x,x';t;T_{FR})\mathcal{H}_t(x') \tag{16.7.7}$$

The Dirac-Feynman formula given in Eq. 15.2.9 relates the Lagrangian $\mathcal{L}(t_n)$ to the Hamiltonian operator and yields

$$\langle f_t | e^{-\epsilon\mathcal{H}_f} | f_{t+\epsilon}\rangle = \mathcal{N} e^{\epsilon L(t)} \tag{16.7.8}$$

where $\mathcal{N}$ is a normalization. The Lagrangian yields the interest rates Hamiltonian via Eq. 16.7.8.

Equation 16.7.7 is re-written using Gaussian integration and (ignoring henceforth irrelevant constants), using notation

$$\int Dp \equiv \prod_x \int_{-\infty}^{+\infty} dp(x)$$

yields

$$\exp\{\epsilon L(t)\}$$
$$= \int Dp \exp\{-\frac{\epsilon}{2} \int_{x,x'} p(x)\mathcal{D}(x,x';t;T_{FR})p(x') + i \int_x p(x)\mathcal{H}_t(x)\}$$

where the propagator $\mathcal{D}(x,x';t;T_{FR})$ is the inverse of $\mathcal{D}^{-1}(x,x';t;T_{FR})$.

Re-scaling the integration variable

$$p(x) \to \sigma(t,x)p(x) \ ; \ \ \mathcal{H}_t(x) \to f_{t+\epsilon}(x) - f_t(x) - \epsilon\alpha(t,x)$$

Equation 16.7.7 yields (up to an irrelevant constant)

$$e^{\epsilon L(t)} = \int Dp \exp\{i \int_x p(x) \left(f_{t+\epsilon} - f_t - \epsilon\alpha_t\right)(x)\}$$
$$\times \exp\{-\frac{\epsilon}{2} \int_{x,x'} \sigma(t,x)p(x)\mathcal{D}(x,x';t;T_{FR})\sigma(t,x')p(x')\} \quad (16.7.9)$$

Hence, the Dirac-Feynman formula given in Eq. 16.7.8 yields the Hamiltonian as follows

$$\mathcal{N}e^{\epsilon L(t)} = \langle f_t | e^{-\epsilon\mathcal{H}_\phi} | f_{t+\epsilon}\rangle \quad (16.7.10)$$
$$= \exp\{-\epsilon\mathcal{H}_t(\delta/\delta f_t)\} \int Dp \exp\{i \int_x p(f_{t+\epsilon} - f_t)\} \quad (16.7.11)$$

For each instant of time, there are infinitely many independent interest rates (degrees of freedom) represented by the collection of variables $f_t(x), x \in [t, t + T_{FR}]$. The Hamiltonian is written in terms of functional derivatives in the co-ordinates of the *dual* state space variables f_t.

To obtain the Hamiltonian, define

$$M_\sigma(x,x';t) = \sigma(t,x)\mathcal{D}(x,x';t;T_{FR})\sigma(t,x')$$

In abbreviated notation, Eqs. 16.7.9, 16.7.10 and 16.7.11 yield

$$\left[1 - \epsilon\mathcal{H}_t(\delta/\delta f_t)\right] \int Dp e^{i\int_x p(f_{t+\epsilon}-f_t)}$$
$$\simeq \left[1 - \frac{\epsilon}{2}\int_{x,x'} p(x)M(x,x';t)p(x') - i\epsilon\int_x p(x)\alpha_t(x)\right] \int Dp e^{i\int_x p(f_{t+\epsilon}-f_t)}$$
$$= \left[1 + \frac{\epsilon}{2}\int_{x,x'} M(x,x';t)\frac{\delta^2}{\delta f_t(x)\delta f_t(x')}\right.$$
$$\left. + \epsilon\int_x \alpha_t(x)\frac{\delta}{\delta f_t(x)}\right] \int Dp e^{i\int_x p(f_{t+\epsilon}-f_t)} \quad (16.7.12)$$

The sign of the drift $\alpha_t(x)$ depend crucially on the sign of t; note that in the defining equation given in Eq. 16.7.6 the time derivatives for $f(t,x)$ are for calendar time and not remaining time, and this determines the sign of the drift in the Hamiltonian.

The degrees of freedom $f_t(x)$ refer to time t only through the domain on which the Hamiltonian is defined. Unlike the action $S[f]$ that spans all instants of time from the initial to the final time, the Hamiltonian is an infinitesimal generator in time, and refers to only the instant of time at which it acts on the state space. This is the reason that in the Hamiltonian the time index t can be dropped from the variables $f_t(x)$ and replaced by $f(x)$ with $t \le x \le t + T_{FR}$.

The Hamiltonian for the forward interest rates, from Eq. 16.7.12, is given by [4]

$$\mathcal{H}(t) = -\frac{1}{2}\int_t^{t+T_{FR}} dxdx' M_\sigma(x,x';t)\frac{\delta^2}{\delta f(x)\delta f(x')} - \int_t^{t+T_{FR}} dx\alpha(t,x)\frac{\delta}{\delta f(x)} \tag{16.7.13}$$

The derivation only assumes that the volatility $\sigma(t,x)$ is deterministic. The drift term $\alpha(t,x)$ in the Hamiltonian is completely *general* and can be any (nonlinear) function of the interest rates.

The quantum field $f(t,x)$ is more fundamental that the velocity quantum field $\mathcal{A}(t,x)$; the Hamiltonian cannot be written in terms of the $\mathcal{A}(t,x)$ degrees of freedom. The reason being that the dynamics of the forward interest rates are contained in the time derivative terms in the Lagrangian, namely the term containing $\partial f(t,x)/\partial t$; in going to the Hamiltonian representation, the time derivative $\partial f/\partial t$ becomes a differential operator $\delta/i\delta f(t,x)$.[5]

16.8 Path Integral: Martingale

Recall the evolution of financial instruments following a *martingale process* is the mathematical realization of the theory of arbitrage-free pricing of financial instruments. A path integral derivation of the martingale process is given for forward interest rates [4, 5].

The path integral calculation can be carried out explicitly because the forward interest rates are linear and hence described by a Gaussian quantum field. A similar calculation for imposing the martingale condition on Libor can be defined, but the path integral is nonlinear and hence to actually carry out the path integral to determine the Libor drift is an intractable problem. However, it will be shown in Sect. 16.15 that the Libor Hamiltonian can be used to exactly determine nonlinear Libor drift.

The martingale condition for a zero coupon bond states that the price of the zero coupon bond $B(t_*,T)$ at some future time $T > t_* > t$, when *discounted* from time t_* to time t, is the equal to the bond price $B(t,T)$. Future cash flows are discounted by the risk free interest rate $r(t) = f(t,t)$; the martingale condition yields the following

$$B(t_0,T) = E[e^{-\int_{t_0}^{t_*} r(t)dt}B(t_*,T)] \tag{16.8.1}$$

where $E[X]$ denotes the average value of X over all the stochastic forward interest rates in the time interval $[t,t_*]$. Note that

$$B(t_0,T) = \exp\{-\int_{t_0}^T dxf(t_0,x)\}\ ;\ B(t_*,T) = \exp\{-\int_{t_*}^T dxf(t_*,x)\} \tag{16.8.2}$$

[5]If one wants to use the velocity degrees of freedom $\mathcal{A}(t,x)$ in the state space representation, one needs to use the formalism of phase space quantization discussed in [7].

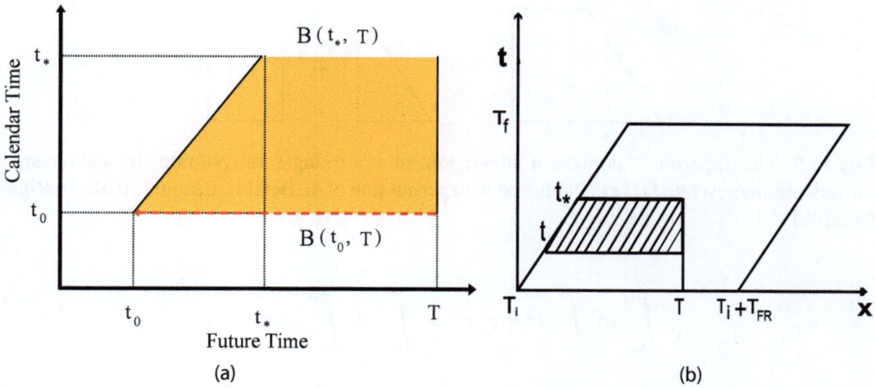

Fig. 16.8 **a** The trapezoid enclosing the zero coupon bonds is the domain for the forward interest rates. The shaded portion represents the domain $\mathcal{T}$. **b** Domains for deriving the martingale condition for zero coupon bonds $B(t_*, T)$. The horizontal lines at t_* and t represent $B(t_*, T)$ and $B(t, T)$ respectively. The vertical line at T represents the maturity time of the zero coupon bonds. Published with permission of ©Belal E. Baaquie 2020. All Rights Reserved

In terms of the Feynman path integral, Eq. 16.8.1 yields

$$B(t_0, T) = \frac{1}{Z}\int Df e^{S[f]} e^{-\int_{t_0}^{t_*} r(t)dt} B(t_*, T) \; ; \; Z = \int Df e^{S[f]} \tag{16.8.3}$$

In the path integral given in Eq. 16.8.3, there are two domains; namely the domain for the zero coupon bond that is nested inside the domain of the forward interest rates. These domains are shown in Fig. 16.8a and b.

From Eq. 16.3.2

$$\frac{\partial f(t,x)}{\partial t} = \alpha(t,x) + \sigma(t,x)\mathcal{A}(t,x)$$
$$\Rightarrow \; f(t,x) = f(t_0,x) + \int_{t_0}^{t} d\tau \alpha(\tau,x) + \int_{t_0}^{t} d\tau \sigma(\tau,x)\mathcal{A}(\tau,x) \tag{16.8.4}$$

As shown in Fig. 16.9, the domain of integration for the spot rate $r(t)$ is over the triangle. For $r(t) = f(t,t)$, Eq. 16.8.4 yields

$$\int_{t_0}^{t_*} dt r(t) = \int_{t_0}^{t_*} dt f(t_0,t) + \int_{t_0}^{t_*} dt \int_{t_0}^{t} d\tau \alpha(\tau,t)$$
$$+ \int_{t_0}^{t_*} dt \int_{t}^{t_*} d\tau \sigma(\tau,t)\mathcal{A}(\tau,t) \tag{16.8.5}$$

Doing the integration over the triangle in two different ways yields

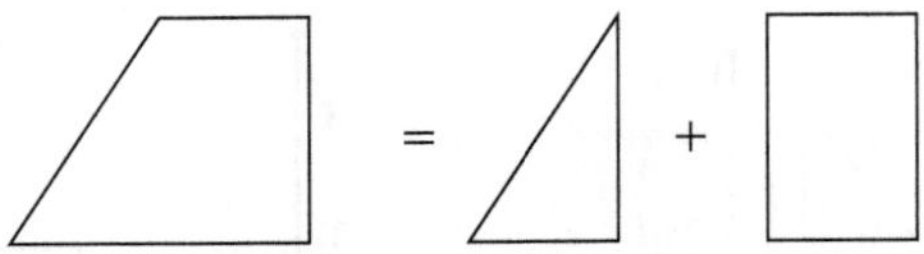

Fig. 16.9 The trapezoid $\mathcal{T}$ domain of integration: sum of triangle for spot rate $r(t)$ and rectangle for forward interest rate $f(t,x)$. Published with permission of © Belal E. Baaquie 2020. All Rights Reserved

$$\int_{t_0}^{t_*} dt \int_{t_0}^{t} d\tau h(t,\tau) = \int_{t_0}^{t_*} dt \int_{t}^{t_*} dx h(t,x)$$

Equation 16.8.5 is re-written as follows

$$\int_{t_0}^{t_*} dt r(t) = \int_{t_0}^{t_*} dt f(t_0,t) + \int_{t_0}^{t_*} dt \int_{t}^{t_*} dx \alpha(t,x) + \int_{t_0}^{t_*} dt \int_{t}^{t_*} dx \sigma(t,x)\mathcal{A}(t,x) \tag{16.8.6}$$

As shown in Fig. 16.9, the domain of integration for the forward interest is over the trapezoid. The zero coupon bond $B(t_*,T)$ yields

$$\int_{t_*}^{T} dx f(t_*,x) = \int_{t_*}^{T} dx f(t_0,x) + \int_{t_*}^{T} dx \int_{t_0}^{t_*} dt \alpha(t,x) + \int_{t_*}^{T} dx \int_{t_0}^{t_*} dt \sigma(t,x)\mathcal{A}(t,x) \tag{16.8.7}$$

Hence, combining Eqs. 16.8.6 and 16.8.7 yields

$$\begin{aligned} \int_{t_0}^{t_*} dt r(t) + \int_{t_*}^{T} dx f(t_*,x) &= \int_{t_0}^{T} dx f(t_0,x) \\ + \int_{t_0}^{t_*} dt \int_{t}^{T} dx \alpha(t,x) &+ \int_{t_0}^{t_*} dt \int_{t}^{T} dx \sigma(t,x)\mathcal{A}(t,x) \\ = \int_{t_0}^{T} dx f(t_0,x) &+ \int_{\mathcal{T}} \alpha(t,x) + \int_{\mathcal{T}} \sigma(t,x)\mathcal{A}(t,x) \end{aligned} \tag{16.8.8}$$

where the integration over a trapezoid domain $\mathcal{T}$, for some arbitrary function $G(t,x)$ is given by

$$\int_{\mathcal{T}} G(t,x) = \int_{t_0}^{t_*} dt \int_{t}^{T} dx G(t,x)$$

The domain $\mathcal{T}$ determining the risk-neutral measure is shown as the shaded domain in Fig. 16.8. Hence

$$e^{-\int_t^{t_*} r(t)dt} B(t_*, T) = \exp\left\{-\int_T \alpha(t,x) - \int_T \sigma(t,x)\mathcal{A}(t,x)\right\} B(t_0, T) \quad (16.8.9)$$

The bond $B(t, T)$ is determined by the boundary condition at time t and is not stochastic; furthermore, the drift $\alpha(t, x)$ for linear forward interest rates is not stochastic. Hence, taking the expectation value of Eq. 16.8.9 yields

$$E[e^{-\int_t^{t_*} r(t)dt} B(t_*, T)] = e^{-\int_T \alpha(t,x)} B(t_0, T) E[e^{-\int_T \sigma(t,x)\mathcal{A}(t,x)}] \quad (16.8.10)$$

From Eqs. 16.8.3 and 16.8.10, the martingale condition yields the following

$$e^{\int_T \alpha(t,x)} = E[e^{-\int_T \sigma(t,x)\mathcal{A}(t,x)}] \quad (16.8.11)$$

Using the results of Gaussian path integration given in Eq. 13.10.5, the martingale condition given in Eq. 16.8.11 yields

$$\exp\int_T \alpha(t,x) = \frac{1}{Z}\int DAe^{-\int_T \sigma(t,x)\mathcal{A}(t,x)} e^{\int_T S[A]} \quad (16.8.12)$$

$$= \exp\left\{\frac{1}{2}\int_{t_0}^{t_*} dt \int_t^T dx dx' \sigma(t,x)\mathcal{D}(x,x';t,T_{FR})\sigma(t,x')\right\} \quad (16.8.13)$$

Differentiating both sides of Eq. 16.8.13 on the time coordinate t_* yields

$$\int_{t_*}^T dx \alpha(t_*, x) = \frac{1}{2}\int_{t_*}^T dx dx' \sigma(t_*, x)\mathcal{D}(x,x';t_*,T_{FR})\sigma(t_*,x') \quad (16.8.14)$$

Differentiating above expression with respect to T yields the drift velocity

$$\alpha(t_*, T) = \sigma(t_*, T)\int_{t_*}^T dx' \mathcal{D}(x,x';t_*,T_{FR})\sigma(t_*,x')$$

$$\Rightarrow \quad \alpha(t,x) = \sigma(t,x)\int_t^x dx' \mathcal{D}(x,x';t,T_{FR})\sigma(t,x') \quad (16.8.15)$$

The forward interest rates having a martingale time evolution, from Eqs.16.8.4 and 16.8.15 are given by

$$f(\tau,x) = f(t_0,x) + \int_{t_0}^{\tau} dt\sigma(t,x)\int_t^x dy \mathcal{D}(x,y;t,T_{FR})\sigma(t,y)$$
$$+ \int_{t_0}^{\tau} dt\sigma(t,x)\mathcal{A}(t,x) \quad (16.8.16)$$

The drift obtained ensures that financial instruments priced using the forward interest rates are free from arbitrage opportunities.

16.9 Numeraire

In the derivation of the martingale condition discussed in Sect. 16.8, the spot rate $r(t)$ was used, with the discounting instrument being the return on the money market account. The following instrument has a martingale evolution

$$\frac{B(t_*, T)}{\exp\{\int_t^{t_*} dt' r(t')\}}$$

Hence, since the discounted instrument has a martingale evolution and Eq. 16.8.1 can be re-written as follows

$$E\left[\frac{B(t_*, T)}{\exp\{\int_t^{t_*} dt' r(t')\}}\right] = \frac{B(t, T)}{\exp\{\int_t^{t} dt' r(t')\}} = B(t, T)\ ;\ t_* > t$$

with the drift, from Eq. 16.8.15, given by

$$\alpha(t, x) = \sigma(t, x) \int_t^x dx' \mathcal{D}(x, x'; t, T_{FR}) \sigma(t, x') \tag{16.9.1}$$

The choice of a numeraire is quite arbitrary and any positive valued instrument satisfying some general requirement is adequate. The two most commonly used numeraires are (a) discounting by the spot interest rate and (b) discounting using the zero coupon bond numeraire. The generality of choosing a numeraire is addressed in [5]. The property of numeraire invariance is central to the concept of martingales.

The martingale derivation given for the zero coupon bonds in Sect. 16.9 can be repeated with a numeraire given by the zero coupon bond $B(t_*, T)$. For a suitably chosen drift $\alpha_*(t, x)$ and $t_* > t_0$, the instrument

$$\frac{B(t_*, T)}{B(t_0, t_*)}$$

is a martingale: the expectation of its future value is equal to its present day value. Hence, since $B(t_0, t_0) = 1$

$$E_*\left[\frac{B(t_*, T)}{B(t_0, t_*)}\right] = \left[\frac{B(t_0, T)}{B(t_0, t_0)}\right] \quad \Rightarrow \quad B(t_0, T) = B(t_0, t_*) E_* \left[B(t_*, T)\right]$$

Similar to the result for the drift for discounting by the spot rate $r(t)$ in Eq. 16.8.15, the drift for the bond numeraire $B(t_0, t_*)$ is given by[6]

$$\alpha_*(t, x) = \sigma(t, x) \int_{t_*}^x dx' \mathcal{D}(x, x'; t) \sigma(t, x') \tag{16.9.2}$$

[6]To simply the notation let $T_{FR} \to \infty$.

To clarify the concept of numeraire and its associated drift, the change of numeraire is derived for the forward interest rates. The expectation values for the two numeraires are given by the following path integrals

$$B(t,T) = \int Df \exp\{S[f]\exp\{-\int_t^{t_*} dt' r(t')\}B(t_*,T) \tag{16.9.3}$$

$$B(t,T) = B(t,t_*)\int Df \exp\{S_*[f]B(t_*,T)\} \tag{16.9.4}$$

The actions for the two different numeraires, from Eq. 16.4.13, are given by

$$S = \int_{t_1}^{t_2} dt \int_{x,x'} \mathcal{L}(t,x,x')\ ; \ S_* = \int_{t_1}^{t_2} dt \int_{x,x'} \mathcal{L}_*(t,x,x') \tag{16.9.5}$$

where the Lagrangian densities, from Eq. 16.4.14 are given by

$$\begin{aligned}&\mathcal{L}(t,x,x') \\ &= -\frac{1}{2}\left[\frac{\partial f(t,x)/\partial t - \alpha(t,x)}{\sigma(t,x)}\right]\mathcal{D}^{-1}(x,x';t)\left[\frac{\partial f(t,x')/\partial t - \alpha(t,x')}{\sigma(t,x')}\right]\end{aligned} \tag{16.9.6}$$

$$\begin{aligned}&\mathcal{L}_*(t,x,x') \\ &= -\frac{1}{2}\left[\frac{\partial f(t,x)/\partial t - \alpha_*(t,x)}{\sigma(t,x)}\right]\mathcal{D}^{-1}(x,x';t)\left[\frac{\partial f(t,x')/\partial t - \alpha_*(t,x')}{\sigma(t,x')}\right]\end{aligned} \tag{16.9.7}$$

It is shown in [4] that one has the following *identity*

$$e^{S_*} = \frac{\exp\{-\int_t^{t_*} dt' r(t')\}}{B(t,t_*)} e^S \tag{16.9.8}$$

Hence

$$\begin{aligned}B(t,T) &= \int Df \exp\{S[f]\exp\{-\int_t^{t_*} dt' r(t')\}B(t_*,T) \\ &= B(t,t_*)\int Df \exp\{S_*[f]B(t_*,T)\}\end{aligned} \tag{16.9.9}$$

The result given in Eqs. 16.9.8 and 16.9.9 above shows the reason why both Eqs. 16.9.3 and 16.9.4 give the same result. Recall the drift for the two choice of numeraire was fixed by imposing the martingale condition: changing the numeraire results in changing the drift that off-sets the change of the numeraire and gives a result independent of the numeraire.

16.10 Zero Coupon Bond Call Option

The price of an option at time t, is the discounted value of the payoff function, which is settled at future time T. The concept of *discounting* requires a discounting factor, or equivalently a numeraire. Discounting by r, the spot interest rate yields the numeraire $\exp\{r(T-t)\}$, which is the money market numeraire. A more suitable choice for evaluating the zero coupon bond optin is to choose the zero coupon bond $B(t, t_*)$ as the numeraire.

For the drift of the forward interest rates $\alpha_*(t, x)$ given by Eq. 16.9.2, the combination $C(t, t_*, K)/B(t, t_*)$ is a martingale. The conditional expectation value of the discounted future (random) price of the payoff is equal to it's present value; hence, one has the following

$$\frac{C(t, t_*, K)}{B(t, t_*)} = E\left[\frac{C(t_*, t_*, K)}{B(t_*, t_*)}\right] = E\left[(B(t_*, T) - K)_+\right] \tag{16.10.1}$$

$$\Rightarrow C(t, t_*, K) = B(t, t_*)E\left[(B(t_*, T) - K)_+\right] \tag{16.10.2}$$

where $C(t_*, t_*, K)$ is the payoff function.

Equation 16.10.2 is the basis of pricing options $C(t, t_*, K)$ written on any financial security.

From Eq. 16.3.1

$$B(t_*, T) = \exp\{-\int_{t_*}^{T} dxf(t_*, x)\} \tag{16.10.3}$$

From the definition of the Dirac delta-function given in Eq. 13.4.1 the payoff function can be written as

$$\begin{aligned}
[B(t_*, T) - K]_+ &= \int dG\delta(G - \ln(B(t_*, T))(e^G - K)_+ \\
&= \int dG\delta(G + \int_{t_*}^{T} dxf(t_*, x))(e^G - K)_+
\end{aligned}$$

Representing the delta-function as in Eq. 13.4.3 yields

$$\begin{aligned}
[B(t_*, T) - K]_+ &= \frac{1}{2\pi}\int_{-\infty}^{+\infty} dGdpe^{ip(G+\int_{t_*}^{T} dxf(t_*,x))}(e^G - K)_+ \\
&= \int_{-\infty}^{+\infty} dG\Psi(G, t_*, T)(e^G - K)_+
\end{aligned}$$

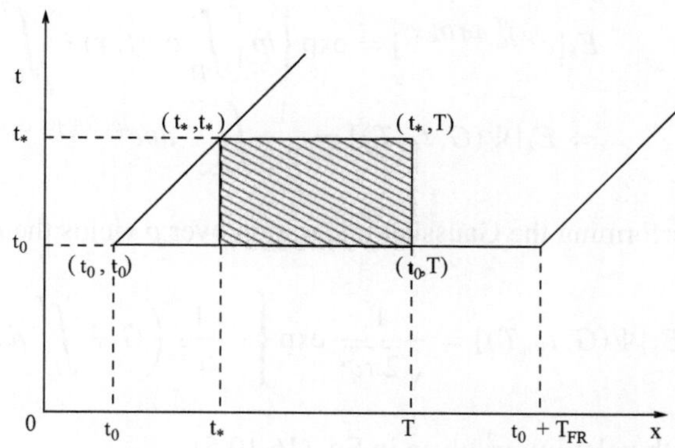

Fig. 16.10 The shaded portion is domain $\mathcal{R}$ for the zero coupon bond option price. Published with permission of © Belal E. Baaquie 2020. All Rights Reserved

The price of the option is obtained by discounting the future value of the payoff function with the bond $B(t_0, t_*)$. From Eq. 16.10.2, the current price of the option, for $t = t_0$, is given by

$$\begin{aligned} C(t_0, t_*, T, K) &= B(t_0, t_*)E_*[(B(t_*, T) - K)_+] \\ &= \int_{-\infty}^{+\infty} dG E_*[\Psi(G, t_*, T)](e^G - K)_+ \end{aligned} \tag{16.10.4}$$

For domain $\mathcal{R}$ defined in Fig. 16.10, Eqs. 16.8.4 and 16.10.3 yield

$$\int_{t_*}^{T} dx f(t_*, x) = \int_{\mathcal{R}} \alpha_*(t, x) + \int_{t_*}^{T} dx f(t_0, x) + \int_{\mathcal{R}} \sigma(t, x)\mathcal{A}(t, x)$$

From the expression above and the payoff function given in Eq. 16.10.4 one needs to compute

$$\begin{aligned} E_*[e^{ip\int_{\mathcal{R}} \sigma(t,x)\mathcal{A}(t,x)}] &= \frac{1}{Z}\int D\mathcal{A}e^{ip\int_{\mathcal{R}} \sigma(t,x)\mathcal{A}(t,x)}e^S = \exp\{-\frac{q^2}{2}p^2\} \\ \Rightarrow q^2 &= \int_{t_0}^{t_*} dt \int_{t_*}^{T} dx dx' \sigma(t, x) D(x, x'; t)\sigma(t, x') \end{aligned} \tag{16.10.5}$$

Using the expression given in Eq. 16.9.2 for $\alpha_*(t, x)$, one obtains

$$\begin{aligned} \int_{\mathcal{R}} \alpha_*(t, x) &= \int_{t_*}^{T} dx \int_{t_0}^{t_*} dt \alpha_*(t, x) \\ &= \int_{t_*}^{T} dx \int_{t_0}^{t_*} dt \sigma(t, x) \int_{t_*}^{x} dx' D(x, x'; t)\sigma(t, x') = \frac{q^2}{2} \end{aligned} \tag{16.10.6}$$

Collecting the results and using Eq. 16.10.6 yields

$$E_*[e^{ip\int_{t_*}^T dxf(t_*,x)}] = \exp\left\{ip\left[\int_{\mathcal{R}} \alpha_*(t,x) + \int_{t_*}^T dxf(t_0,x)\right] - \frac{q^2}{2}p^2\right\}$$

$$\Rightarrow E_*[\Psi(G,t_*,T)] = \frac{1}{2\pi}\int_{-\infty}^{+\infty} dp e^{ip(G+\int_{\mathcal{R}}\alpha_*(t,x)+\int_{t_*}^T dxf(t_0,x))-\frac{q^2}{2}p^2}$$

Performing the Gaussian integration over p yields the result that

$$E_*[\Psi(G,t_*,T)] = \frac{1}{\sqrt{2\pi q^2}}\exp\left\{-\frac{1}{2q^2}\left(G + \int_{t_*}^T dxf(t_0,x) + \frac{q^2}{2}\right)^2\right\} \tag{16.10.7}$$

with volatility q^2 given in Eq. (16.10.5).

The European bond call option is given by

$$C(t_0,t_*,T,K) = \int_{-\infty}^{+\infty} dG E_*[\Psi(G,t_*,T)](e^G - K)_+ \tag{16.10.8}$$

The final answer can be read off by a direct comparison with the case of the option price for a single equity given in Eq. 11.10.3. The zero coupon bond call option is given by [4]

$$C(t_0,t_*,T,K) = B(t_0,t_*)[F(t_0,t_*,T)N(d_+) - KN(d_-)] \tag{16.10.9}$$

$$F \equiv F(t_0,t_*,T) = \exp\{-\int_{t_*}^T dxf(t_0,x)\}\ ;\ d_\pm = \frac{1}{q}[\ln\frac{F}{K} \pm \frac{q^2}{2}]$$

The expression for the price of the European option for a zero coupon bond is very similar to the one for equity derived by Black and Scholes. The major difference arises in the expression for the volatility q as this contains the correlation of the volatilities $\sigma(x,t)$ due to the nontrivial correlation of all the forward interest rates.

16.11 Libor: Simple Interest Rate

Libor is a simple interest rate instrument that was launched on 1 January 1986 by British Bankers' Association. Libor is a daily quoted rate based on the interest rates at which *commercial banks* are willing to lend funds to other banks in the London interbank money market. Euribor (Euro Interbank Offered Rate) is similar to Libor and is the benchmark rate of the Euro money market, which has emerged since 1999. Euribor is a simple interest rate that is paid for fixed deposits in the Euro currency. The market widely uses Libor and Euribor this instrument and market data on interest rates is largely based on Libor.

The UK Financial Conduct Authority has announced that Libor is going to replaced by 2022.[7] The discussions in this and subsequent sections analyze Libor. However, the mathematics developed in this and subsequent sections is independent of Libor and is equally applicable to any interest rates instrument.

Cash deposits can earn simple interest rates for a given period of time. For example, one can lock-in at time t a *simple interest rate* for a fixed deposit from future time T_1 to T_2, which is denoted by $L(t; T_1, T_2)$. The period of the deposit, namely $T_2 - T_1$, is the *tenor* of the simple interest rate.

A deposit of \$ 1, made from time T_1 to T_2 and paying simple interest, will increase to an amount $1 + (T_2 - T_1)L(t; T_1, T_2)$. Conversely, the present day value of a Libor zero coupon bond $B(t, T)$, paid at future time T, is given by

$$B(t, T) = \frac{1}{1 + (T - t)L(t; t, T)}$$

and more generally

$$B(t, T_2) = B(t, T_1)\frac{1}{1 + (T_2 - T_1)L(t; T_1, T_2)} \tag{16.11.1}$$

From the definition of zero coupon bonds given in Eq. 16.3.1 the simple interest rates are given in terms of the instantaneous forward interest rates by Eq. 16.11.1

$$\exp\{-\int_{T_1}^{T_2} dx f(t, x)\} = \frac{1}{1 + (T_2 - T_1)L(t; T_1, T_2)}$$

$$\Rightarrow L(t; T_1, T_2) = \frac{1}{T_2 - T_1}[\exp\{\int_{T_1}^{T_2} dx f(t, x)\} - 1] \tag{16.11.2}$$

For $f(t, x) > 0$, Eq. 16.11.2 yields

$$L(t; T_1, T_2) > 0 \tag{16.11.3}$$

Libor for a $\ell =$ three months tenure is widely used for pricing interest rate options and other instruments; henceforth Libor for only three months tenure is considered. Libor time lattice is defined by $T_n = T_0 + \ell n$ and is shown in Fig. 16.11a. The following notation is used

$$L(t, T_n) \equiv L(t; T_n, T_n + \ell)$$

From Eqs. 16.3.1 and 16.11.2

$$L(t, T_n) = \frac{1}{\ell}\left[\exp\{\int_{T_n}^{T_n+\ell} dx f(t, x)\} - 1\right] \tag{16.11.4}$$

[7] Regulator calls on banks to replace Libor by 2022. https://www.ft.com/content/04dd3316-72ab-11e7-aca6-c6bd07df1a3c.

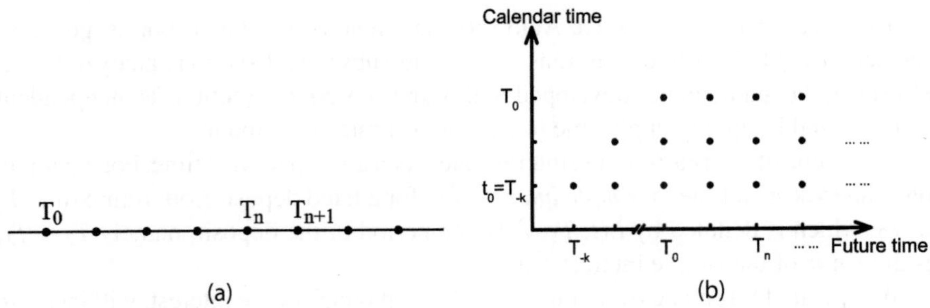

Fig. 16.11 **a** Libor future time lattice $T_n = T_0 + \ell n$; the tenor (future time lattice spacing) is given by $\ell = 90$ days. **b** Libor future and calendar time lattice. The zero coupon bond is issued at T_0 and expires at $T_n + \ell$. It's future price is given at (present) Libor time $t_0 = T_{-k}$

In terms of zero coupon bonds, Eq. 16.11.4 yields

$$L(t, T_n)B(t, T_{n+1}) = \frac{1}{\ell}[B(t, T_n) - B(t, T_{n+1})] \tag{16.11.5}$$

$$\Rightarrow \quad L(t, T_n) = \frac{B(t, T_n) - B(t, T_{n+1})}{\ell B(t, T_{n+1})}. \tag{16.11.6}$$

The Libor rates are related to the zero coupon bonds by Eq. 16.11.5, namely that

$$B(t, T_n + \ell) = \frac{B(t, T_n)}{1 + \ell L(t, T_n)} \tag{16.11.7}$$

Equation 16.11.7 provides a recursion equation that allows one to express $B(t, T)$ solely in terms of $L(t, T)$. Note that Libors are only defined for discrete future time given by *Libor future time* $T = T_n = n\ell,\ n = 0, 1, 2, \ldots \infty$. Libor future and calendar time lattice is shown in Fig. 16.11.

Hence, from Eq. 16.11.7

$$B(t, T_{k+1}) = \frac{B(t, T_k)}{1 + \ell L_k(t)} = B(t, T_0) \prod_{n=0}^{k} \frac{1}{1 + \ell L_n(t)}; \quad L_n(t) = L(t, T_n)$$

Bonds $B(t, T_0)$ that have time t not at a Libor time ℓk cannot be expressed solely in terms of Libor rates. Consider zero coupon bonds that are issued at Libor time, say T_0, and mature at another Libor time T_{k+1}; the Libor calendar and future time lattice is shown in Fig. 16.11b. Since $B(T_0, T_0) = 1$, zero coupon bond $B(T_0, T_{k+1})$ can be expressed entirely in terms of Libor as follows

$$B(T_0, T_{k+1}) = \prod_{n=0}^{k} \frac{1}{1 + \ell L(T_0, T_n)} = \prod_{n=0}^{k} \frac{1}{1 + \ell L_n(T_0)} \tag{16.11.8}$$

The forward bond $F(t_0, T_0, T_n + \ell)$ can be expressed solely in terms of Libor rates as follows

$$F(t_0, T_0, T_n + \ell) = \frac{B(t_0, T_n + 1)}{B(t_0, T_0)} = \prod_{i=0}^{n} \frac{1}{1 + \ell L(t_0, T_i)}. \quad (16.11.9)$$

16.12 Libor Market Model

The continuously compounded instantaneous forward interest rates have been discussed in Sects. 16.4–16.10. The instantaneous forward rates are not directly observable in the market, and need to be extracted from zero coupon bonds [8]. Hence, it is difficult to use forward interest rates to price traded instruments such as caps.

A major intrinsic shortcoming of the HJM model of forward interest rates, including its quantum finance generalization discussed given in Eqs. 16.3.1 and 16.3.2, respectively, is that the forward interest rates have a finite probability of being negative. This does not lead to any problems in the modeling of bonds using the forward interest rates since the bonds can never go negative. However, as can seen from Eq. 16.11.4, negative forward interest rates leads to Libor that is *negative*, and which leads to the failure of the model.

Instead of starting from forward interest rates with the evolution equation given by Eq. 16.3.2, one can directly model Libor in terms of $L(t, T_n)$. The BGM formulation of LMM is defined by [15]

$$\frac{1}{L(t, T_n)} \frac{\partial L(t, T_n)}{\partial t} = \xi_n(t) + \gamma_n(t) R(t)$$

$\xi_n(t)$ is the drift which is a function of Libor rates $L(t, T_n)$ and hence this makes the model nonlinear. $\gamma_n(t)$ is a deterministic function and $R(t)$ is Gaussian white noise given by

$$E[R(t)] = 0\,; \quad E[R(t)R(t')] = \delta(t - t')$$

The quantum formulation of LMM, which generalizes the BGM-formulation, is given by [5]

$$\frac{1}{L(t, T_n)} \frac{\partial L(t, T_n)}{\partial t} = \xi(t, T_n) + \int_{T_n}^{T_n+\ell} dx \gamma(t, x) \mathcal{A}(t, x) \quad (16.12.1)$$

$\gamma(t, x)$ is a *deterministic volatility* function and $\xi(t, T_n)$ is a nonlinear function of $\mathcal{A}(t, x)$. The quantum generalization of Libor Market Model gives a full description of correlation functions and accurately models the imperfect correlations of Libors that exist in the market data [11].

Equation 16.12.1 is well defined and yields convergent solutions for Libor because $\gamma(t, x)$ is a deterministic volatility function. In particular, $\gamma(t, x)$ is determined from market data and it is shown later how to completely fix the drift $\xi(t, T_n)$ as a function of $L(t, T_n)$ and $\gamma(t, x)$ using the martingale condition. Determining the drift $\xi(t, T_n)$, in effect, shows that Eq. 16.12.1 is consistent with Eq. 16.3.2 [5].

16.13 Libor Forward Interest Rates

Recall in studying the modeling coupon bonds and options thereof, one starts with the forward interest rates $f(t, x)$, as in Eq. 16.3.2, and then all the bond instruments are expressed in terms of $f(t, x)$. Since Libor is given by a set of forward interest rates, as in Eq. 16.11.4, why does one model $L(t, T_n)$ directly, as in LMM, instead of starting from the forward interest rates? One can do that and soon one finds that there are insurmountable instabilities in the Libor forward interest rates, denoted by $f_L(t, x)$–to distinguish it from the bond forward interest rates $f(t, x)$.

From Eq. 16.11.4, Libor is given by

$$L(t, T_n) = \frac{1}{\ell}\left[\exp\{\int_{T_n}^{T_n+\ell} dx f_L(t, x)\} - 1\right] \tag{16.13.1}$$

Similar to Eq. 16.3.2, Libor forward interest rates are defined by

$$\frac{\partial f_L(t, x)}{\partial t} = \mu(t, x) + v(t, x)\mathcal{A}_L(t, x) \tag{16.13.2}$$

$$f_L(t, x) = f_L(t_0, x) + \int_{t_0}^{t} dt\mu(t, x) + \int_{t_0}^{t} dt v(t, x)\mathcal{A}_L(t, x) \tag{16.13.3}$$

For Eq. 16.12.1 to be consistent with Eq. 16.13.1, it is shown in [5] that

$$v(t, x) = \frac{\ell L(t, T_n)}{1 + \ell L(t, T_n)}\gamma(t, x)\ ;\quad x \in [T_n, T_{n+1}) \tag{16.13.4}$$

A key *assumption* of the Libor Market Model is that the Libor volatility function $\gamma(t, x)$ is a deterministic function that is independent of the Libor rates. Market data for Libors or for interest rates caplets can be used for determining the empirical value of $\gamma(t, x)$ [30].

In Libor Market Model, $v(t, x)$ yields a model of the Libor forward interest rates with stochastic volatility. Equation 16.13.4 can be viewed as fixing the volatility function $v(t, x)$ of the forward interest rates $f_L(t, x)$ so as to ensure that all Libor are strictly positive.

To have a better understanding of $v(t, x)$ consider the limit of $\ell \to 0$, which yields $\int_{T_n}^{T_{n+1}} dx f_L(t, x) \simeq \ell f_L(t, x)$. From Eqs. 16.11.4 and 16.13.4

$$v(t, x) \simeq [1 - e^{-\ell f_L(t,x)}]\gamma(t, x)$$

The following are the two limiting cases

$$v(t, x) = \begin{cases} \ell\gamma(t, x) f_L(t, x) & ; \quad \ell f_L(t, x) << 1 \\ \gamma(t, x) & ; \quad \ell f_L(t, x) >> 1 \end{cases} \tag{16.13.5}$$

For small values of $f_L(t, x)$, the volatility $v(t, x)$ is proportional to $f_L(t, x)$. It is known [31] that Libor forward interest rates $f_L(t, x)$ with volatility $v(t, x) \simeq f_L(t, x)$ are unstable and diverge after a finite time. However, in Libor Market Model, when the Libor forward rates become large, that is $\ell f_L(t, x) >> 1$, the volatility $v(t, x)$ becomes deterministic and equal to $\gamma(t, x)$. It is shown in [5] that Libor forward interest rates $f_L(t, x)$ are never divergent and Libor dynamics yields finite $f_L(t, x)$ for all future calendar time.

16.14 Libor Lagrangian

Libor $L(t, T_n) > 0$; the fact that Libor is always positive naturally leads to an exponential representation for the fundamental degrees of freedom. *Define* the *dimensionless* log Libor two dimensional quantum field $\varphi(t, x)$

$$\ell L(t, T_n) = \exp\{\int_{T_n}^{T_{n+1}} dx \varphi(t, x)\} \equiv e^{\varphi_n} \tag{16.14.1}$$

To find the evolution equation for $\varphi(t, x)$ requires $\partial \ln L(t, T_n)/\partial t$. From the definition of the Libor evolution equation given by Eq. 16.12.1,

$$\frac{1}{L(t, T_n)} \frac{\partial L(t, T_n)}{\partial t} = \xi(t, T_n) + \int_{T_n}^{T_n+\ell} dx \gamma(t, x) \mathcal{A}(t, x) \tag{16.14.2}$$

The differential of log Libor is given by

$$\begin{aligned} \frac{\partial \ln L(t, T_n)}{\partial t} &= \lim_{\epsilon \to 0} \frac{1}{\epsilon} [\ln L(t + \epsilon, T_n) - \ln L(t, T_n)] \\ &= \frac{1}{L(t, T_n)} \frac{\partial L(t, T_n)}{\partial t} - \frac{\epsilon}{2} \left[\frac{1}{L(t, T_n)} \frac{\partial L(t, T_n)}{\partial t} \right]^2 + O(\epsilon) \end{aligned} \tag{16.14.3}$$

Note

$$\left[\frac{1}{L(t, T_n)} \frac{\partial L(t, T_n)}{\partial t} \right]^2 = \int_{T_n}^{T_n+\ell} dx \gamma(t, x) \mathcal{A}(t, x) \int_{T_n}^{T_n+\ell} dx' \gamma(t, x) \mathcal{A}(t, x')$$

The Wilson expansion, given in Eq. 16.5.2, yields

$$\mathcal{A}(t,x)\mathcal{A}(t,x') = \frac{1}{\epsilon}D(x,x';t)$$

and hence

$$\left[\frac{1}{L(t,T_n)}\frac{\partial L(t,T_n)}{\partial t}\right]^2 = \frac{1}{\epsilon}\Lambda_{nn}(t)$$

where the Libor autocorrelator $\Lambda_{nn}(t)$ is given by

$$\begin{aligned}\Lambda_{nn}(t) &= \int_{T_n}^{T_{n+1}} dx \int_{T_n}^{T_{n+1}} dx'\gamma(t,x)D(x,x';t)\gamma(t,x') \\ &\equiv \int_{T_n}^{T_{n+1}} dx\Lambda_n(t,x) \qquad (16.14.4)\end{aligned}$$

with

$$\Lambda_n(t,x) = \gamma(t,x)\int_{T_n}^{T_{n+1}} dx' D(x,x';t)\gamma(t,x') \qquad (16.14.5)$$

The **definition** of density $\rho_n(t,x)$ of Libor drift $\xi(t,T_n)$—as in Eq. 16.14.2—is given by the following

$$\xi(t,T_n) = \int_{T_n}^{T_{n+1}} dx\rho_n(t,x) \qquad (16.14.6)$$

Hence, from Eqs. 16.14.2, 16.14.3, 16.14.4 and 16.14.6

$$\begin{aligned}\frac{\partial \ln L(t,T_n)}{\partial t} &= \int_{T_n}^{T_{n+1}} dx\rho_n(t,x) + \int_{T_n}^{T_{n+1}} dx\gamma(t,x)\mathcal{A}_L(t,x) \\ &- \frac{1}{2}\int_{T_n}^{T_{n+1}} dx\Lambda_n(t,x) \qquad (16.14.7)\end{aligned}$$

The density of drift $\rho_n(t,x)$ is derived later in Eq. 16.15.11, and for completeness is given below

$$T_n \le x < T_{n+1} \ : \ \rho_n(t,x) = \begin{cases} \sum_{m=I+1}^{n} \frac{\ell L(t,T_m)}{1+\ell L(t,T_m)}\Lambda_m(t,x), & T_n > T_I \\ 0, & T_n = T_I \\ -\sum_{m=n+1}^{I} \frac{\ell L(t,T_m)}{1+\ell L(t,T_m)}\Lambda_m(t,x). & T_n < T_I \end{cases}$$

Note that unlike Eq. 16.14.6 where, for a given time t, there is a single value for the drift given by $\xi(t, T_n)$, the drift $\rho_n(t, x)$ is a function of x, which lies in the interval $x \in [T_n, T_{n+1})$.

Integrating Eq. 16.14.7 over time yields

$$L(T_0, T_n) = L(t_0, T_n)e^{\beta(t_0,T_0,T_n)+W_n} \tag{16.14.8}$$

$$\beta(t_0, T_0, T_n) = \int_{t_0}^{T_0} dt\xi(t, T_n) \ ; \ q_n^2 = \int_{t_0}^{T_0} dt\Lambda_{nn}(t) \tag{16.14.9}$$

$$W_n = -\frac{1}{2}q_n^2 + \int_{t_0}^{T_0} dt \int_{T_n}^{T_{n+1}} dx\gamma(t, x)\mathcal{A}_L(t, x) \tag{16.14.10}$$

Libor dynamics leads to *positive* Libor, as given in Eq. 16.14.8; Libor is proportional to the exponential of real quantities, namely a real-valued drift $\xi(t, T_n) - q_n^2/2$ and a real-valued (Gaussian) quantum field $\mathcal{A}_L(t, x)$.

The defining Eq. 16.14.1 for log-Libor gives

$$\frac{\partial \ln L(t, T_n)}{\partial t} = \int_{T_n}^{T_{n+1}} dx \frac{\partial \varphi(t, x)}{\partial t}$$

From Eq. 16.14.7, dropping the integral $\int_{T_n}^{T_{n+1}} dx$, yields

$$\frac{\partial \varphi(t, x)}{\partial t} = \tilde{\rho}_n(t, x) + \gamma(t, x)\mathcal{A}_L(t, x) \ ; \ \ T_n \le x < T_{n+1} \tag{16.14.11}$$

with

$$\tilde{\rho}_n(t, x) = \rho_n(t, x) - \frac{1}{2}\Lambda_n(t, x) \ ; \ \ T_n \le x < T_{n+1}$$

Note that the drift density $\rho_n(t, x)$ depends on $\varphi(t, x)$ and is stochastic whereas $\Lambda_n(t, x)$ is deterministic. Integrating Eq. 16.14.11 yields

$$\phi(T, x) = \phi(T_0, x) + \int_{T_0}^{T} dt \left\{ -\frac{1}{2}\Lambda(t, x) + \rho(t, x) + \gamma(t, x)\mathcal{A}(t, x) \right\} \tag{16.14.12}$$

To explicitly write out the drift terms in the Lagrangian, consider the Heaviside step function $H_k(x)$, defined in Eq. 16.14.13 and shown in Fig. 16.12; $H_k(x)$ has the value 1 in the Libor range $T_k \le x < T_{k+1}$ and is equal to 0 when outside this range. More precisely

$$H_k(x) = \begin{cases} 1 \ T_k \le x < T_{k+1} \\ 0 \ x \notin [T_k, T_{k+1}) \end{cases} \tag{16.14.13}$$

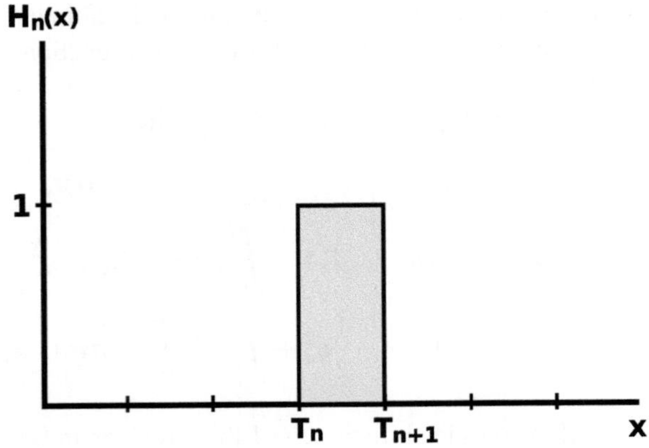

Fig. 16.12 The characteristic Heaviside function $H_n(x)$ for the Libor interval $[T_n, T_{n+1})$. Published with permission of © Belal E. Baaquie 2020. All Rights Reserved

The Heaviside function has the following two important properties.

$$f(x) = \sum_{n=0}^{\infty} H_n(x) f_n(x) \;\; \Rightarrow \;\; f(x) = f_n(x) \text{ for } T_n \leq x < T_{n+1}$$

and

$$\int_{T_n}^{T_{n+1}} dx H_n(x) = \ell \delta_{n-m}$$

The deterministic drift term is written as

$$\Lambda(t,x) = \sum_{n=0}^{\infty} H_n(x) \Lambda_n(t,x) \; ; \;\; \Lambda_n(t,x) = \int_{T_n}^{T_{n+1}} dx' M_\gamma(x,x';t)$$

In later calculations, the following representation is used

$$\Lambda(t,x) = \sum_{n=0}^{\infty} \int_{x'} H_n(x) M_\gamma(x,x';t) H_n(x')$$

The nonlinear term in the Hamiltonian is the stochastic drift $\rho(t,x)$, which depends of the field $\varphi(t,x)$ and is written as

$$\rho(t,x) = \sum_{n=0}^{\infty} H_n(x) \rho_n(t,x)$$

The martingale condition is used in Sect. 16.15 for an exact derivation of $\rho_n(t,x)$ using the Hamiltonian formalism.

With the definitions above, one has the evolution equation for log-Libor—without any restriction on x—given by

$$\frac{\partial\varphi(t,x)}{\partial t}=\tilde{\rho}(t,x)+\gamma(t,x)\mathcal{A}_L(t,x) \tag{16.14.14}$$

Similar to Eq. 16.4.5, define a change of variables relating two quantum fields $\varphi(t,x)$ and $\mathcal{A}_L(t,x)$ given by[8]

$$\mathcal{A}_L(t,x)=\frac{\partial\varphi(t,x)/\partial t-\tilde{\rho}(t,x)}{\gamma(t,x)} \tag{16.14.15}$$

The drift term $\tilde{\rho}(t,x)$ in Eq. 16.14.15, for future notational convenience, has been written as a sum of two terms using

$$\tilde{\rho}(t,x)=\frac{1}{2}\Lambda(t,x)-\rho(t,x)$$

Libor drift $\tilde{\rho}(t,x)$ is *nonlinear*, depending on the field $\varphi(t,x)$. Nonlinear drift $\tilde{\rho}(t,x)$ makes the Libor field $\varphi(t,x)$ a nonlinear two dimensional Euclidean quantum field. The defining feature of the LMM is that the volatility function $\gamma(t,x)$ is a deterministic function. This yields an evolution equation for Libor $L(t,T_n)$ that does not diverge and ensures a strictly positive Libor: $L(t,T_n)>0$. Nonlinear drift $\tilde{\rho}(t,x)$ is fixed by the martingale condition. The deterministic volatility function $\gamma(t,x)$ has been evaluated from market data [11].

The Lagrangian and action for the Gaussian quantum field $\mathcal{A}_L(t,x)$ is given by the following

$$\mathcal{L}[\mathcal{A}_L]=-\frac{1}{2}\mathcal{A}_L(t,x)\mathcal{D}_L^{-1}(t,x,x')\mathcal{A}_L(t,x')\ ;\ S[\mathcal{A}_L]=\int_{\mathcal{T}}\mathcal{L}[\mathcal{A}_L] \tag{16.14.16}$$

with the semi-infinite trapezoidal domain $\mathcal{T}$ given in Fig. 16.6b. The partition function, from Eq. 16.4.11, is given by

$$Z=\int D\mathcal{A}_L e^{S[\mathcal{A}_L]} \tag{16.14.17}$$

The Lagrangian and action for logarithmic Libor quantum field $\varphi(t,x)$ is given by

$$\mathcal{L}[\varphi]=-\frac{1}{2}\left[\frac{\partial\varphi(t,x)/\partial t-\tilde{\rho}(t,x)}{\gamma(t,x)}\right]\mathcal{D}_L^{-1}(t,x,x')\left[\frac{\partial\varphi(t,x')/\partial t-\tilde{\rho}(t,x')}{\gamma(t,x')}\right]$$

$$S[\varphi]=\int_{t_0}^{\infty}dt\int_t^{\infty}dxdx'\mathcal{L}[\varphi] \tag{16.14.18}$$

$$\tilde{\rho}(t,x)=\sum_{n=0}^{\infty}H_n(x)\left[-\frac{1}{2}\Lambda_n(t,x)+\rho_n(x)\right] \tag{16.14.19}$$

[8]The subscript L in $\mathcal{A}_L(t,x)$ is to differentiate it from the field defined in Eq. 16.4.5.

The Neumann boundary conditions $\mathcal{A}_L(t, x)$ given in Eq. 16.4.8 yields the following boundary conditions on $\varphi(t, x)$

$$\frac{\partial}{\partial x}[\frac{\partial\varphi(t,x)/\partial t - \tilde{\rho}(t,x)}{\gamma(t,x)}]\Big|_{x=t} = 0 \tag{16.14.20}$$

It is shown in [5] that the Jacobian of the transformation given in Eq. 16.14.15 is a constant, independent of $\varphi(t, x)$; note that the Jacobian is a constant in spite of the fact that the transformation in Eq. 16.14.15 in nonlinear due to the nonlinearity of Libor drift $\rho(t, x)$. A constant Jacobian leads to $\varphi(t, x)$ being *flat variables*, with no measure term in the path integral. Flat variables have a well defined leading order Gaussian path integral that generates a Feynman perturbation expansion for all financial instruments, thus greatly simplifying all calculations that are based on $\varphi(t, x)$.

In summary, up to an irrelevant constant, the log Libor path integral measure is given by

$$\int D\mathcal{A}_L = \int D\varphi = \prod_{t=t_0}^{\infty}\prod_{x=t}^{\infty}\int_{-\infty}^{+\infty} d\varphi(t,x)$$

The partition function for φ is given by

$$Z = \int D\varphi e^{S[\varphi]} = \int D\mathcal{A}_L e^{S[\mathcal{A}_L]}$$

The expectation value of a financial instrument $\mathcal{O}$ is given by

$$E[\mathcal{O}] = \frac{1}{Z}\int D\mathcal{A}_L\mathcal{O}[\mathcal{A}_L]e^{S[\mathcal{A}_L]} = \frac{1}{Z}\int D\varphi\mathcal{O}[\varphi]e^{S[\varphi]} \tag{16.14.21}$$

16.15 Libor Hamiltonian: Martingale

The nonlinear Libor Hamiltonian is similar to the one in Eq. 16.7.13 and is given by [5, 8]

$$\mathcal{H}_\varphi(t) = -\frac{1}{2}\int_{x,x'} M_\gamma(x,x';t)\frac{\delta^2}{\delta\varphi(x)\varphi(x')} + \frac{1}{2}\int_x \Lambda(t,x)\frac{\delta}{\delta\varphi(x)} - \int_x \rho(t,x)\frac{\delta}{\delta\varphi(x)}$$

where

$$M_\gamma(x,x';t) = \gamma(t,x)\mathcal{D}_L(x,x';t)\gamma(t,x')\ ;\quad \int_x \equiv \int_t^{+\infty} dx$$

$\Lambda(t, x)$ is a deterministic function and $\rho(t, x)$ is a nonlinear stochastic term.

As mentioned in Sect. 16.8, the martingale condition for Libor, unlike the case for the forward interest rates, cannot be solved using the path integral since the path integral is nonlinear. However, using the Libor Hamiltonian, the martingale condition can be solved exactly, and results in a nonlinear drift for Libor. This calculation shows the advantage of having both the path integral and Hamiltonian formulations for studying Libor. In particular, a nonlinear problem that is intractable using the path integral can be solved exactly using the Hamiltonian.

Recall from Sect. 16.9 that any positive valued financial instrument is equally suitable as a numeraire, with the drift changing with the change of numeraire. For Libor, it is most suitable to choose the zero coupon bond $B(t, T_{I+1})$ as the numeraire—having a fixed maturity at future time T_{I+1}.

The nonlinear drift for LMM, using the Hamiltonian approach, has been determined exactly in [5, 8]. Choose the numeraire to be the zero coupon bond $B(t, T_{I+1})$; for all n, the drift is fixed by the following instrument

$$\chi_n(t) \equiv \frac{B(t, T_{n+1})}{B(t, T_{I+1})} \tag{16.15.1}$$

obeying the *martingale* condition given by

$$E[\frac{B(t_*, T_{n+1})}{B(t_*, T_{I+1})}] = \frac{B(t, T_{n+1})}{B(t, T_{I+1})} \ ; \ t_* > t \ : \ \text{Martingale}$$

The Hamiltonian condition for martingale, obtained in Eq. 14.6.6, yields

$$\mathcal{H}_\varphi(t)\chi_n(t) = \mathcal{H}_\varphi(t)\left[\frac{B(t, T_{n+1})}{B(t, T_{I+1})}\right] = 0 : \ \text{for all } n \tag{16.15.2}$$

Let $t = \ell k$, for some k, be a Libor time; from the definition of the zero coupon bond given in Eq. 16.11.8

$$B(t, T_{n+1}) = \prod_{k=0}^{n}\left(\frac{1}{1+\ell L_k}\right) \ ; \ L_k \equiv L(t, T_k)$$

From Eq. 16.15.1, the following are the three cases for $\mathcal{X}_n(t)$.

- $n = I$

$$\mathcal{X}_I(t) = 1$$

- $n > I$

$$\mathcal{X}_n(t) = \prod_{k=I+1}^{n}\left(\frac{1}{1+\ell L_k}\right) = \exp\{-\sum_{k=I+1}^{n}\ln(1+\ell L_k)\} \tag{16.15.3}$$

- $n < I$

$$\mathcal{X}_n(t) = \prod_{k=n+1}^{I} (1 + \ell L_k) = \exp\{+\sum_{k=n+1}^{I} \ln(1 + \ell L_k)\} \tag{16.15.4}$$

To evaluate the action of the Hamiltonian on χ_n, the functional derivatives, discussed in Noteworthy 13.9, have to be evaluated. Recall

$$\ell L(t, T_k) = \exp\{\varphi_k\} \; ; \; \varphi_k = \int_{T_k}^{T_{k+1}} dx \varphi(t, x)$$

Functional differentiation by $\varphi(x)$, using Eq. 13.9.3, yields

$$\begin{aligned} \frac{\delta L(t, T_k)}{\delta\varphi(x)} &= \frac{\partial L(t, T_k)}{\partial\varphi_k} \frac{\delta\varphi_k}{\delta\varphi(x)} = L(t, T_k)\frac{\delta\varphi_k}{\delta\varphi(x)} \\ &= L(t, T_k) \int_{T_k}^{T_{k+1}} dy \delta(\varphi(x) - \varphi(y)) = H_k(x) L(t, T_k) \end{aligned} \tag{16.15.5}$$

Whenever the index of time t is not required, it will be suppressed.

The following are the three cases for the martingale condition given in Eq. 16.15.2.

Case (i): $n > I$

$$\begin{aligned} \chi_n(t) &= \prod_{k=I+1}^{n} \frac{1}{1 + \ell L(t, T_k)} = \exp\{-\sum_{k=I+1}^{n} \ln(1 + \ell L(t, T_k))\} \\ &= \exp\{-\sum_{k=I+1}^{n} \ln(1 + e^{\varphi_k})\} \end{aligned}$$

Hence, from Eq. 16.15.5

$$\frac{\delta\chi_n(t)}{\delta\varphi(x)} = -\sum_{k=I+1}^{n} \frac{e^{\varphi_k} H_k(x)}{1 + e^{\varphi_k}} \tag{16.15.6}$$

The summation term above is due to the discounting by the forward numeraire $B(t, T_{I+1})$. The second derivative yields

$$\begin{aligned} \frac{\delta^2\chi_n(t)}{\delta\varphi(x)\delta\varphi(x')} &= -\sum_{k=I+1}^{n} (\frac{e^{\varphi_k}}{1 + e^{\varphi_k}})^2 H_k(x) H_k(x') - \sum_{k=I+1}^{n} \frac{e^{\varphi_k} H_k(x) H_k(x')}{1 + e^{\varphi_k}} \\ &+ \sum_{j,k=I+1}^{n} \frac{e^{\varphi_j + \varphi_k} H_k(x) H_k(x')}{(1 + e^{\varphi_j})(1 + e^{\varphi_k})} \end{aligned}$$

Applying the log Libor Hamiltonian on $\chi_n(t)$ yields

$$\mathcal{H}_\varphi \chi_n(t) = \tfrac{1}{2} \int_{x,x'} M_\gamma(x,x') \Big\{ \sum_{k=I+1}^{n} (\frac{e^{\varphi_k}}{1+e^{\varphi_k}})^2 H_k(x) H_k(x')$$

$$+ \sum_{k=I+1}^{n} \frac{e^{\varphi_k} H_k(x) H_k(x')}{1+e^{\varphi_k}} - \sum_{j,k=I+1}^{n} \frac{e^{\varphi_j+\varphi_k} H_k(x) H_j(x')}{(1+e^{\varphi_j})(1+e^{\varphi_k})} \Big\}$$

$$-\tfrac{1}{2} \sum_{p=0}^{\infty} \int_{x,x'} H_p(x) M_\gamma(x,x') H_p(x') \sum_{k=I+1}^{n} \frac{e^{\varphi_k} H_k(x)}{1+e^{\varphi_k}}$$

$$+ \int_x \rho(t,x) \sum_{k=I+1}^{n} \frac{e^{\varphi_k} H_k(x)}{1+e^{\varphi_k}} \tag{16.15.7}$$

The Heaviside functions obey the identity

$$H_p(x) H_k(x) = \delta_{p-k} H_k(x) \tag{16.15.8}$$

and leads to the cancellation of the second last term in Eq. 16.15.7 above with the second term in Eq. 16.15.7.

Note the remarkable identity

$$\frac{1}{2} \sum_{j,k=I+1}^{n} A_{jk} = \sum_{j=I+1}^{n} \sum_{k=I+1}^{j} A_{jk} - \frac{1}{2} \sum_{k=I+1}^{n} A_{kk}$$

Taking

$$A_{jk} = \frac{e^{\varphi_j+\varphi_k} H_k(x) H_j(x')}{(1+e^{\varphi_j})(1+e^{\varphi_k})}$$

yields, after some cancellations, the following

$$\mathcal{H}_\phi \chi_n(t) = \int_x \rho(t,x) \sum_{k=I+1}^{n} \frac{e^{\varphi_k} H_k(x)}{1+e^{\varphi_k}}$$

$$- \int_{x,x'} M_\gamma(x,x') \sum_{k=I+1}^{n} \sum_{j=I+1}^{k} \frac{e^{\varphi_j+\varphi_k} H_k(x) H_j(x')}{(1+e^{\varphi_j})(1+e^{\varphi_k})} \tag{16.15.9}$$

Choose the Libor drift to be given by

$$\rho_n(t,x) = \sum_{j=I+1}^{n} \frac{e^{\varphi_j}}{1+e^{\varphi_j}} \int_{T_j}^{T_{j+1}} dx' M_\gamma(x,x';t)$$

$$= \sum_{j=I+1}^{n} \frac{e^{\varphi_j}}{1+e^{\varphi_j}} \Lambda_j(x);\quad T_n \le x < T_{n+1}$$

Inserting the value chosen for $\rho_n(t,x)$ into Eq. 16.15.9 and using Eq. 16.15.8 leads to the cancellation of the two terms and yields the final result

$$\mathcal{H}_\varphi \chi_n(t) = 0 \ : \ \text{martingale} \tag{16.15.10}$$

Case (ii): $n < I$

The derivation for Case (ii) similar to Case (i).

$$\chi_n(t) = \prod_{k=n+1}^{I} (1 + \ell L(t, T_k)) = \exp\{\sum_{k=n+1}^{I} \ln(1 + \ell L(t, T_k))\}$$

Choose

$$\rho_n(t,x) = -\sum_{j=n+1}^{I} \frac{e^{\varphi_j}}{1+e^{\varphi_j}} \Lambda_j(x); \quad T_n \le x < T_{n+1}$$

and this leads to the fulfillment of the martingale condition.

Case (iii): $n = I$

$\chi_n(t) = \chi_I(t) = 1$ and yields

$$\mathcal{H}_\varphi \chi_I(t) = 0 \Rightarrow \rho_I(t,x) = 0$$

The full drift is given by

$$\rho(t,x) = \sum_{n=0}^{\infty} H_n(x)\rho_n(t,x)$$

The three cases can be summarized as follows

$$T_n \le x < T_{n+1} \ : \ \rho_n(t,x) = \begin{cases} \sum_{m=I+1}^{n} \frac{e^{\varphi_m(t)}}{1+e^{\varphi_m(t)}} \Lambda_m(t,x) & T_n > T_I \\ 0 & T_n = T_I \\ -\sum_{m=n+1}^{I} \frac{e^{\varphi_m(t)}}{1+e^{\varphi_m(t)}} \Lambda_m(t,x) & T_n < T_I \end{cases}$$

Since

$$e^{\varphi_m(t)} = \ell L(t, T_m)$$

the drift can also be written as the following

$$T_n \le x < T_{n+1} \ : \ \rho_n(t,x) = \begin{cases} \sum_{m=I+1}^{n} \frac{\ell L(t,T_m)}{1+\ell L(t,T_m)} \Lambda_m(t,x) & T_n > T_I \\ 0 & T_n = T_I \\ -\sum_{m=n+1}^{I} \frac{\ell L(t,T_m)}{1+\ell L(t,T_m)} \Lambda_m(t,x) & T_n < T_I \end{cases} \tag{16.15.11}$$

The drift $\rho(t,x)$ for the LMM model with different numeraires has been studied numerically in [12].

16.16 Black's Caplet Price

Recall from Eq. 16.12.1

$$\frac{1}{L(t,T_n)} \frac{\partial L(t,T_n)}{\partial t} = \xi(t,T_n) + \int_{T_n}^{T_n+\ell} dx \gamma(t,x) \mathcal{A}(t,x) \tag{16.16.1}$$

For numeraire $B(t,T_{I+1})$, the drift $\rho_I(t,x) = 0$ for $T_n = T_I$ due to Eq. 16.15.11. Hence, $\xi(t,T_n) = 0$ due to Eqs. 16.14.6 and 16.16.1 yields the following (martingale) evolution for $L(t,T_I)$

$$\frac{\partial L(t,T_I)}{\partial t} = L(t,T_I) \int_{T_I}^{T_{I+1}} dx \gamma(t,x) \mathcal{A}_L(t,x)$$

and from Eq. 16.14.7

$$L(t_*,T_I) = L(t_0,T_I) \exp\{-\frac{1}{2} q_I^2 + \int_{t_0}^{t_*} dt \int_{T_I}^{T_{I+1}} dx \gamma(t,x) \mathcal{A}_L(t,x)\} \tag{16.16.2}$$

The payoff for a midcurve caplet on Libor $L(t_*,T_I)$, for a notional sum of V and maturing at time $t_* < T_I$, similar to the bond option as in Eq. 16.10.1, is given by [24]

$$\text{Caplet}(t_*,t_*,T_I) = \ell V B(t_*,T_I+\ell) \left[L(t_*,T_I) - K\right]_+$$

A caplet that matures when Libor $L(t_*,T_I)$ becomes operational is obtained by setting $t_* = T_I$. The caplet is a traded instrument and follows a martingale evolution for numeraire $B(t,T_{I+1})$; hence, the price of caplet at present time t_0 is given by

$$\frac{\text{Caplet}(t_0,t_*,T_I)}{B(t_0,T_{I+1})} = E\left[\frac{\text{Caplet}(t_*,t_*,T_I)}{B(t_*,T_{I+1})}\right] = \ell V E\left[L(t_*,T_I) - K\right]_+$$
$$\Rightarrow \text{Caplet}(t_0,t_*,T_I) = \ell V B(t_0,T_{I+1}) E\left[L(t_*,T_I) - K\right]_+ \tag{16.16.3}$$

The payoff can be re-expressed, from Eq. 16.6.2, as follows

$$[L(t_*, T_I) - K]_+ = \int_{-\infty}^{+\infty} dQ \frac{d\eta}{2\pi} e^{i\eta(\int_{t_0}^{t_*} dt \int_{T_I}^{T_{I+1}} dx \gamma(t,x) \mathcal{A}_L(t,x) + Q)} \times \left[L(t_0, T_I) e^{-\frac{1}{2}q_I^2 - Q} - K \right]_+ \tag{16.16.4}$$

To obtain the caplet price one evaluates Eq. 16.16.3. The action for $\mathcal{A}_L(t, x)$ is given in Eq. 16.14.16; using Gaussian integration, as in Eq. 13.10.5, yields

$$\begin{aligned} E\left[e^{i\eta \int_{t_0}^{t_*} dt \int_{T_I}^{T_{I+1}} dx \gamma(t,x) \mathcal{A}_L(t,x)} \right] &= \frac{1}{Z} \int D\mathcal{A}_L e^S e^{i\eta \int_{t_0}^{t_*} dt \int_{T_I}^{T_{I+1}} dx \gamma(t,x) \mathcal{A}_L(t,x)} \\ &= \exp\{-\frac{1}{2}\eta^2 \int_{t_0}^{t_*} dt \int_{T_I}^{T_{I+1}} dx dx' \gamma(t, x) \mathcal{D}_L(x, x'; t) \gamma(t, x')\} \\ &= \exp\{-\frac{1}{2} q_I^2 \eta^2\} \end{aligned} \tag{16.16.5}$$

where

$$q_I^2 = \int_{t_0}^{t_*} dt \int_{T_I}^{T_{I+1}} dx \int_{T_I}^{T_{I+1}} dx' M_\gamma(x, x'; t)$$

Hence, from Eqs. 16.16.3, 16.16.4 and 16.16.5, the caplet price is given by

$$\begin{aligned} &\frac{\text{Caplet}(t_0, t_*, T_I)}{B(t_0, T_{I+1})} \\ &= \ell V \int_{-\infty}^{+\infty} dQ \frac{d\eta}{2\pi} e^{-\frac{1}{2}q_I^2 \eta^2} e^{i\eta Q} \left[L(t_0, T_I) e^{-\frac{1}{2}q_I^2 - Q} - K \right]_+ \\ &= \ell V \int_{-\infty}^{+\infty} \frac{dQ}{\sqrt{2\pi q_I^2}} e^{-\frac{1}{2q_I^2} Q^2} \left[L(t_0, T_I) e^{-\frac{1}{2}q_I^2 - Q} - K \right]_+ \\ &= \ell V \left[L(t_0, T_I) N(d_+) - K N(d_-) \right] \; ; \; d_\pm = \frac{1}{q_I} \ln[\frac{L(t_0, T_I)}{K}] \pm \frac{q_I}{2} \end{aligned} \tag{16.16.6}$$

Equation 16.16.7 is the well known Black's formula for a Libor caplet [16, 24]. The additional information that the quantum LMM yields is that Black's volatility $\sigma_B^2(t_* - t_0)$ is in fact equal to q_I^2, which in turn is given by Libor volatility $\gamma(t, x)$ and the correlator $\mathcal{D}_L(x, x'; t)$. More precisely

$$\sigma_B^2 = \frac{q_I^2}{t_* - t_0} = \frac{1}{t_* - t_0} \int_{t_0}^{t_*} dt \int_{T_I}^{T_{I+1}} dx \int_{T_I}^{T_{I+1}} dx' M_\gamma(x, x'; t) \tag{16.16.7}$$

One can either choose to calibrate the quantum LMM from caplet data and ascertain $\gamma(t, x)$ or else obtain $\gamma(t, x)$ independently from the correlation of Libor rates. Knowing $\gamma(t, x)$ and the Libor auto-correlation function allows one to predict the value of the caplet [29].

Appendix
Mathematics of Numbers

A.1 Introduction

Starting from an elementary and intuitive definition of numbers, the integers and the real and complex numbers are defined. One is naturally led to the set of all integers $\mathbb{Z}$, which has infinitely many elements. The set of integers $\mathbb{Z}$ opens the path to mathematical logic, with many paradoxes and puzzles embedded in the concept of a set with infinitely many elements. Two famous and central ideas of mathematical logic are briefly discussed, that of Cantor's diagonal construction and of Gödel's incompleteness theorem.

A.2 Integers

A mathematical set $\mathcal{S}$ is a collection of objects, called elements and that are labeled by some index u. An element belonging to the set is denoted by $u \in \mathcal{S}$; and an object not in the set is denoted by $u \notin \mathcal{S}$. A set containing elements $u, v, w, \ldots$ is written as

$$\mathcal{S} = \{u, v, w, \ldots\}$$

An operation $*$ can be defined on elements of the set given by $u * v$. If the result of $*$ yields an element also in the set, for example, if $u * v \in \mathcal{S}$, then the set is said to be closed under the operation $*$.

Consider the finite set of $\mathbb{N}$ integers, also called *natural numbers*, is given by

$$\mathbb{N} = \{1, 2, 3 \ldots, n\}$$

The operation of addition, denoted by '+', is defined on the elements of the set $\mathbb{N}$. From the 'minimum' element 1 one can generate the entire set by the operation

B. E. Baaquie, *Mathematical Methods and Quantum Mathematics for Economics and Finance*, https://doi.org/10.1007/978-981-15-6611-0

$$1+1=2; 2+1=3; \ldots; (n-1)+1=n$$

What about $n+1$? It does not belong to $\mathbb{N}$ and hence $\mathbb{N}$ is not closed under the operation of +.

The set $\mathbb{N}$ can be *extended* by including *all* positive integers and yields

$$\mathbb{N}=\{1, 2, \ldots, n, \ldots, +\infty\}$$

Note ∞ is not an ordinary integer since

$$\infty+n=\infty$$

and with this definition of ∞, the set $\mathbb{N}$ is closed under addition. The element zero 0 is defined by

$$0+n=n$$

The negative integers $-N$ are then defined by

$$n+(-n)=0$$

Collecting all positive and negative integers, including 0 and $\pm\infty$ yields the *set of integers*

$$\mathbb{Z}=\{-\infty, \ldots, -n, \cdots, -2, -1, 0, 1, 2, \ldots, +n, \ldots, +\infty\}$$

The integers are ordered > (greater than) defined by

$$m>n \;\Rightarrow\; m=n+1+1\cdots+1$$

Since integers are ordered, they can be arranged in a *straight line*, with the smallest integers to the left. This is the foundation of the real line, as will be seen later.

The operation $\cdot$ stands for multiplication and is a short-hand for addition since

$$m \cdot n=\underbrace{m+\cdots+m}_{\text{n-times}}$$

Similarly, m raised to the power n is a short hand for multiplication since

$$m^n=\underbrace{m \ldots m}_{\text{n-times}}$$

The operation of division can be defined for only a certain set of integers. For m, n, p, r integers

$$m=n+pr \;\Rightarrow\; \frac{m-n}{p}=r$$

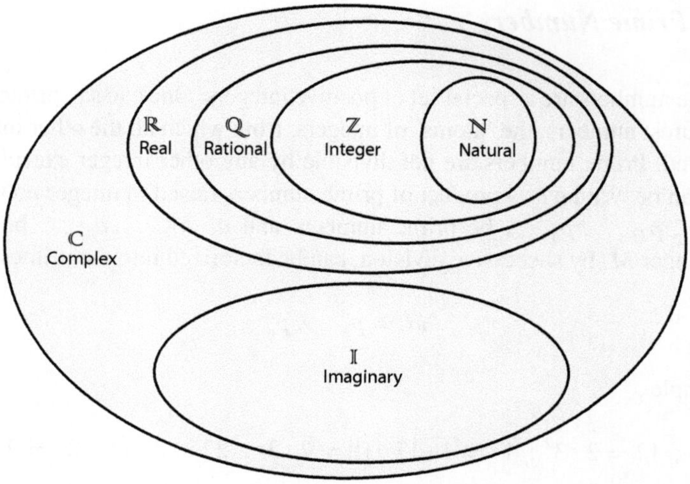

Fig. A.1 The Venn diagram of the sets of numbers. Published with permission of © Belal E. Baaquie 2020. All Rights Reserved

For example

$$20 = 4 + 2 \cdot 8 \;\Rightarrow\; \frac{20-4}{2} = 8$$

What about division of integers by other integers that does not result in an integer? Such as

$$1/3, 4/5, 27/7, 35/6, \ldots$$

The numbers above are fractions; all fractions can be expressed as

$$\frac{m}{n} = r + f$$

where r is an integer and f is a *proper fraction* given by

$$f = \frac{k}{s} < 1 \;\Rightarrow\; k < s;\;\; k, s \neq 0 \;:\; \text{integers}$$

If one includes *all fractions* in a set with $\mathbb{Z}$, then one obtains the set of all rational numbers $\mathbb{Q}$ given by (Fig. A.1)

$$\mathbb{Q} = \{\mathbb{Z};\; \frac{m}{n} \;:\; m, n \in \mathbb{Z};\, n \neq 0\}$$

A.2.1 Prime Numbers

The prime numbers are a special set of positive integers. One can say prime numbers are the purest numbers, the 'atoms' of integers, from which all the other integers are constructed. Prime numbers are not divisible by any other integer except 1. Every integer can be written as a product of prime numbers raised to integer powers.

Let $p_1, p_2, \ldots, p_n, \ldots$ be prime numbers and $a_1, a_2, \ldots, a_n, \ldots$ be integers. *Every* integer M, by successive division, can be factorized into the primes to yield

$$M = p_1^{a_1} \ldots p_n^{a_n}$$

For example

$$3 = 1 \cdot 3\ ;\ 12 = 2 \cdot 3^2\ ;\ 17 = 1 \cdot 17\ ; 18 = 2 \cdot 3^2\ ;\ 22 = 2 \cdot 11\ ;\ 28 = 2^2 \cdot 7\ ;\ \ldots$$

There are infinitely many primes. To see this, suppose that p_k, for some k, is the largest prime; then we can obtain an integer given by

$$1 + p_1 p_2 \ldots p_k$$

which is also a prime: contradiction. Hence, to avoid the contradiction, one concludes that there cannot be a largest prime.

It is known that the number of primes less than N, for large N, is less than $N/\ln(N)$. In general, very little is known about the primes. For instance, there is no formula that can generate all the primes.

A.2.2 Irrational Numbers

We have obtained the set of all rational numbers $\mathbb{Q}$, of which the integers are a special case. Are there any other numbers? Consider a right triangle with base a height b and hypotenuse given by c; its area is given by $ab/2$ as can be seen by embedding the triangle in a rectangle of sides a, b. Pythagoras theorem states that

$$a^2 + b^2 = c^2$$

Consider a right triangle with $a = 1 = b$; then

$$c^2 = 1 + 1 \ \Rightarrow\ c = \sqrt{2}$$

$\sqrt{2}$ is an example of an irrational number and it can be shown that $\sqrt{2}$ is not equal to any rational number. In fact, none of the irrational numbers can be represented by a fraction.

So what is an irrational number? To see the connection of $\sqrt{2}$ with the rational numbers one first needs to study infinite series, which is discussed in Sect. 1.5. For now, note that irrationals are the *limiting points* of rational numbers.

To understand a limiting point, consider the sequence of rational numbers

$$1, 1/2, 1/3, 1/4, \ldots, 1/N, \ldots$$

As N becomes larger and larger, the sequence value approaches 0, but does not reach it in any finite number of steps. So one writes

$$\lim_{N\to\infty} \frac{1}{N} \to 0$$

and 0 is the *limiting value* of the sequence $1/N$. There are infinitely many points near zero. The convergence of an infinite sequence is discussed in Sect. 1.6.

There is a question of cardinality, which is a statement about how many rational numbers are there, and how many numbers are irrational. As one can imagine, all rationals can be put in a one to one mapping with the integers, and hence are said to be *countable*. In contrast, there are far more irrationals than rationals, and the irrational numbers are said to be *uncountable*.

A.2.3 Transcendental Numbers

An irrational number can be the solution of an algebraic equation. For instance, $\sqrt{2}$ is the solution of the equation

$$x^2 = 2 \;\Rightarrow\; x = \sqrt{2}$$

The irrational numbers that are the solution of polynomial equations with integer coefficients are called algebraic numbers. Transcendental numbers are numbers that are irrational numbers but are not algebraic numbers. Although it is known from general results that there are uncountably infinite transcendental numbers, it is difficult to prove that any one given number is transcendent. Two well known transcendent numbers are the exponential e and the number π. It can be shown that if a is an algebraic number and b is an irrational algebraic number, then a^b is a transcendental number.[1]

[1]Note a is not zero or one, since then $a^b = a$.

A.3 Real Numbers

Consider the collection of all rational numbers as well as all the irrational numbers. The combined set yields the real numbers $\mathbb{R}$ and is given by

$$\mathbb{R} = \mathbb{Q} \oplus \text{irrationals}$$

The real numbers $\mathbb{R}$ form a *continuous* set, and can be represented by a continuous line. By continuous is meant that arbitrarily near every point, there are infinitely many other distinct points; there are no 'gaps' between points on the real line. This was already encountered in considering the sequence $1/N$. One says that points infinitely close to say 0, but not equal to 0, are at a 'distance' of *epsilon* (ϵ) from 0. The entire field of calculus is based on the idea of ϵ. Crudely speaking

$$\epsilon \simeq \lim_{N \to \infty} \frac{1}{N}$$

ϵ is the 'inverse' of infinity. ϵ is not equal to 0 and ϵ is also *not* finite. It is 'in-between' zero and a finite number. ϵ is the 'smallest' real number. ϵ cannot be assigned a numerical value since any real number, if we divide by say 10, will produce a smaller number.

A.3.1 Decimal Expansion

Every real number has a decimal expansion and can be written in the decimal notation; for example

$$\begin{aligned}
\frac{1}{2} &= 0.500000 = 0.5 \\
\frac{1}{3} &= 0.3333333\ldots \\
\frac{1}{6} &= 0.171717\ldots \\
\frac{1}{7} &= 0.\underbrace{142857}\,\underbrace{142857}\ldots
\end{aligned}$$

The decimal expansion for rational fractions has a finite number of terms—a pattern of a finite length—that repeats infinitely many times. In contrast, for irrational numbers the decimal expansion is infinitely long, with no pattern being repeated. For example

$$\sqrt{2} = 1.41421356237\ldots$$

In decimal notation, one can heuristically write

$$\epsilon = \underbrace{0.0000\ldots 1}_{\text{infinitely many zeros}}$$

Note the apparent paradox that

$$1 = 3 \times \frac{1}{3} = 0.999999999\ldots$$

Finite valued infinitesimals are not allowed in the real number system: the infinitesimal ϵ is *not* a finite number. Hence, there is no inconsistency since 1 is infinitesimally different from 0.99999. . . .[2] One can heuristically write

$$1 = \lim_{\epsilon \to 0}[0.999999999 \cdots + \epsilon] = 0.999999999\ldots$$

A.3.2 Complex Numbers

Negative numbers lead to the following possibility

$$z^2 = -r^2$$

What is z? No real number, when squared, can be negative. A new class of 'numbers', called imaginary numbers, are required for describing z. Define

$$i = \sqrt{-1}$$

One solution to $z^2 = -r^2$ is then given by $z = ir$ (the other being $z = -ir$).

In general, any complex numbers can be written as follows

$$z = x + iy \ ; \ \ x, y \in \Re$$

z can be thought of as a pair of real numbers $z = (x, y)$. One can represent the complex number z as an element of two copies of the real numbers, and is written

$$z \in \mathbb{C} = \mathbb{R} \times \mathbb{R} = \mathbb{R}^2$$

The space $\mathbb{R}^2$ is two dimensional and z can be thought of as a 'vector' in this space.

Complex numbers obey the same rules of arithmetic as real numbers and are closed under these rules. Let

$$z = x + iy \ ; \ \ w = u + iv$$

[2] Reference. https://en.wikipedia.org/wiki/0.999.

Then

$$z + w = (x + u) + i(v + y) = a + ib$$

and

$$wz = (ux - vy) + i(vx + uy) = a + ib$$

Complex conjugation is defined by changing i to $-i$ wherever it occurs. In particular

$$z^* = x - iy$$

Also

$$z + z^* = 2x \quad : \text{Real} \quad z - z^* = 2iy \quad : \text{Imaginary}$$

Hence

$$|z|^2 \equiv zz^* = x^2 + y^2 : \quad \text{(real) length of } z$$

A.4 Cantor's Diagonal Construction

Consider the set of all positive integers $\mathbb{N} = \{1, 2, 3, \ldots, +\infty\}$. There are infinitely many subsets of $\mathbb{N}$ that can be put into one-one correspondence with $\mathbb{N}$. For example

$$1 \longleftrightarrow 1^2, 2 \longleftrightarrow 2^2, 3 \longleftrightarrow 3^2, \ldots, n \longleftrightarrow n^2, \ldots$$

Also

$$1 \longleftrightarrow 2, 2 \longleftrightarrow 4, 3 \longleftrightarrow 6, \ldots, n \longleftrightarrow 2n, \ldots$$

One of the defining properties of an infinite set, such as $\mathbb{N}$, that its subset can equal the set itself. From above, we can see that, for example, the even numbers, a subset of $\mathbb{N}$, can be put into a one-one correspondence to all the elements of $\mathbb{N}$.

As a warm-up to the study of Gödel's incompleteness theorem, consider the set of all integers $\mathbb{Z}$ and the set of all real numbers $\mathbb{R}$. The question Cantor asked is the following:

Can the elements (points) of $\mathbb{R}$ be put into a one-one correspondence with the elements (numbers) of $\mathbb{Z}$?

To simplify our discussion consider only the decimal expansion of real numbers less than 1; this can always be achieved by dividing any number by factors of 10 to make it less than 1. For example, a typical decimal expansion of a real number is

$$0.97653320288863598231747420\ldots$$

There is no pattern for irrational numbers, and the decimal expansion goes on indefinitely.

Table A.1 The diagonal entries are s_{ii}

Number of sequence	Decimal representation
S_1	0.**9**7653320288863598231747420...
S_2	0.3**2**1074741765332729853972354949....
S_3	0.76**0**0598237524363989783334...
...	...
S_n	0.674020288784664...859377983773645561209484...
...	...
$\mathcal{M}$ (Missing)	**0.455...4...**

The set of all positive integers $\mathbb{Z}$ is a set with infinite many elements; this set is said to be countably infinite since one can recursively enumerate all the integers, even though the counting will take forever. In particular, in a finite number of steps, one can keep adding the number 1 to reach any specific element. A set is said to be uncountably infinite if all the elements of the set cannot be enumerated—even in one takes infinitely many steps.

We can now address the question: can all the real numbers in the set $\mathbb{R}$ be put into a one-one correspondence with the integers in the set $\mathbb{Z}$?

Let us assume that $\mathbb{R}$ is countable. Since the integers form a called a countable infinite set, if $\mathbb{R}$ forms a countable infinite set, one should be able to create a countably infinite sequence of real numbers, one entry for every positive integer in $\mathbb{Z}$. Since all real numbers have a decimal expansion, one can generate all possible decimal expansions and put each element into a one-one correspondence with a single integer.

Each decimal number is denoted by S_n and one enumerates an infinite collection of sequences $\{S_n | n = 1, 2, \ldots, +\infty\}$. An example of such an infinite list is given in the Table A.1.

The sequence $\mathcal{M}$ in Table A.1 is constructed as follows. The sequence S_n shown in the Table A.1 is labeled by n. Writing only the entries after the decimal point for convenience, the sequence S_n can be represented by

$$S_n = \{s_{n1}, s_{n2}, \ldots, s_{nn}, \ldots\} \ ; \ \ s_{ni} = 0 \text{ or } 1 \text{ or } \ldots 9 \ : \ i = 1, 2, \ldots, \infty$$

Consider a sequence (the last line in the Table A.1), writing only the entries after the decimal point, given by

$$\mathcal{M} = \{m_i \ ; \ i = 1, 2, \ldots, \infty\}$$

Define sequence $\mathcal{M}$ by requiring $m_i \neq s_{ii}$ and given by the following:

$$\mathcal{M} = \{m_i \ ; \ i = 1, 2, \ldots, \infty\} \Rightarrow \begin{cases} m_i = 5 \text{ if } s_{ii} < 5 \\ m_i = 4 \text{ if } s_{ii} \geq 5 \end{cases}$$

Comparing $\mathcal{M}$ with all the sequences S_n, starting from S_1, we see that $\mathcal{M}$ does not match any of the sequences since the diagonal element s_{ii} always differs from m_i. Cantor's argument is called the diagonal construction since it uses the diagonal elements of the entries of the Table A.1.

We conclude that the $\mathcal{M}$, which represents an actual real number that in Table A.1 is given by 0.454. . .4· · · , *does not exist* in the infinite list in the Table A.1, a list that exhausts all the positive integers. Table A.1 shows that the elements of $\mathbb{R}$ cannot be put into a one-one correspondence with the elements of $\mathbb{Z}$.

Hence, we conclude that the set of real numbers $\mathbb{R}$ has more elements than $\mathbb{Z}$—and that $\mathbb{R}$ is a set which has uncountably infinite many number of elements. Irrational and transcendent numbers are also uncountable [33].

Every point in the real line has infinitely many points in its neighborhood. In the notation of calculus, there are infinitely many points that are at a distance of ϵ from any given point. If the continuous real line is considered to be a set of uncountably many points, then integration theory assigns a 'measure zero' to a single point of the real line: in other words, a single point has no intrinsic significance in the real line, it is the connection of every point with its neighboring points that forms the continuous line.

This is what is meant by considering the real line to be a continuous geometric manifold, since each point is so interconnected with the other neighboring points that the exact isolation a point from each other is impossible. The Dirac δ-function achieves this 'impossible' task by introducing a new mathematical structure, not an ordinary function, that picks out a single point of the continuum inside an integration.

A.5 Higher Order Infinities

From this rather shocking result, Cantor concluded that there exists (in a mathematical and not physical sense) a hierarchy of infinities, with each infinity having a cardinality, which is the number of elements in that infinite set—and with each level of infinity having more elements than the preceding level of infinity. The most familiar being the countable infinity of the integers $\mathbb{Z}$ and an infinity with a higher cardinality is the uncountable infinite of $\mathbb{R}$, which is the infinity of the set constituting the continuous real line.

The reasoning of Cantor can be extended to the conclude that there cannot be a complete list of all the subsets $S_1, S_2, \ldots$ of the positive numbers $\mathbb{Z}$. Similar to the diagonal construction, one can define a subset S of $\mathbb{Z}$ such that the integer n is a label of S if $n \notin S_n$. Then S is not in the countable list of the subsets of $\mathbb{Z}$ and hence $\mathbb{Z}$ does not contain all of its own possible subsets. In particular, the number of subsets of $\mathbb{Z}$ are greater than the number of elements of $\mathbb{Z}$ and have the cardinality of the real numbers.

One can go even further: the diagonal argument shows the number of all subsets of any set X, called the power set of X, is larger than the number of elements that the set X contains. To see this, let there be a pairing for every $x \in X$ with a subset of X denoted by S_x. Using Cantor's diagonal argument, one can construct a set S such that it does not contain the element x and hence is distinct from all the subsets of X.

We will see that the diagonal construction of Cantor has a deep connection with Gödel's incompleteness theorem.

A.6 Meta-mathematics

Meta-mathematics, or equivalently of mathematical logic, is the study of the logical basis of *all* mathematical systems. It consists primarily in the drive to make all of mathematics a system of symbols with well defined rules on how to manipulate and calculate with these symbols. The aim is to reduce all mathematical proofs to a number of mechanical steps that would lead from the axioms to the theorems and lemmas.

Let the mathematical system be denoted $\mathcal{G}$. The key ideas underlying meta-mathematics is that all mathematical systems (set of axioms) must have the following properties [33].

- *Completeness*. All valid statements in $\mathcal{G}$ are theorems that can be proven.
- *Decidability*. A list of permissible operations are defined on the symbols in $\mathcal{G}$ that lets one decide whether a statement allowed by the rules in $\mathcal{G}$ is true or false.
- *Consistency*. A theorem cannot be true and false at the same time.

In the effort to completely axiomatize arithmetic, in 1905 Russell and Whitehead wrote their monumental work *Principia Mathematica* (PM) in which they provided a complete set of symbols—using which all possible statements in arithmetic could be made. They also claimed to show that all theorems of arithmetic could proven by a set of rules being applied mechanically, thus removing the possibility of human error. An example of meta-mathematics is that Gödel devised a mapping of statements in PM into the integers. He then used the properties of integers to study the logical structure of the axioms of arithmetic. The Gödel mapping is a meta-mathematical construction that is valid for all axiomatic systems.

A.7 Gödel's Incompleteness

Kurt Gödel's proof in 1931 of the incompleteness of all mathematical systems that contain the integers was an epoch-making event, the first advance in mathematical logic after over 2,000 years of Aristotelian logic [29]. The seminal result of Gödel has led to a view that mathematics, instead of being the science of patterns, is the subject of a self-consistent set of axioms.

We take recourse to the Turing to approach the concept of incompleteness. The property of a Turing machine is that given a list of positive numbers that is called *computably enumerable*, there is a Turing machine that lists all its elements. In particular, we are able to observe how and when a Turing machine generates a number n.

For example, $\mathbb{Z}$ is computably enumerable as are many of its subsets like all even integers, all primes and so on.

Let a Turing machine T_n generate a computably enumerable set W_n. According to the Church-Turing theorem, all the sets W_n generated by the Turing machines are in a computably enumerable master list W generated by all the T_n's. Hence

$$W = \{W_n | n = 1, 2, \ldots, \infty\}$$

Then it follows from Cantor's diagonal construction that there must exist a set $\mathcal{D}$ such that

$$\mathcal{D} = \{d_i | i = 1, 2, \ldots, \infty\} \notin W$$

is not computably enumerable and hence it is not in the master list of computably enumerable sequences.

The result for Turing machines carries over to any formal mathematical system $\mathcal{G}$. Russell and Whitehead in PM provided a complete set of symbols, with the associated rules for the manipulation of these symbols, using which all syntactically correct statements in arithmetic can be written down. For example, a statement such as $0 = 5$ is allowed by the syntax of arithmetic; furthermore, using the rules for the manipulation of the symbols given in PM, it can be shown, in a few mechanical steps, that the statement $0 = 5$ is false.

The key takeaway from PM is that all theorems of $\mathcal{G}$, contained in master list W, are computably enumerable. Hence, from Cantor's diagonal construction, there is a theorem, similar to $\mathcal{D}$, that cannot be proven or disproven since (it will be shown that) it is *not* in the computably enumerable list of theorems that is generated by PM. Note that $\mathcal{D}$ is the proof of a syntactically allowed sequence in PM.

Suppose $\mathcal{G}$ is consistent. Let a sequence be enumerated as follows

$$\mathcal{W}_n = \{w_{n1}, w_{n2}, \ldots, w_{nn}, \ldots\} \; : \; n = 1, 2, \ldots, \infty$$

Consider the sequence $\mathcal{D}$ that is defined as follows

$$\mathcal{D} = \{\tilde{w}_{11}, \tilde{w}_{22}, \ldots, \tilde{w}_{nn}, \ldots\} = \{d_1, d_2, \ldots, d_n, \ldots\}$$

where, using the Cantor diagonal approach

$$d_1 = \tilde{w}_{11} \neq w_{11} \; ; \; d_2 = \tilde{w}_{22} \neq w_{22} \; ; \; \ldots \; ; \; d_n = \tilde{w}_{nn} \neq w_{nn} \; ; \; \ldots$$

Consider a sequence W_m; from its definition

$$\mathcal{W}_m = \{w_{m1}, w_{m2}, \ldots, w_{mn}, \ldots, w_{mm}, \ldots, \}$$

Suppose that W generates W_m that is the sequence $\mathcal{D}$. We then have

$$\mathcal{W}_m = \{\tilde{w}_{m1}, \tilde{w}_{m2}, \ldots, \tilde{w}_{mn}, \ldots, \tilde{w}_{mm}, \ldots, \} = \mathcal{D}$$

This leads to a contradiction since the definition of $\mathcal{D}$ demands that

$$w_{mm} = d_m = \tilde{w}_{mm} \neq w_{mm} \quad : \text{ Contradiction}$$

In effect, $w_{mm} \in W_m$ implies that '$w_{mm} \notin W_m$'—in other words, both the statement and its opposite are correct if W_m exists. The only way out of this conundrum is that the sequence W_m does not exist in the computably enumerable list of theorems.

Hence, consistency demands that the computably enumerable sequence W_m—which is a proof of a syntactically allowable string in PM—*cannot* be generated by PM, and hence PM is *incomplete*.

- The first proof of the incompleteness of PM given by Gödel in 1931 was based on explicitly constructing a logical statement and then mapping it to the Gödel number G. Gödel then showed that if the logical statement he generated was true, then the negation of the logical statement was also true. This is similar to the proof given above that $w_{mm} \in W_m$ implies that '$w_{mm} \notin W_m$'.
- It was then shown by Gödel that the consistency of PM requires that the number G does not exist in the set of all Gödel numbers that are generated by PM.
- Gödel, in effect, generated the missing number G, analogous to the Cantor diagonal construction that shows that a real number is missing in the countable list. Apparently, John von Neumann had sent a proof to Gödel of the arithmetical derivation of Gödel's incompleteness theorem.

Gödel explicitly generated a logical statement in PM that is not provable, thus demonstrating the incompleteness of any formal system $\mathcal{G}$. The *halting problem* is the question of whether, given some input, can one predict that an algorithm will or will not terminate in a finite number of steps? This is another unprovable statement, where it cannot be predicted that an algorithm will or will not halt. In other words, one cannot write a computer program that can ascertain whether a specific computer program for an arbitrary input will stop or not stop in a finite number of steps.

Unlike the case of mathematical logic (Gödel's example) and computer algorithms (the halting problem), no unprovable statements have so far been found in arithmetic. There are a number of conjectures that so far that have defied all attempts to prove them. These include (a) the Goldbach conjecture, that all even numbers are the sum of two primes, (b) the Riemann conjecture that the Riemann zeta function—function of a complex variable—has complex zeros that lie on a line that is parallel to the imaginary axis, and (c) the twin prime conjecture that there are infinitely many primes that come in pairs that differ by 2.

The reason that these conjectures cannot be classified as unprovable is because no proof has, as yet, been given that they are unprovable. It may be the case that the conjectures have not been proven simply because a proof has not yet been found.

References

1. Anthony, M., Biggs, N. L., & Biggs, N. (1996). *Mathematics for economics and finance: Methods and modelling*. UK: Cambridge University Press.
2. Asano, A. (2003). *An introduction to mathematics for economics*. UK: Cambridge University Press.
3. Baaquie, Belal Ehsan (2019). Merton's equation and the quantum oscillator: Option pricing. *Physica A: Statistical Mechanics and its Applications*, 532 (121792)
4. Baaquie, Belal E. (2004). *Quantum finance: Path integrals and Hamiltonians for options and interest rates*. Cambridge: Cambridge University Press.
5. Baaquie, Belal E. (2010). *Interest rates and coupon bonds in quantum finance* (1st ed.). UK: Cambridge University Press.
6. Baaquie, Belal E. (2013). *The theoretical foundations of quantum mechanics*. UK: Springer.
7. Baaquie, Belal E. (2014). *Path integrals and Hamiltonians: Principles and methods*. Cambridge: Cambridge University Press.
8. Baaquie, Belal Ehsan. (2018). *Quantum field theory for economics and finance*. Cambridge: Cambridge University Press.
9. Baaquie, Belal Ehsan. (2020). *Lattice quantum field theory of the Dirac and Gauge fields*. Singapore: World Scientific.
10. Baaquie, Belal Ehsan. (2020). Merton's equation and the quantum oscillator: Pricing risky corporate coupon bonds. *Physica A: Statistical Mechanics and its Applications*, *541*(123367)
11. Baaquie, Belal E., & Yang, Cao. (2009). Empirical analysis of quantum finance interest rate models. *Physica A*, *388*(13), 2666–2681.
12. Baaquie, Belal E., & Pan, Tang. (2012). Simulation of nonlinear interest rates in quantum finance: Libor market model. *Physica A*, *391*, 1287–1308.
13. Baaquie, Belal E., Corianó, C., & Srikant, M. (2004). Hamiltonian and potentials in derivative pricing models: Exact results and lattice simulations. *Physica A: Statistical Mechanics and its Applications*, *334*(3), 531–557.
14. Bergin, J. (2016). *Mathematics for economists with applications*. UK: Springer.
15. Brace, A., Gatarek, D., & Musiela, M. (1996). The market model of interest rate dynamics. *Mathematical Finance*, *7*, 127–154.
16. Cairns, A. J. G. (2004). *Interest rate models*. Princeton: Princeton University Press.
17. Campolieti, G., & Makarov, R. N. (2014). *Financial mathematics: A comprehensive treatment*. UK: CRC Press.
18. Chiang, K., & Wainwright, A. C. (2005). *Fundamental methods of mathematical economics*. UK: McGraw Hill.

B. E. Baaquie, *Mathematical Methods and Quantum Mathematics for Economics and Finance*, https://doi.org/10.1007/978-981-15-6611-0

19. Curtis, C. W. (1984). *Linear algebra an introductory approach*. UK: Springer.
20. Eichhorn, W., & Gleibner, W. (2016). *Mathematics and methodology for economics*. Germany: Springer.
21. Feynman, R. P., & Hibbs, A. R. (1965). *Quantum mechanics and path integrals*. USA: McGraw-Hill.
22. Harrison, M., & Waldron, P. (2011). *Mathematics for economics and finance*. USA: Routledge.
23. Heath, D., Jarrow, R., & Morton, A. (1992). Bond pricing and the term structure of interest rates: A new methodology for contingent claim valuation. *Econometrica, 60*, 77–105.
24. Hull, J. C. (2000). *Options, futures, and other derivatives* (4th ed.). New Jersey: Prentice Hall.
25. Krichene, N., & Mirakhor, A. (2014). *Introductory mathematics and statistics for Islamic finance*. USA: Wiley.
26. Liu, S., Lu, J. C., Kolpin, D. W., & Meeker, W. Q. (1997). Multiperiod corporate default prediction with stochastic covariates. *Environmental Science and Technology, 31*, 635–665.
27. Mavron, V. C., & Phillips, T. N. (2007). *Elements of mathematics for economics and finance*. UK: Science and Business Media.
28. Merton, R. C. (1974). On the pricing of corporate debt: The risk structure of interest rates. *Journal of Finance, 29*, 449–470.
29. Nagel, E., & Newman, J. R. (1973). *Gödel's proof*. USA: M. Dekker.
30. Rebonata, R. (1996). *Interest-rate option models*. USA: Wiley.
31. Rebonata, R. (2002). *Modern pricing of interest-rate derivatives*. Princeton: Princeton University Press.
32. Shreve, S. E. (2008). *Stochastic calculus for finance II: Continuous-time models*. Germany: Springer Finance.
33. Stillwell, John. (2019). *A concise history of mathematics for philosophers*. UK: Cambridge University Press.
34. Strang, G. (1986). *Introduction to applied mathematics*. USA: Wellesley Cambridge Press.
35. Sundaresan, S. (2013). A review of merton's model of the firm's capital structure with its wide applications. *Annual Review of Financial Economics, 5*(1), 21–41.
36. Suzuki, M., Kashiwa, T., & Ohnuki, Y. (1997). *Path integral methods*. UK: Claredon Press.
37. Ummer, E. K. (2012). *Basic mathematics for economics, business and finance*. USA: Routledge.
38. Vali, S. (2014). *Principles of mathematical economics*. France: Atlantis Press.
39. Yeh, J. (2001). *Stochastic processes and the Wiener integral*. USA: New York University Press.
40. Zastawniak, T., & Capinski, M. (2003). *Mathematics for finance an introduction to financial engineering*. UK: Springer.

Index

B. E. Baaquie, *Mathematical Methods and Quantum Mathematics for Economics and Finance*, https://doi.org/10.1007/978-981-15-6611-0

MIX
Papier aus verantwortungsvollen Quellen
Paper from responsible sources
FSC® C105338

If you have any concerns about our products,
you can contact us on
ProductSafety@springernature.com

In case Publisher is established outside the EU,
the EU authorized representative is:
Springer Nature Customer Service Center GmbH
Europaplatz 3, 69115 Heidelberg, Germany

Printed by Libri Plureos GmbH
in Hamburg, Germany